Studies in the History of Philosophy of Mind

Volume 18

More information about this series at http://www.springer.com/series/6539

Jack P. Cunningham • Mark Hocknull
Editors

Robert Grosseteste and the pursuit of Religious and Scientific Learning in the Middle Ages

Springer

Editors
Jack P. Cunningham
Bishop Grosseteste University
Lincoln, United Kingdom

Mark Hocknull
University of Lincoln
Lincoln, United Kingdom

Studies in the History of Philosophy of Mind
ISBN 978-3-319-33466-0 ISBN 978-3-319-33468-4 (eBook)
DOI 10.1007/978-3-319-33468-4

Library of Congress Control Number: 2016947425

Printed on acid-free paper

This Springer imprint is published by Springer Nature
The registered company is Springer International Publishing AG Switzerland

This volume is dedicated to Professor Pietro B. Rossi in gratitude for his outstanding contribution to our understanding of Grosseteste's pursuit of Religious and Scientific learning.

Preface

In July 2014 scholars from all over the globe met in Lincoln for Bishop Grosseteste University's third international conference on Robert Grosseteste which took as its title, *Robert Grosseteste and the pursuit of Religious and Scientific Learning in the Middle Ages*. The group made up an eclectic body of academics from a wide range of disciplines including theology, physics, cosmology, history, philosophy and experimental psychology. Quite possibly the whole exercise should have failed since academics from such different subject groupings usually have little to say to one another when it comes to their work. It was instead a resounding success as colour scientists explained to medievalists Grosseteste's colour theories, historians described to modern cosmologists the inner workings of the medieval scientific mind and physicists provided profound insights into what all this meant in terms of the relationship between faith and science. Two questions emerged above all others as the 3 days of conference progressed. Firstly, how might we best place the Bishop of Lincoln in the history of science after the bold assertions of Alistair Crombie in the 1950s and the new understandings that are emerging from the tremendously important work of the *Ordered Universe Project* at Durham University? Secondly, what if anything, might all this say to us in the twenty-first century about the relationship between science and religion? This volume does not pretend to present a single answer to either of these questions; indeed, our two final chapters represent quite opposing points of view. What it does hope to do is present fifteen contributions to the answering of these and related questions from scholars with a wide range of expertise who might combine their learning to produce something that is able, in a small way, to approach the inner workings of a mind as staggeringly intelligent as the medieval polymath that was Robert Grosseteste.

When the Archbishop of Canterbury, Randall Thomas Davidson, asked Einstein what effect his theory of relativity would have on religion, Einstein is reported to have replied 'None. Relativity is purely a scientific matter and has nothing to do with religion' (Eddington 1939). On the face of it, this is a simple statement, well supporting the common view of the separation of the sciences from religion with the popular aphorism of science dealing with the how questions and that of religion dealing with the question of why. Yet this statement of Einstein belies both the historical complexity of the relationship between science and religion and their interconnectedness in Einstein's own scientific work and religious belief structure. One of the reasons that Einstein rejected the Copenhagen statement of 1927 on quantum indeterminacy, and the possibility of only statistical accounts of the quantum world, was his deterministic view of the universe drawn from a religious view of the world as the creation of Mind. It would seem that attempts to compartmentalise human thought are not so simple and straightforward as we might sometimes wish. Such a separation makes for interesting analytical schemes but belies the complexity of historical and personal realities. Einstein himself in subsequent writings seems to have discarded this separation thesis. Whilst this could be explained away as a change of mind, it is perhaps better understood in a different way. In his response to the Archbishop, Einstein had in mind institutional or organised religion: he was after all replying to the head of a religious institution. In his subsequent reflections on the relationship between science and religion, he was more interested in ideas and the impact science might have on religion or theology as a systematic discipline and personal belief system. Such apparent contradictions within the reported output of one modern scientist indicate the great difficulty the historian faces in analysing the relationships between the many different areas of the thought and work of historical figures. If the historian faces such problems with a modern, twentieth-century figure where the sources are plentiful and well attested, how much more difficult in the case of a medieval figure such as the thirteenth-century Bishop of Lincoln.

The middle of the twentieth century saw an explosion of interest in the ideas of Robert Grosseteste as a significant figure in the development of medieval science and thus as a pioneer and forerunner of the developments which lead to modern experimental science. This expansion of interest was no doubt related to the 700th anniversary of the death of Grosseteste, but also must be connected to the discovery of the connection between Grosseteste and Roger Bacon, who had been somewhat lionised by historians of the nineteenth century as the persecuted harbinger of experimental science. Earlier in the century, Ludwig Baur made a decisive move in arguing for the importance of Grosseteste in the development of both experimental method and the mathematical description of the physical world (Baur 1917). Though as one might expect with such a development of interest, there was no consensus about the importance of Grosseteste, and there was considerable debate about the precise nature of that influence and indeed the essence of Grosseteste's scientific identity. In this regard, the most widely known work to examine the place of Grosseteste in the history of science is that of Alistair Crombie (1953). Crombie's central thesis is not merely that experimental method was developed within Grosseteste's school at Oxford but that this development stands in direct continuity to modern experimental science. The experimental method, and its allied term empirical observation, in modern science means something that is along the lines of the contrived or controlled observation of the effects of different variables. It is this contrived means of manipulating and observing the natural world that it is claimed Grosseteste developed in his writings and reflections on the physical world. Whilst there may be little or no difficulty in demonstrating Grosseteste's insistence that the physical world be described mathematically and on the basis of observation (see his remarks in *De Lineis, angulis et figuris*, for example), demonstrating that the observations which he refers to constitute experimental observation is quite a different matter. The term *experimentia* understood in its proper thirteenth-century context means nothing other than the observations made from experience, normal, everyday, common experience, as Bruce Eastwood convincingly demonstrates (Eastwood 1968). Moreover, it is not always clear that when Grosseteste refers to *experimentia*, he always means his own direct observation, for he also uses the term to enlist the support of observations recorded in his sources. McEvoy draws our attention to Grosseteste's *Notes on Physics* in this respect (1986). As such, Grosseteste's method remains firmly Aristotelian and bears no relationship to the controlled experiment central to modern science. On this view, Crombie has gone beyond the limits of his sources in claiming for Grosseteste the development of controlled experimental observation. It is far from clear, however, that considerations such as these settle the question of Grosseteste's place in the history of science—far from it, in fact, since criticisms of Crombie have merely increased and broadened the discussion. Alexandre Koyré, for example, whilst deeply critical of Crombie's assessment of Grosseteste's practice of *experimentium*, nevertheless sees in Grosseteste the beginnings of the mathematical description of the physical world which has been one of the distinguishing marks of modern science since at least the time of Newton (1957). For Koyré, it is Grosseteste's turning to mathematics that is the defining moment in determining his place in the history of science, for this love of mathematical or geometrical description marks the decisive turn from Aristotelian empiricism. Grosseteste's *De Lineis* begins with the opening sentence 'The value of considering lines, angles and figures is very extensive, since it is impossible to understand natural philosophy without them.' We might paraphrase this view as 'it is impossible to understand the physical world without mathematics'. Herein then, perhaps, lies Grosseteste's place in the history of science, not as the progenitor of experimental method but in making a decisive step towards the naturalistic, mathematical description of the physical world. For Grosseteste, mathematics is no mere abstraction from the world; it is rather the very nature of that world—not for Grosseteste the distinction that became so important during the Reformation between an abstract mathematical description of the, in this case heliocentric, universe used as an aid to calculation and that same mathematics claimed as an actual description of the physical world. Grosseteste anticipates Kepler's sentiment *Ubi materia, ibi geometria* by some four centuries.

This still leaves open the question of the relationship between Grosseteste's Christian faith and this mathematical description of the world. Is it possible that this mathematical innovation is connected with Grosseteste's faith? McEvoy believe that it is. According to him, the step towards mathematical description of the world derives directly from Grosseteste's belief in a creator God who orders the universe according to precise calculations (1986). According to McEvoy, Grosseteste's conception of God removed him from the conceptual world of ancient Greece, allowing him to conceive of the unity of the world in the service of humanity. This faith is nothing more and nothing less than a belief in the account of Creation given in Genesis and expounded through the Church fathers, most prominently St Augustine, but it led Grosseteste to further develop the naturalistism that informed the Greek conception of the heavens. In a sense Grosseteste's conception of God was deeply traditional, laying stress on the infinite power and wisdom of God as Creator, but in the context of his mathematical developments this old idea is given new content and the conviction of the rationality of the world is worked out for the first time in terms of mathematics and geometry. Perhaps, it is this new grounding for the conviction of the rationality of the world, a *sine qua non* for the development and practice of experimental science that marks Grosseteste's real significance in the history of science.

Lincoln, OR — Mark Hocknull
August 2015 — Jack P. Cunningham

References

Baur, L. (1917). *Die Philosophie des Robert Grosseteste, Bischofs von Lincoln (1253)* Münster iW: Verlag der Aschendorffschein Buchhandlung.

Crombie, A. C. (1953). *Robert Grosseteste and the Origins of Experimental Science 1100–1700.* Oxford: Clarendon Press.

Eastwood, B. S. (1968). Medieval empiricism. The case of Robert Grosseteste's optics. *Speculum, 43*, 306–321.

Eddington, A. S. (1939). *The philosophy of physical science.* Cambridge: Cambridge University Press.

Koyré, A. (1957). *Die Urspünge der modernen Wissenschaft.* Deutungsversuch Kiepenheuer et Witsch.

McEvoy, J. (1986). *The philosophy of Robert Grosseteste.* Oxford: Oxford University Press.

Acknowledgements

The editors of this book would like to thank Bishop Grosseteste University, Lincoln, for hosting the conference in July 2014, from which the idea for this volume first emerged, and for their continuing support for the pursuit of knowledge of all things Grossetestian. They would also like to acknowledge and thank *The Society for the Study of Medieval Languages and Literature* for their generous sponsorship of this event which allowed a number of postgraduate students to attend and provide us with their much valued contributions. We are also indebted to the *Montgomery Trust* who facilitated the attendance of Dr. Christopher Southgate whose paper proved to be one of the highlights of the proceedings. Our debt of gratitude to the *Ordered Universe Project* is, of course, enormous. Not only did they inspire the scientific and religious theme of the conference but papers from members of their team provided at the time a highly stimulating input and now some exceedingly important contributions to the volume that we have before us. It is thanks to the academic bravery and imagination of the progenitors of this venture and of the numerous scholars who have contributed to their undertakings that we are now entering an exciting and, we are certain, tremendously fruitful episode in the study of Robert Grosseteste.

Contents

Part I Rainbows, Light and Optics

1 **Unity and Symmetry in the *De Luce* of Robert Grosseteste** 3
Brian K. Tanner, Richard G. Bower, Thomas C.B. McLeish, and Giles E.M. Gasper

2 **Grosseteste's Meteorological Optics: Explications of the Phenomenon of the Rainbow After Ibn al-Haytham** 21
Nader El-Bizri

3 **Robert Grosseteste and the Pursuit of Learning in the Thirteenth Century** 41
Jack P. Cunningham

4 **All the Colours of the Rainbow: Robert Grosseteste's Three-Dimensional Colour Space** 59
Hannah E. Smithson

Part II Purity: Physical and Spiritual

5 **Medicine for the Body and Soul: Healthy Living in the Age of Bishop Grosseteste c. 1100–1400** 87
Christopher Bonfield

6 **The Corruption of the Elements: The Science of Ritual Impurity in the Early Thirteenth Century** 103
Sean Murphy

Part III Robert Grosseteste and Roger Bacon

7 **From *Sapientes antiqui* at Lincoln to the New *Sapientes moderni* at Paris c. 1260–1280: Roger Bacon's Two Circles of Scholars** 119
Jeremiah Hackett

8 The Theological Use of Science in Robert Grosseteste and Adam Marsh According to Roger Bacon: The Case Study of the Rainbow 143
Cecilia Panti

9 Laying the Foundation for the Nomological Image of Nature: From Corporeity in Robert Grosseteste to Species in Roger Bacon 165
Yael Kedar

Part IV Infinities and Transcendentals

10 Robert Grosseteste on Transcendentals 189
Gioacchino Curiello

11 A Theoretical Fulcrum: Robert Grosseteste on (Divine) Infinitude 209
Victor Salas

12 The Fulfillment of Science: Nature, Creation and Man in the *Hexaemeron* of Robert Grosseteste 221
Giles E.M. Gasper

Part V Science and Faith: Some Lessons from the Thirteenth Century?

13 *Intelligo ut credam, credo ut intelligam*: Robert Grosseteste Between Faith and Reason 245
Angelo Silvestri

14 Can Science and Religion Meet Over Their Subject-Matter? Some Thoughts on Thirteenth and Fourteenth-Century Discussions 263
Dónall McGinley

15 Medieval Lessons for the Modern Science/Religion Debate 281
Tom McLeish

Index 301

Part I
Rainbows, Light and Optics

Chapter 1
Unity and Symmetry in the *De Luce* of Robert Grosseteste

Brian K. Tanner, Richard G. Bower, Thomas C.B. McLeish, and Giles E.M. Gasper

1.1 Introduction

The treatise on light (*De luce*) of Robert Grosseteste, was written sometime between 1200 and 1225, the latter date having gained most recent consensus (Panti 2011). A date as late as 1240 was suggested by Sir Richard Southern, following Servus Gieben (Panti 2013a; Southern 1992). If this attribution is correct, it is amongst Grosseteste's mature scientific treatises, written at or around the same time as the *Commentary on Posterior Analytics* and as such the *De luce* reflects a significant influence of Aristotle's scientific thinking. It is arguably the best known of Grosseteste's works. Its model of an expanding universe stimulated speculation as to whether Georges Lemaître in 1927, who was a Catholic priest, was aware of Grosseteste's thinking when he introduced the modern 'Big Bang' model of cosmology (Lemaitre 1927; Panti 2011). In the *De luce* Grosseteste develops the consequences of his metaphysics of light towards a physics of light, introduced to explain the stability of solids, into a complete cosmogony. This connection between the perfect heavens and the imperfect earth is an astonishing intellectual feat, rooted on the premise that there exists a unity in the fundamental explanations of the causes of natural phenomena. It is underpinned by the principle of the uniformity of nature (Crombie 1953). The principle forms the basis of the predictability of nature, contrasting the Platonic view in which the observed world is a shifting incomprehensible shadow of an ideal, perfect world.

B.K. Tanner (✉) • R.G. Bower • T.C.B. McLeish • G.E.M. Gasper
Durham University, Durham, UK
e-mail: b.k.tanner@durham.ac.uk; r.g.bower@durham.ac.uk; t.c.b.mcleish@durham.ac.uk; g.e.m.gasper@durham.ac.uk

J.P. Cunningham, M. Hocknull (eds.), *Robert Grosseteste and the pursuit of Religious and Scientific Learning in the Middle Ages*, Studies in the History of Philosophy of Mind 18, DOI 10.1007/978-3-319-33468-4_1

1.2 Unity and Breakdown of Complex Problems into Testable Components

His belief in the unity underlying all physical phenomena enabled Grosseteste to exploit the technique, ubiquitous among modern physicists, of breaking down a complex problem into small, testable components. For example, in his treatise 'On the Rainbow' (*De iride*), he subdivides the passage of light through a cloud and the associated mist of rain into the passage of light across the boundaries between these various regions. After discussing the phenomenon of refraction in some detail, he goes on to state very specifically:

> Therefore, in accordance with what was said before about the refraction of rays and the size of the angle of refraction at the interface between two transparent media, solar rays must be refracted first at the interface between the air and the cloud and then at the interface between the cloud and the mist (Lindberg 1974; Baur 1912).[1]

He is then able to reason how the light is refracted at these interfaces, based on observations of refraction of light at boundaries between air and dense materials. The complex problem of formation of the rainbow is broken down into discussion of observable phenomena. Grosseteste understood the principles of geometrical optics and perspective very well, as illustrated in the first section of *De Iride*. He divided the subject into three areas, saying that;

> The first part [of perspective] is exhausted by the science we say deals with sight; the second by that which we call 'on mirrors'. But the third part has remained untouched and unknown among us until the present time (Baur 1912).

He then asserts that, 'it is to this third part of optics that the study of the rainbow is subordinated' (Ibid.).

1.2.1 Symmetry and Tests of Refraction

In the subsequent detailed discussion of the phenomenon of refraction, Grosseteste makes an assertion about the magnitude of refraction at a boundary between different materials. His law of refraction, namely that, with respect to the interface normal, the angle of refraction is half of the angle of incidence, despite being extremely elegant and symmetric does not withstand detailed examination. About 50 years after Grosseteste was writing, Witelo followed Ptolemy (Smith 1999) and Alhazen (Smith 2010) in recording precise measurements of the refraction of light between air and water (Fig. 1.1). Witelo, who described the experiments in great detail in his *Perspectiva* (or *Optica*), was not able to express his results in simple mathematical terms but nevertheless the data of Ptolemy (Smith 1982) and Witelo (Risner 1572; Baeumker 1908; Crombie 1953) are astonishingly good, even by

[1] The translations from the *De iride* by Sigbjørn Sønnesyn: private communication.

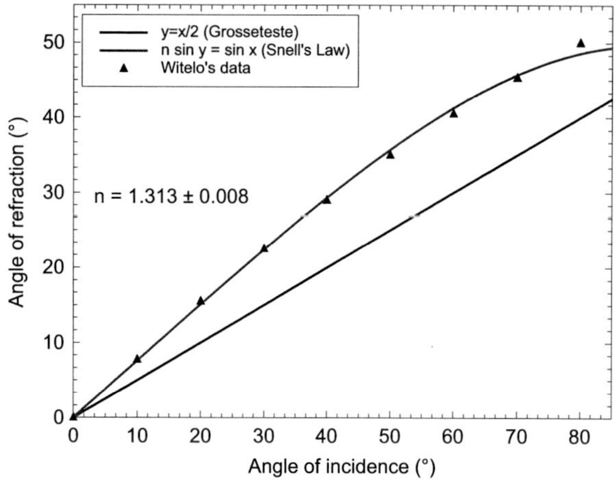

Fig. 1.1 Witelo's thirteenth century measurements of refraction at an air/water boundary, compared with Grosseteste's law of refraction and Snell's Law

modern standards. As seen in Fig. 1.1, there is excellent agreement with the modern theory known as Snell's Law.[2] We have numerically-fitted Witelo's data to a smooth curve generated using Snell's Law, enabling a refractive index n of water to be extracted from the measurements of Witelo and Ptolemy. The value obtained is $n = 1.313 \pm 0.008$. Considering the limitations on light sources and machining precision in the period at which these measurements were taken, the precision, determined by the spread of data points about the smooth numerically generated curve and given as the '$\pm$' number, is impressively high. The accuracy is also high, a modern value for the refractive index of water being 1.33299 ± 0.00001.[3] These data do suggest that the type of careful systematic experiment which we would recognise as the hallmark of modern laboratory science, was being conducted in the second half of the thirteenth century.

Examination of Fig. 1.1 might suggest that Grosseteste never made measurements to test his assertion that the angle of refraction bisects the angle of incidence, as the straight line predicted by this law diverges very substantially from Ptolemy and Witello's experimental data. In assessing Grosseteste's apparent lack of experimental measurement, it is, however, salutary to consider a similar measurement of the refraction between air and glass. Here the refractive index is higher, and the discrepancy becomes rather less (Fig. 1.2). The measurements were made using a glass block standing on a sheet of paper with a pencil and a ruler. The numerical fit to Snell's Law yields a refractive index of 1.53 ± 0.02, typical of a modern optical glass.

[2] Snell's Law, based on the wave nature of light, states that the sine of the incidence angle i and the sine of the refracted angle r are related to a property of the medium (called the refractive index n) by $\frac{\sin i}{\sin r} = n$. In Francophone countries Descarte's name is attached to this relationship, although it was first described by Ibn Sahl of Baghdad in 984.

[3] Quoted value is at 20 °C and a light wavelength of 589.3 nm; *Handbook of Chemistry and Physics* 52nd Edition (1972) The Chemical Rubber Co. p E203.

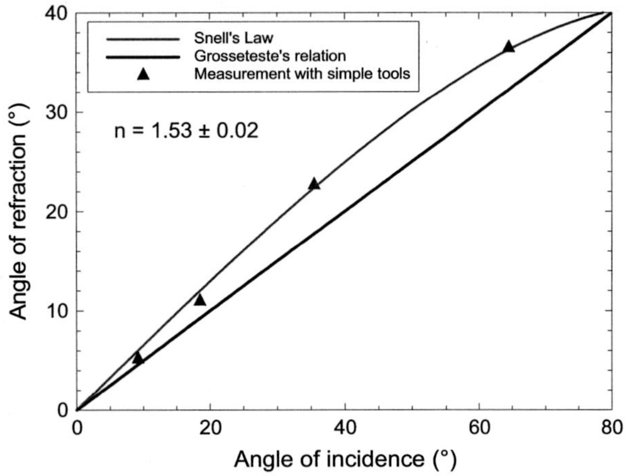

Fig. 1.2 Modern measurement, with simple experimental tools, of refraction at an air/glass boundary compared with Grosseteste's law of refraction and Snell's Law

Although it is probable that Grosseteste arrived at his rule for the refractive angle bisecting the incidence angle by appealing to the essential simplicity and symmetry of natural phenomena, less than careful measurements will give credence to the model if the measurements are made for the air/glass interface. (Although it is uncertain that a glass block whose sides were sufficiently parallel to perform this experiment will have been available in the early thirteenth century, rock crystal [quartz] of sufficient size with naturally parallel faces will certainly have been available. As the refractive index of quartz is 1.54, the suggestion that the discrepancy may not have been recognised remains credible.) In the *De Iride* Grosseteste hints at both approaches, arguing:

> However, what in this way determines the size of the angle in the fraction of the ray is shown to us through experiences [or experiments] similar to those through which we learn that the reflection of a ray on a mirror is at an angle equal to the angle of incidence. This same fact is made manifest to us by that principle of natural philosophy, that 'all operations of nature are in the most complete, most ordered, shortest, and best way possible for it' (Baur 1912).[4]

The rule had the simplicity of that of Ptolemy which stated in effect that the ratio of the incidence to refracted angles was constant (Crombie 1953). Grosseteste does not mention Alhazen's caveat that this ratio is not in fact constant and that Ptolemy's measurements do not support simple proportionality.

This appeal to the simplicity of natural laws is illustrative of Grosseteste's approach to the economy of premise that is found by invoking principles of symmetry. While the credit often goes to William of Ockham (Maurer 1978), we

[4] It is noteworthy that this statement falls into a long-running development of this idea. In modern optics, Fermat's Principle makes a similar claim, namely that light follows the path between two points for which it takes the shortest time. From Fermat's Principle, it is easy to prove Snell's Law (see above). The extension of the principle from classical physics ideas into quantum mechanics leads to the Feynman path integral, a standard tool in particle physics.

find that about 100 years earlier, Robert Grosseteste was propounding the principle that where there are several possible explanations, all of which save the appearances, the preferred explanation is the one that invokes fewest assumptions.[5] In the commentary on Aristotle's *Posterior Analytics* we find him arguing that:

> That is better and more valuable which requires fewer, other circumstances being equal, just as that demonstration is better, other circumstances being equal, which necessitates the answering of a smaller number of questions for a perfect demonstration or requires a smaller number of suppositions and premises from which the demonstration proceeds... Similarly in natural science, in moral science and in mathematics the best is that which needs no premises [i.e. immediate perception of truth without the need for discursive reasoning] and the better that which needs the fewer, other circumstances being equal.[6]

This approach was a development of Aristotle's view of the efficiency of operation of natural phenomena, Grosseteste, in his treatise On Lines (*De lineis*) quotes Aristotle as saying, '...in Book V of the Physics, because nature operates in the shortest way possible. But the straight line is the shortest of all, as he says in the same place.'[7]

1.3 Spherical Symmetry of the Universe Arising from Light as the First Form

The concept of simplicity of physical laws lies behind Grosseteste's insistence on the role of mathematics, particularly geometry, in understanding the physical world. He saw in mathematics a tool to describe observations and correlate variations in the observed effects. This insistence on the role of mathematics resulted in his arguments being mathematically structured even though he had no mathematical notation at his disposal beyond rudimentary numerals. In the case of the *De luce* we have shown that it is indeed possible to translate his arguments into modern mathematical symbols (symbolic language) and solve numerically the resultant equations (Bower et al. 2014). We have found that his model of the role and behaviour of light does lead quantitatively to the remarkable conclusions that he reaches from a very simple set of premises.

It is noteworthy that Grosseteste begins his treatise on *light* with an analysis of a problem concerning the theory of *matter*. The property of extension, or alternatively the 'stability' of matter is an old, but not necessarily obvious, problem to

[5] But even Aristotle writes in his *Posterior Analytics*, I.25 'Let that demonstration be better which, other things being equal, depends on fewer postulates or suppositions or propositions.' (Barnes 1984).

[6] Translated (with emendation) in Crombie (1953). See also (Rossi 1981).

[7] 'Aristoteles V Physicorum, quia natura operatur breviori modo, quo potest. Sed linea recta omnium est brevissima, ut ibidem dicit' (Baur 1912). Translated in Crombie (1953).

explain.[8] The opening section of *De luce* contains a strong, if implied, critique of the pure classical atomism of Democritus and Lucretius. He rejects the continuum description of matter of Aristotle and Plato but, in identifying first matter as 'a simple substance without any dimension,' Grosseteste points out that, in the absence of much more complexity, a theory of matter that has it consisting of however large a number of infinitesimal, indivisible atoms cannot account for extension (he requires an infinite number to do this).[9] We might illustrate his point by a classical thought-experiment: solids composed of however many billions of point-like particles would simply pass through each other.[10] At this point Grosseteste appeals to a mathematical argument as a vehicle for his physics. This in itself was something of an innovation (though of course central to the way physics works today). He observed, in a quite detailed argument, that an infinite sum of infinitesimals may indeed result in a finite magnitude. To obtain a three dimensional solid from matter without dimension, he sought a corporeity or 'first form' that multiplied itself infinitely. This he identified as light.

The key to the rest of the treatise lies in this infinitely self-replicating property of light. After the mathematical justification of how matter can be stabilised, Grosseteste resumes by:

> Returning to my topic, then, I say that light by the infinite multiplication of itself made uniformly in every direction extends matter uniformly on all sides into a spherical form, with the necessary result that the outermost parts of this extension of matter are more extended and more rarefied than the innermost parts near the centre (Panti 2013b; Lewis 2013).

This conceptual leap is only made possible by Grosseteste's notion of the underlying unity of natural phenomena and the possibility of a unity of explanation, from the stability of matter to the formation of the whole observable universe. Specifically he states that:

> So light, which is the first form in created first matter, by its nature infinitely multiplying itself everywhere and stretching uniformly in every direction, at the beginning of time, extended matter [which it could not leave], drawing it out along with itself into a mass the size of the world-machine. . .Light, then [which in itself is simple] must, when infinitely multiplied, extend matter [which is equally simple] into dimensions of finite size.

---

[8] The use of the notion of 'stability' goes one step beyond the phenomenon of the 'extension' of matter. To explain the solid state in a modern paradigm, for example, it is required not only that fixed molecular positions are an equilibrium solution to their mutual force laws, but that this solution is stable with respect to small external perturbations. For example, classical point charges under electrostatic forces do not satisfy this requirement.

[9] Robert Grosseteste, *De luce*: '. . .cum tamen utraque, corporeitas scilicet et material, sit substantia in se ipsa simplex carens omni dimensione. . .[. . .despite the fact that both corporeity and matter are in themselves simple substances lacking any dimension]' (Panti 2013b; Lewis 2013).

[10] An analogy of a physical process we know about today might be found in the propagation of neutrinos, tiny sub-atomic particles produced in prodigious quantities during nuclear fusion processes in the sun. Neutrinos interact with normal matter only very weakly and as a result pass through the earth almost unimpeded.

In attempting, to identify general explanations by induction and from them arrive by logical deduction at new observable conclusions,[11] Grosseteste was able to argue coherently from his initial postulates relating to the structure and stability of matter to the structure of the cosmos.

A particularly beautiful feature of this scheme is illustrated in Grosseteste's statement that, '. . .by its nature light spreads itself in every direction in such a way that as large as possible a sphere of light is instantaneously generated from a point of light [provided nothing opaque stands in the way]' (Panti 2013b; Lewis 2013). The generation of the universe by this mechanism automatically results in spherical symmetry. Although nowhere does Grosseteste explicitly refer to the importance of this, the highest degree of symmetry possible, it underpins the Aristotelean concept of the sphericity of the celestial orbits. The issue of the behaviour of the wandering stars (planets) caused Grosseteste considerable puzzlement. He knew of the theory of Ptolemy, which introduced epicycles to explain both retrograde planetary motion and the apparent changes in the diameter of the Moon, and although he regarded this as being possible, he also believed that such motion could not correspond to physical reality[12] because the Aristotelean celestial spheres were concentric. Grosseteste did not appear to have ever resolved this conflict between observation, the associated mathematical theory needed to explain it and the elegance of Aristotle's model incorporating a single prime mover.

Nevertheless, his model in the *De luce* creates just the spherical symmetry required by the Aristotelean universe. Grosseteste realised that, as light drags matter outwards, the density must decrease as the radius increases. He did, implicitly, invoke a conservation law, many centuries before the concept of conservation laws became a fundamental tenet of science.[13] In order to make sense of his model of light, of form inextricably linked to matter, dragging matter outwards, he made the assumption that there is no new creation of matter during the process. In other words, he assumed conservation of mass (*molem*). Indeed, there is no new creation of matter at all in the whole cosmological model. Presumably, as a Christian, Grosseteste will have regarded matter, together with its first form, light, as being created by God 'in the beginning' in Genesis 1:3. The expanding universe model of Grosseteste just describes how this matter comes to be distributed through the universe. The spherical expansion, Grosseteste realised, could not go on for ever, although light was itself capable of infinite multiplication. Without making an explicit statement, he invoked the Aristotelian concept of the impossibility of a vacuum. If a vacuum is impossible, there must be a minimum density beyond which

---

[11] We note, incidentally, that to think in such a way now comes as second nature to modern scientists.

[12] Referring to *Almagest* explicitly in his *De sphere* (Baur 1912).

[13] In the mid-nineteenth century, Rudolf Clausius stated the First Law of thermodynamics thus: 'In all cases in which work is produced by the agency of heat, a quantity of heat is consumed which is proportional to the work done; and conversely, by the expenditure of an equal quantity of work an equal quantity of heat is produced' (Clausius 1850; trans: Truesdell 1980).

matter cannot be rarefied and this sets the boundary of the universe. Grosseteste asserted that at this minimum density, there is a 'phase change' (in modern parlance, or 'perfection' in his) of matter-plus-light and that this perfect state can undergo no further change, forming the first celestial sphere of the cosmos—the firmament. Outside of this nothing existed. The question of what is outside of the observable universe still remains impossible to answer today. Scientists avoid the conundrum by stating, correctly, that it is not a scientific question. A scientific question is one which is testable by physical observation. As we cannot make observations outside of the limits of the universe, we cannot know anything about what may be outside, a logic that has not changed between the thirteenth and the twenty-first centuries.

Once this first, spherically symmetric shell had formed, Grosseteste then argued that, as light must continually multiply, the perfected sphere must itself emit light, but of a new, different, kind (*lumen*). The perfected outer shell, consisting only of first form (*lux*) and first matter, emits light which propagates instantaneously towards the centre of the Universe. As it propagates, it sweeps up the (imperfect) matter, or body (*corpus*) and because light and matter are interconnected, the matter is compressed. Because the first sphere is perfect and cannot change its status, and because there cannot be space that is empty, the *lumen* it emits sweeps up and compresses the matter inside the sphere until the matter behind it reaches a critical density. At this point it becomes perfected, cannot undergo change and becomes the second of the heavenly spheres, which we take to be that of the fixed stars. Despite the celestial matter being capable of being perfected, *lumen* is intrinsically less subtle than the first form, *lux*. Even if in other texts, not concerned with first form and matter, he does use *lux* and *lumen* more interchangeably, Grosseteste is consistent within the *De luce* on his use of terms. First matter and form are the most rarefied, and result in the subtlest of bodies. *Lumen* is naturally less subtle and becomes less so as it propagates inwards, eventually becoming incapable of perfecting matter.

Grosseteste seems to have been aware that the critical density at which perfection occurred could not be the same in subsequent spheres and explicitly stated that the light (*lux*) present in the first sphere is doubled in the second (Bower et al. 2014). Although he does not give the exact expression for subsequent spheres, we have followed Grosseteste's mathematical introduction and interpreted his text as requiring that the density must exceed one of a series of quantized thresholds (that is, that the critical density in subsequent shells is a factor 1, 2, 3, 4…greater than the lowest possible density) and that the combined *lux* and *lumen* must be sufficient to perfect the matter. The spheres are perfected until the ninth sphere, that of the moon, whose *lumen* emission is not sufficient to completely perfect the spheres below which comprise the four elements (fire, air, water, earth).

In the treatise it is possible to identify seven physical laws, which although not formally stated, provide the basis for writing down Grosseteste's model using modern mathematical symbols. These include the interaction of light and matter, the critical criteria for perfection, and the re-radiation and absorption of *lumen*. As

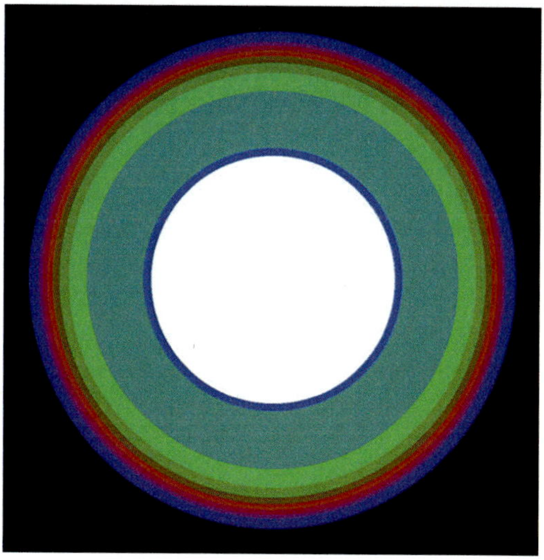

Fig. 1.3 Two-dimensional representation of the three-dimensional nature of Grosseteste's universe numerically simulated under conditions where nine perfected spheres are formed in addition to the imperfect sub-lunar sphere

Grosseteste is at pains to state, the lower celestial spheres, although perfected, are not as pure as the outer ones. He states:

> And the species and perfection of all bodies is light (though that of higher bodies is more spiritual and simple, while that of lower bodies is more corporeal and multiplied). Nor are all bodies of the same species, although they have been perfected by simple or multiplied light. . .(Panti 2013b; Lewis 2013).

Numerical calculations under conditions of spherical symmetry show that, subject to tight restrictions on the values associated with the above concepts and including a term relating to the partial opacity of the perfected spheres, it is possible to find solutions which contain nine perfected celestial spheres and one imperfect sub-lunar sphere (Fig. 1.3).

The spherical symmetry of the model is critical to the next step in Grosseteste's argument, his explanation of vertical motion. The four spheres of the elements, which Grosseteste treats as a single entity, are not completely actualised or perfected and hence they are subject to compression or rarefaction. The *lumen* in them thus inclines them to move towards or away from the centre of the universe (earth), movement away from the centre results in rarefaction and motion towards the centre results in condensation. The elements can thus be moved upwards and downwards, in contrast to the celestial material. Grosseteste says:

> But because the elements are incomplete, having a capacity for rarefaction and condensation, the luminosity that is in them either inclines away from the centre so as to rarefy or toward the centre so as to condense, and this is why the elements are naturally able to move upward or downward (Lewis 2013; Panti 2013b).

Objects made of the elements move because they are naturally moving to their proper place and they are moved through the change in the light within them. Such a model also explains the relative densities of the four spheres of fire air, water and earth in that order. Referring to the sub-lunar material, it follows that:

> . . .the highest part of this mass, though made fire by its dispersal, was not dispersed as much as possible and still remained elemental matter. And this element too, begetting luminosity from itself and concentrating the mass contained below itself, dispersed the outermost parts of it, though with a dispersal less than of fire itself, and in this way produced air (Ibid.).

In this scheme, the spherical symmetry with the sphere of the earth as the centre of the cosmos provides an explanation for motion. Dense objects fall towards the centre of the earth as the higher density of *lumen* in them moves them towards the centre of the universe.[14] Finally, we note that Grosseteste's idea of actualisation or perfection enables him very simply to explain why the motion of the celestial spheres is tangential and not radial. Because such matter is not any longer receptive to rarefaction or compression, the light in them does not cause rarefaction or compression of the matter towards or away from the centre of the universe. The only motion possible of the celestial spheres is circular motion, driven by the 'intellectual moving power' (*motiva intellectiva*).

1.3.1 Unity of Celestial and Terrestrial Matter

Grosseteste's unifying explanation of cosmic structure does not distinguish between the origins of heavenly and terrestrial matter. It is all the same but, in modern terminology, it is in a different phase.[15] The phase change results in different properties, but it is essentially the same constituent. Unlike Aristotle, he does not need to invoke the concept of different types of matter. Matter is intrinsically the same, but because of its differing light content, it behaves differently. Since, '. . .it is clear that every higher body in respect of the luminosity begotten from it [every higher body] is the species and perfection of the following body.' Then, he reasons, 'Earth, in contrast, is all higher bodies by the collection in it of the higher luminosities' (Panti 2013b; Lewis 2013).

Aristotle had built a simple geometrical model of the Universe as a means of explaining observable phenomena, but he did not attempt a grand unity of synthesis such as Grosseteste envisions. Grosseteste's inspiration for the physical hierarchy

[14] Grosseteste was aware of the rotundity of the earth; for example, in *De Sphera* he quotes observations to this effect by Ptolemy and Thabit. In that work, he notes that all the stars revolve around the Pole Star in circles, the circle diameters being smaller the nearer the stars are to the fixed Pole Star, thus demonstrating the sphericity of the universe.

[15] The concept of actualisation or perfection resonates with modern ideas of phase changes, sometimes structural, sometimes electronic, which result in markedly differing behaviour of matter as a result of, for example, a small change in temperature. The first order phase transition in water, which will have been as easily observable in the thirteenth century as now, results in the dramatically different properties of ice and liquid water when the temperature changes by an infinitesimally small amount.

of inherited properties may well have included his reading of Avicenna (who in turn was strongly influenced by al-Farabi). In his *Metaphysics*, following Aristotle, Avicenna associated an 'intelligence' with each sphere, but in a departure from the Philosopher, described a sequential begetting of these intelligences 'from the outermost one in' (Morewedge 1972). Just as in his identification of Avicenna's 'first form' with light itself, the development of a hierarchy of sequentially generated intelligences into a hierarchy of sequentially generated material spheres exemplifies Grosseteste's movement from metaphysics to physics, and in this case too his debt to Islamic commentators on Aristotle.

By inverting the argument, Grosseteste reaches the remarkable conclusion that,

> ...the luminosity of any celestial sphere [may be] drawn out from earth into act and operation, and so from earth, as if from a kind of mother, any god will be procreated (Panti 2013b; Lewis 2013).

He argues that within the earth all necessary constituents exist to recreate the whole cosmos, another example of his search for unification. This 'holistic' feature enables the curious passage preceding this statement to be understood simply in terms of poetic embellishment. Indeed, it is joyful word-play, showing his dexterity in the manipulation of and quotation from his sources. When he notes that the earth 'is named Cybele as if cubele from a cube [that is from solidity],' (Ibid.) the words *as if* are crucial and remove any suggestion that the subsequent comment concerning the earth's density is a logical deduction. On the basis of his model, Grosseteste is playing with words in the context of concepts from classical poetry. Behind the poetic imagery, however, is an extraordinary claim. This is the statement that the perfected heavenly spheres are not just of the same substance as the imperfect terrestrial material that we find around us, but that perfected bodies could, in principle, be created out of that earthly substance. Everything needed to create the celestial spheres is available on earth. Here is a unity of concept that represents a high point in Grosseteste's thought.

1.4 Symmetry and the Number of Celestial Spheres

At first sight, the final section of the treatise does not connect well to the earlier text. Suddenly the discussion turns from the consequences of light being the first form into an argument, apparently ad hoc and based on numerology, concerning the number of celestial spheres. In an otherwise extraordinarily tautly argued exposition, which follows so logically that we have been able to translate his statements into mathematics and numerically compute the consequences of his model (Bower et al. 2014), this final section may be seen to fade away into speculation. Sir Richard Southern, for whom the treatise was, 'one of the most lucid and brilliantly conceived pieces of writing of Grosseteste's later years,' continued to comment. 'Yet it must also be observed that, like much else that he wrote, it tails away into a rather chaotic and unintelligible sequel in its final paragraphs' (Southern 1992).

After building a magnificent model of the observable universe, the use of numerical arguments to show the necessity of just nine celestial Aristotelian spheres in addition to the imperfect sub-lunar sphere, might appear to be an afterthought that is not of the same intellectual rigour. However, through our mathematically-assisted reading it becomes seen as a natural part of the flow of the text, completing Grosseteste's attempt to provide a fundamental framework for the properties of matter by accounting for the number of the celestial spheres. Careful inspection reveals that it exhibits all the taut analysis characteristic of the earlier sections of the treatise. Indeed, it is based on principles of symmetry that find resonance in fundamental physics to this day.

With the support of numerical simulation, we have shown that Grosseteste's model predicts a different number of celestial spheres, depending on the initial starting conditions and the optical properties of the perfected and unperfected material. By selecting appropriate parameters, we were able to demonstrate a stable numerical calculation resulting in nine perfected spheres plus the one imperfect sphere of the elements. Without such tools, Grosseteste was unable to make the model predict the number of spheres at which the process of perfection would stop. He knew that there needed to be ten spheres but the model could not determine that number. Therefore he had to take a different approach. He appealed once again to the simplicity of structure underlying natural phenomena. Interestingly, Grosseteste starts his argument by discussing the properties of the highest of the celestial spheres on the grounds that this is the simplest. 'Now, in the highest body—which is the most simple of bodies—we can find four things; namely, form, matter, composition, and the composite' (Panti 2013b; Lewis 2013). The identification of four components of this the most simple of bodies draws on ideas from the Arabic scholar Albumasar (Abu Ma'shar al-Balkhi) and Daniel of Morley, cited by Panti (2013b), Abū Ma'săr (1995) and Maurach (1979).

Grosseteste sees form as the most simple of these qualities and allots to it the number one. The unity is of attribute, not dimension. Matter has two characteristics, namely 'the capacity to take on impressions and its capacity to retain them.' Capacity to take on, or receptiveness of, impressions is a concept that does not map directly onto modern scientific thinking. The nearest scientific equivalent is 'hardness',[16] which relates to a variety of tests used in materials engineering and geology. Receptiveness to, or capacity to retain, impressions may refer to viscoplastic behaviour of substances such as wax.[17] These yield on the application

[16] There are three common types of hardness test. The earliest to be developed was the scratch hardness test of Moh, used extensively in geology. Indentation tests involve driving a sharp pointed object into the specimen surface with a constant load and measuring the dimension of the indent produced. The rebound test measures the 'bounce' of a diamond-tipped hammer dropped from a fixed height onto a material. The various tests all have their own unique hardness scales.

[17] A viscoelastic material returns to it initial state very gradually over time. A viscoplastic material never returns to its initial state. The image of wax is perhaps the most common analogy used by medieval authors for reception of images.

of external pressure and substantial change of shape is possible, after which the substance does not return immediately to its original dimensions. As a result of the twofold nature of the characteristics of matter, Grosseteste ascribes to matter the number two. He then proceeds to discuss the components of composition, showing them to be three:

> The composition, however, has in itself the number three, because in it are formed matter, enmattered form, and the very attribute of composition (which is found to be a third item other than matter and form). In the composite also, form, matter, and composition, and that which belongs to the composite besides these three, are included under the number four (Panti 2013b; Lewis 2013).

In his association of composition with three and composite with four, there is an inconsistency with his earlier argument that ascribes the number two to matter. He invokes, for each of composition and composite, an additional attribute that is unique to the quality itself, in contrast to his discussion of matter where no such attribute is present. Thus, he arrives at the numerical sequence 1, 2, 3 and 4 relating to the four fundamental characteristics of the most simple of bodies.

If they ever existed, no diagrams associated with the *De luce* survive and he does not refer to a geometric representation, the sequence does have underlying symmetry. If we consider close packed circles (or a two-dimensional representation of spheres on a flat surface, we find that for close packing, a triangle such as shown in Fig. 1.4 is formed.[18] Working upwards from the bottom, the first circle touches two circles in the second row. These two circles touch three in the third row and they subsequently touch four in the top row. This is the sequence that Grosseteste is trying to rationalise by invoking the concept of the 'very attribute of composition' and 'that which belongs to the composite'.[19] We note that the sequence forms a hexagonal array with sixfold symmetry in the plane.[20] When the attributes are arranged (Fig. 1.5) in a similar array as are the circles of Fig. 1.4, the apparently forced sequence attains a simplicity which is characteristic of Grosseteste's approach to explanation. It is then only a small step to argue that $1+2+3+4=10$ and that therefore ten must be the perfect number for the universe. There is then a rationale behind the nine perfected celestial spheres plus the one sub-lunar imperfect sphere.

[18] An inverted form of Fig. 1.4 is to be found in Kepler's 1611 paper on the snowflake, *Strena Seu de Nive Sexangula* in which he discusses close packing of spheres. Kepler's Conjecture states that hexagonal and its related face centred cubic close packing results in the highest packing density possible.

[19] What makes various types of material different is a property which is different to those of form and matter but is required to describe composition. The nearest modern analogy might be allotropes of elements, such as diamond and graphite where the same type of atoms are bound together in different configurations, resulting in hugely different physical properties.

[20] The scheme shown is exactly that of hexagonal close packing of spheres which results in the crystal structures of the elements tin and zinc. The crystallographic symmetry arises simply from the modelling of each atom as a rigid sphere and requiring close packing. It is an identical sequence to that discussed by Kepler.

Fig. 1.4 Geometric representation of close packing of circles showing the sequence of 1, 2, 3 and 4 rows

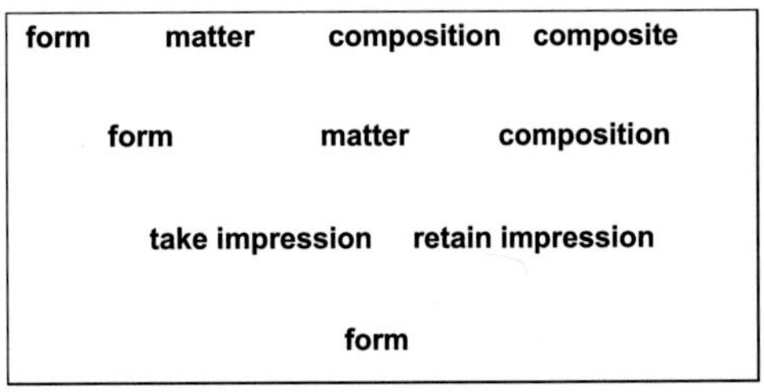

Fig. 1.5 Grosseteste's characteristics set out in the geometric representation of Fig. 1.4

As if to try and reinforce the point, Grosseteste then appeals to characteristics of music. In order to persuade the reader that this approach is reasonable he connects the five ratios generated by the numbers 1, 2, 3, and 4 with the mathematical relations underlying music. He asserts: '. . .only the five ratios found in these four numbers [one, two, three and four] . . . . are harmonious in musical measures, gestures and rhythms'(Panti 2013b; Lewis 2013). The five ratios that can be generated by the four numbers are $^{1}/_{2}:^{1}/_{3}:^{1}/_{4}:^{2}/_{3}$: and $^{3}/_{4}$. In the sixth century BCE, the Pythagoreans showed that there was a fundamental relation between musical pitch and the length of a vibrating string. When the string is divided into the proportions of the five ratios, the octave, the twelfth, the double octave, the fifth and the fourth musical intervals are generated. These were the perfect intervals of the quadrivium described by Boethius, who was the main transmitter of Pythagorean ideas to the late antique and medieval West (Boethius 1989). Similarly, the fundamental rhythmic metres found in music are these four numbers. There is an urgency and forcefulness about a rhythm consisting of a series of single strong beats of equal emphasis. A more dainty, tripping rhythm is associated with a strong pulse followed by a single weak one and much folk music has this cheerful duple time. A strong pulse followed by two weak ones is a dance rhythm, and a strong beat followed by three weaker ones is a strong martial rhythm with a driving sense of purpose and energy. These four rhythmic patterns dominate Western music to this day.[21] (Five or seven beat rhythms are mentally subdivided into groups of $2+3$, and $3+4$, beats respectively.) Alternatively, it has been suggested that Grosseteste may not have been considering musical rhythms, but rather have been referring to the rhythmic meters in poetry. It is not clear what he meant by the claim that bodily movements are also in the same proportions. In the much earlier work on the liberal arts *De artibus liberalibus*, Grosseteste says:

> And since the proportions of the human voice and the gesticulations of the human body are regulated by the same modulation as that by which sound and the motion of other bodies are, musical thought is subalternated not only to the harmony of human voice and gesticulation, but also of instruments and of those whose delectation consist in motion or sound and with these the harmony of celestial and non-celestial. And since the concordance of times and the composition and harmony of the lower world and of all things composed of four elements come from celestial motions, and, moreover, since it is necessary to find the harmony of causes in their effects, the study of music also extends to knowing the proportions of times and the constitution of the elements of the lower world, and even the composition of all the elements (Sønnesyn et al. forthcoming; Baur 1912).

The connection between the celestial and terrestrial motions is evident, the planets influencing the most appropriate time to undertake certain actions. Grosseteste's reasoning in the *De luce* will have been based on the connection of celestial properties, through the harmony of music, with motion on earth. Whatever is meant, he finds parallels between number associated with music and number associated with the heavenly spheres.

1.4.1 Modern Analogues

Despite presenting such arguments in different forms, modern physicists continue to invoke such aesthetic considerations in developing theoretical models. Indeed, a similar framework based on symmetry underpins the whole of the Standard Model of Particle Physics. In the 1960s experimental physicists discovered several families of high mass, extremely short-lived sub-atomic particles in the products of the collisions of energetic protons in particle accelerators. Within a family of heavy particles, named using Greek letters and collectively called baryons, the masses and electrical charges differed but the other properties were remarkably similar. Using principles of symmetry, initially called the 'Eight Fold Way', and the concept of 'strangeness' the Nobel laureate Murray Gell-Mann grouped certain of these particles (Fig. 1.6) in arrays of 4:3:2:1 in the manner demonstrated above. The four $\mathbf{\Delta}^*$ particles differ only in their charge and are placed in the top row. The three $\mathbf{\Sigma}^*$ particles, with negative, neutral and positive charge respectively, have a different mass to the $\mathbf{\Delta}^*$ particles and are grouped in a row of three below. Below them, again with different mass come the two $\mathbf{\Xi}^*$ particles. The interesting feature of the

[21] Bar notation in music is a recent attribute and did not exist in the thirteenth century. However the association of 1, 2, 3 and 4 beats to a bar maps directly onto the four rhythmic attributes in the text.

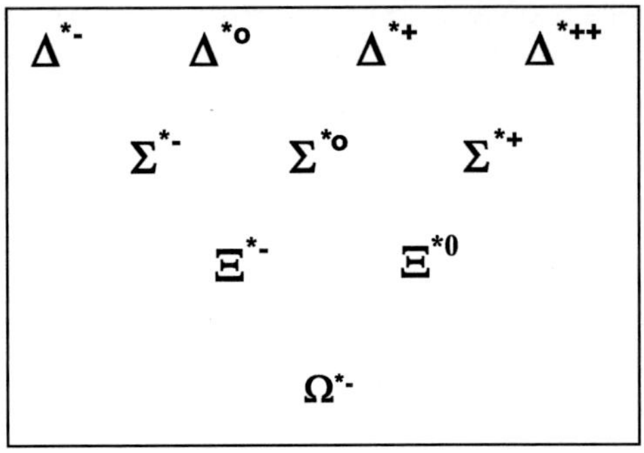

Fig. 1.6 Strangeness-charge diagram of the baryon decuplet in Gell-Mann's Eightfold Way format. Strangeness is zero on the *top row*, 1 on the second, 2 on the third and 3 for the *bottom* ($\mathbf{\Omega}^{*-}$) particle. Charge increases diagonally to the *right*

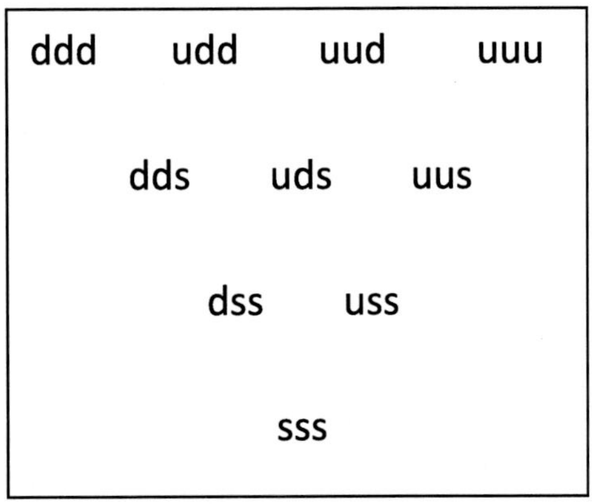

Fig. 1.7 Equivalent symmetry diagram of the particle properties constructed on the basis of composition of three types of quark. These are *up* (u), *down* (d) and *strange* (s)

family shown in the Fig. 1.6 was that to complete it, there should be the singlet state particle, the $\mathbf{\Omega}^{*-}$, at the bottom of the diagram. At that time, such a particle had not been found experimentally. However, from the symmetry of the diagram, the mass of particle could be predicted and indeed a detailed search in that mass region duly revealed the presence of the $\mathbf{\Omega}^{*-}$ particle, with the expected properties. The success in predicting the presence and properties of the $\mathbf{\Omega}^{*-}$ cemented the model, on which is based the Standard Model of particle physics, predicated on combinations of more fundamental particles called quarks. These quarks are of six 'flavours', three of which are named *up*, *down* and *strange*. These have charges of $+\frac{2}{3}$, $-\frac{1}{3}$ and $-\frac{1}{3}$ respectively. The equivalent diagram can then be constructed assuming that each of the baryons consists of three of these quarks and arranging them in a similar symmetric arrangement (Fig. 1.7). The recent (twenty-first century) experimental discovery of the Higgs boson at the CERN's Large Hadron Collider was a similar triumph of the power of prediction of the whole Standard Model based on an underlying symmetry.[22]

[22] It is somewhat ironic that the apparent symmetry in Fig. 1.6 is not reflected by the more fundamental underlying symmetry of three families of two quarks. Figure 1.7 shows that the

Modern physicists still appeal to aesthetic principles as an integral part of physical arguments. Elegance in mathematical formulation and physical statement is an underlying feature of the most powerful of scientific theories. Grosseteste was applying a similar criterion of elegance and simplicity to explain the number of spheres in the known universe. Viewed in this light, the problematic last section of the *De Luce* falls into proper perspective within an extraordinarily powerful cosmological exposition. Southern's assessment could not be further from the truth.

References

Abū Ma'sār. (1995). In R. Lemay (Ed.), *Liber introductorii maioris ad scientiam judiciorum astrorum, Traduction latine de Hermann de Carinthie*. Napoli: Istituto Universitario Orientale.

Baeumker, C. (1908). *Witelo, ein Philosoph und Naturforscher des XIII Jahrhunderts, Beiträge zur Geschichte der Philosophie des Mittelalters*. Munster: Aschendorff.

Barnes, J. (1984). *The complete works of Aristotle* (The revised Oxford translation, Vol. 2). Princeton: Princeton University Press.

Baur, L. (1912). *Die Philosophischen Werke des Robert Grosseteste, Bischofs von Lincoln*. Münster i. W: Aschendorff.

Boethius. (1989). *De institione Musicae* (C. M. Bower, Trans.). New Haven, CT: York University Press.

Bower, R. G., McLeish, T. C. B., Tanner, B. K., Smithson, H. E., Panti, C., & Lewis, N. (2014). A medieval multiverse?: Mathematical modelling of the thirteenth century universe of Robert Grosseteste. *Proceedings of the Royal Society of London A, 470*, 20140025.

Clausius, R. (1850). Über die bewegende Kraft de Wärme und dir Gesetze, welche sich daraus für die Wärmelehre selbst ableiten lassen. Annalen der Physik und Chemie, *155*: 368–395.

Crombie, A. C. (1953). *Robert Grosseteste and the origins of experimental science 1100–1700*. Oxford: Oxford University Press.

Lemaitre, G. (1927). Un Univers homogène de masse constante et de rayon croissant rendant compte de la vitesse radiale des nébuleuses extra-galactiques. *Annales de la Société Scientifique de Bruxelles, 47*, 49.

Lewis, N. (2013). Robert Grosseteste's De luce (An english translation). In J. Flood, J. Ginther, & J. W. Goering (Eds.), *Robert Grosseteste and his intellectual milieu* (pp. 239–247). Toronto: Pontifical Institute of Medieval Studies.

Lindberg, D. C. (1974). Robert Grosseteste: On the rainbow. In E. Grant (Ed.), *A source book in medieval science* (Vol. 1, pp. 385–388). Cambridge, MA: Harvard University Press.

Maurach, G. (1979). Daniel von Morley, Philosophia. *Mittellateinisches Jahrbuch, 14*, 204–255.

---

diagram is built up from two quarks of one family (u and d) plus one other (s). Nevertheless in the search for a deeper understanding of the operation of the Universe, use of symmetries plays an important role, even if these symmetries have persistently been found to be broken at some level. (The more fundamental symmetries break at low energy, but experimental particle physics is always working from conditions of low to higher energy as particle accelerators become more powerful, so we initially see the broken symmetry.) This, of course, does not negate the power of symmetry within Group Theory in the understanding of crystal structures; indeed the whole discipline of crystallography is based on principles of symmetry. However, great excitement is always generated when symmetry rules appear to be broken, as was the case for quasi-crystals, whose apparent fivefold symmetry seemed to contravene the symmetries permitted by Group Theory.

Maurer, A. (1978). Method in Ockham's Nominalism. *The Monist, 61*, 426–443.
McLeish, T. C. B., Bower, R. G., Tanner, B. K., Smithson, H. E., Panti, C., Lewis, N., & Gasper, G. E. M. (2014). A medieval multiverse. *Nature, 507*, 161–163.
Morewedge, P. (1972). The Logic of Emanationism and Ṣūfism in the Philosophy of Ibn Sīnā (Avicenna), Part II. *Journal of the American Oriental Society, 92*, 1–18.
Panti, C (Ed.). (2011). *Robert Grosseteste. La Luce: Introduzione, Testo Latino, Traduzione E Commento* (N. Lewis, Trans.). Pisa: Edizioni Plus.
Panti, C. (2013a). Robert Grosseteste and Adam of Exeter's physics of light, remarks on the transmission, authenticity, and chronology of Grosseteste's scientific *Opuscula*. In J. Flood, J. R. Ginther, & J. W. Goering (Eds.), *Robert Grosseteste and his intellectual milieu* (pp. 165–190). Toronto: Pontifical Institute of Medieval Studies.
Panti, C. (2013b). Robert Grosseteste's De Luce (A critical edition). In J. Flood, J. R. Ginther, & J. W. Goering (Eds.), *Robert Grosseteste and his intellectual milieu* (pp. 219–238). Toronto: Pontifical Institute of Medieval Studies.
Risner, F. (Ed.). (1572). *Opticae thesaurus*. Basel: Per Epicopios (contains Alhacen's and Witelo's *Perspectivae*. Facsimile reprint with forward by David Linberg, 1972. New York: Johnson Reprint Corp).
Rossi, P. (Ed.). (1981). *Robertus Grosseteste, Commentarius in Posteriorum Analyticorum Libros*. Florence: Leo S. Olschki.
Smith, A. M. (1982). Ptolemy's search for a law of refraction: A case-study in the classical methodology of "saving the appearances and its limitations.". *Archive for History of Exact Sciences, 26*, 221–240.
Smith, A. M. (1999). Ptolemy and the foundations of ancient mathematical optics: A source based guided study (Trans.). *American Philosophical Society New Series, 89*, 1–172.
Smith, A. M. (2010). Alhacen on refraction (Trans.). *American Philosophical Society New Series, 100*, 213–331.
Sønnesyn, S. (2015). *De artibus liberlibus—On the liberal arts* (Forthcoming publication with others).
Southern, R. W. (1992). *Robert Grosseteste: The growth of an English mind in medieval Europe* (2nd ed.). Oxford: Oxford University Press.
Truesdell, C. A. (1980). *The tragicomical history of thermodynamics 1822–1854*. New York: Springer.

Chapter 2
Grosseteste's Meteorological Optics: Explications of the Phenomenon of the Rainbow After Ibn al-Haytham

Nader El-Bizri

2.1 The Meteorological Optics of Robert Grosseteste

This study focuses on the meteorological optics of Robert Grosseteste with an emphasis on examining his explication of the phenomenon of the rainbow, while also aiming at situating his investigations comparatively in-between the optical inquiries of the polymaths Ibn al-Haytham (Alhazen; eleventh century CE) and Kamāl al-Dīn al-Fārisī (thirteenth century CE).

One principal influence on the scientific endeavours of Grosseteste is attributed to his adaptive reception and assimilation of Latin renditions of Aristotle's *Physics* (Ross 1936) and of transmitted commentaries on it. This is reflected in inquiries on motion (*kinêsis*), the influence of the four elements (*stoikheia*), and in investigations of the nature and comportment of physical light. Grosseteste's take on Aristotelian natural philosophy was nonetheless Neoplatonist in orientation and also impacted by teachings attributed to Pseudo-Dionysius, St. Augustine and Boethius. Grosseteste developed a metaphysical account of light that grasped divine illumination as the presence to the mind of eternal ideas and necessary truths. Human understanding is receptive of this illuminative providence, and sensory perception is a catalyst in activating the workings of reason. In consequence, reasoning also arrives at universals and indubitable truths through analyses of sense data. The metaphysical attributes of illumination evoke the effects of a *spiritual* light as an onto-theological phenomenon, which is intertwined with the *physical* light that is studied in optics (Grant 1974; Zemplén 2005) and that has the capacity to multiply and spread instantaneously in all directions (Zemplén 2005).

N. El-Bizri (✉)
American University of Beirut, Beirut, Lebanon
e-mail: Nb44@aub.edu.lb

J.P. Cunningham, M. Hocknull (eds.), *Robert Grosseteste and the pursuit of Religious and Scientific Learning in the Middle Ages*, Studies in the History of Philosophy of Mind 18, DOI 10.1007/978-3-319-33468-4_2

Based on a Neoplatonic interpretation, such illuminative phenomena are dimensionless in their multiplication as entailed by the process of emanation.

To elucidate the connection between the spiritual and the physical regions of being, Grosseteste did not consider matter (*hulê*) as mere potency and form (*eidos*) as the sole actuality. He argued that matter possessed some reality in itself, and that its potentiality was not mere passivity. Form completes and actualizes matter by granting extension to its utmost capacity, which is a necessary characteristic of corporeity as it is associated with *species* in a hierarchy of perfection. Nevertheless, he followed Aristotle by holding that the essence (*to ti ên einai*) of the hylomorphic compound that constitutes a given thing as an individuated primary substance (*ousia*) is its form (*Metaphysics* Z.7, 1032b1-14). While Grosseteste used an Aristotelian vocabulary of potentiality (*dunamis*) and actuality (*energeia*) as set primarily in Book Θ of Aristotle's *Metaphysics*, he nonetheless did so in a non-Aristotelian way as set out in his *Tractatus de potentia et actu* (Grosseteste 1912a). Despite the nuancing of the language in which potentiality is not strictly distinct from actuality, Grosseteste still emphasizes the causal motion of the potential to the actual (*Omnem autem potentiam praecedit actus naturaliter*), while distinguishing between passive and active potencies (*potentia activa et passiva* [*Tractatus de potentia et actu* 128–129]). Grosseteste followed accordingly a hermeneutic method of interpretation that was not uncommon amongst the lineage of commentators on Aristotle, and in particular as witnessed with the scholastic adaptations of the oeuvres of Averroës (Ibn Rushd) and of the legacy of Avicennism.

In *De luce*, Grosseteste stated that the first corporeal form (*formam primam corporalem* [*De luce* 51]) which some call 'corporeity' is a light (*lucis*) that diffuses itself in every direction instantaneously via spherical (*sphaera*) irradiation, and that it consequently extends matter via such diffusion (Grosseteste 1912e, 1942, 2011; Panti 2013). He also reiterated this view in his *De motu corporali et luce* (Grosseteste 1912c) by arguing that corporeal motion is a phenomenon of the multiplication of light (*motio corporalis est vis multiplicativa lucis* [*De motu corporali et luce* 92]).

Grosseteste grasped *lux* as the first corporeal form of a physical entity that corresponds to its corporeity and its three-dimensional materiality, while he took *lumen* to be a luminosity that is akin to a substance emitted by the celestial spheres and incorporated in physical bodies (*De luce*).[1] He noted in *De lineis angulis et figuris*[2] (Grosseteste 1912b) that the figure that is suitable for describing the propagation of the power of light is the sphere, since every agent emanates its power spherically and all around in every direction. He goes on by stating that this is shown by the manner in which it is possible to draw a line in a certain direction from an agent located at the centre, and in all directions from all the different

[1] Refer also to the authoritative interdisciplinary group publication of the Latin critical edition, annotated English translation, and analytic commentary of *De colore* as set in Dinkova-Bruun et al. (2013).

[2] This tract focused on the reflection and refraction of the ray of light (*De fractionibus et reflexionibus radiorum*) see also Miccoli 2001; Turbayne 1959.

positions, and therefore it is proper to use a spherical figure. He also holds that this aspect is in agreement with what the commentator Averroës (Ibn Rushd) says with regard to Aristotle's *De anima*, '*Ita dicit Commentator super secundum de anima*' (*De lineis angulis et figuris* 64).

In *De Luce*, light is taken to be corporeity (*corporeitas*) itself, and as the first corporeal form '*lux est ergo prima forma corporalis.*' It multiplies itself in an infinite number of times on all sides and spreads itself out uniformly in every direction as a finite (*finitae*) and proportional (*proportione*) extension of matter, '*extensio materiae*' (*De luce* 52). The form (species) and perfection of all bodies are therefore phenomena of light, '*et species et perfectio corporum omnium est lux*' (*De luce* 56–57).

Light in itself is a pure form that is neither extended nor corporeal or spatial. Through its natural tendency, it reproduces itself via the same species that are distinguishable by their different positions in space, and extends in three dimensions instantaneously into a finite sphere that produces extended matter. The science of matter qua physics is hence understood through the science of extension qua geometry and the science of light qua optics (Longeway 2007).

An inquiry about the properties of physical light touches upon themes in optical meteorology, such as the explication of the colouration of the rainbow. It is rather unclear what the disciplinary boundaries were for Grosseteste in determining the distinction and connection between an Aristotelian natural philosophy, a Neoplatonist theology of light, and a Euclidean-Ptolemaic science of optics. Grosseteste had at his disposal Euclid's *Optica* and *Catoptrica* (Ver Eecke 1959) Aristotle's *Meterologica*, al-Kindī's *De aspectibus*, but not necessarily Ptolemy's *Optica* (Lejeune 1956; Crombie 1953). There is no evidence that he had access to any of the Latin renditions of the influential book of optics of Ibn al-Haytham (Alhazen, d. c. 1041 CE), even though these would have been in circulation under the Latinate titles *Perspectiva* or *De aspectibus* since the twelfth century via channels of transmission from Toledo and Sicily.[3] In evoking the commentators on Aristotle in natural philosophy, Grosseteste cites Avicenna (Ibn Sīnā) and Averroës (Ibn Rushd) in various places.

The science of optics was onto-theological in scope for Grosseteste. This penchant in thinking left its mark on his mentoring of Franciscan scholars at Oxford (c. 1229–30 CE), (Little 1926) and possibly extended through the office of his Bishopric at Lincoln (c. 1235–53 CE). His transitional influence reached a next generation of opticians such as Roger Bacon and then Witelo (Grant 1974; El-Bizri 2010), albeit both had access to the fundamentals of Graeco-Arabic optics as embodied in the legacy of Ibn al-Haytham.

Grosseteste advocated what translates into a *species-theory* of light, which was akin to the atomist take on *eidola* (phantasms)[4] while also holding a Platonist thesis about the emission of images from the eye as these are coupled with irradiations from light-sources. His fascination with dioptrics carried a concern over practical

[3] The impact of Ibn al-Haytham's *Optics* reaches eventually the Renaissance perspectivists through its reception by Biagio Pelacani da Parma (d. c. 1416) as witnessed in the latter's *Questiones super perspectiva communi* (Biard et al. 2009; El-Bizri 2010).

[4] See Lucretius (1975).

applications, such as when he noted that a correct understanding of the principles of the refraction of light would help in reading minute letters from large distances. As if he is signalling an awareness of the potentials of the principles of this science to assist in generating lenses that facilitate vision at great distances. However, this is not surprising since the geometric modelling of lenses has had a longstanding history that stretched back to antiquity and found systemic refinements in dioptrics at the hands of polymaths such as Ibn Sahl from the tenth century in 'Abbasid Baghdad, and was further elaborated by Ibn al-Haytham in Book VII of the *Optics*.

Grosseteste partly relied in meteorological optics on Aristotle's meteorology in describing the formation of the rainbow in his *De iride* treatise (Grosseteste 1912d),[5] albeit without restricting his explanation to the Aristotelian claim that the rainbow results from reflections of sunlight rays on droplets of rain (*Meteorologica* III, 4, 373a35–375a8), or the manner this also figured in Avicenna's (Ibn Sīnā) commentary on this phenomenon in *Kitāb al-Shifā'* (*Book of Healing*, Part V, 5 cf.; Ibn Sīnā 1965).

Grosseteste reveals his geometric interest in conics besides spherics in explaining the use of instruments that facilitate vision at large distances. This might be due to the manner conic sections have been systematically deployed since the tenth century in dioptrics in terms of studying the geometrical properties of lenses. There is no indication in this regard that Grosseteste was aware of the theories that experimented with large glass models of individual rain droplets and the way they refracted and reflected light when placed in a *camera obscura* (Zemplén 2005) as for instance entailed by the research of Ibn al-Haytham and the critical and analytic commentaries this experimental work (referred to as: *al-i'tibār*) received later within Arabic and Latin scholarship.

2.2 The Rainbow

Grosseteste considered the phenomenon of the rainbow as a topic of research that concerned the optician qua perspectivist and the physicist qua natural philosopher, '*et perspectivi et physici est speculatio de iride. Sed ipsum quid physici est scire, propter quid vero perspectivi*' (*De iride* 72). However, while the optician/perspectivist seeks explications, the physicist focuses on natural facts. Grosseteste pictures himself as an optician more than being a physicist in his explanation of the occurrence of the natural phenomenon of the rainbow. He states that perspective is the science based on visual figures and that this is subordinate to the science based on figures containing radiant lines and surfaces, whether the irradiation is emitted by the sun, the stars, or some other radiant body, '*et haec subalternat sibi scientiam, quae erigitur super figuras, quas continent lineae et superficies radiosae, sive proiecta sint illa radiosa ex sole, sive ex stellis, sive ex aliquo corpore radiante*' (*De iride* 72–73).

[5] Additional studies on *De iride (De fraccionibus radiorum)* figure in Boyer (1954), Boyer (1958), Eastwood (1966), Lindberg (1966) and Turbayne (1959).

Grosseteste held that the emission of visual rays from the eyes is not imagined and is not lacking in reality. Rather, visual species issuing from the eye are akin to a substance that irradiates like a subtle non-consuming fire when coupled with the irradiation from an external shining body that completes the actualization of natural vision. Whilst the Aristotelian natural philosopher posits an intromission physical theory, the Platonist and the mathematician maintains that vision occurs by way of the emission of subtle light rays from the eyes, '*extramissionem*' (*De iride* 73). It is clear that Grosseteste follows the mathematical thesis that is attributed to Euclidean and Ptolemaic theories about the nature of vision, which also refers back to the emission thesis that is highlighted in Plato's *Timaeus* (45a–47a; esp. 45b–c). However, Grosseteste seems to aim at reconciling this non-Aristotelian position with a tangential commentary on Aristotle's *De generatione animalium* (V.l.781a.I-2. 23; V.l.781b.2-13). On his view, optics (qua natural perspective) has three principal divisions that correlate with the way the rays of light are transmitted. If the propagation of light was along a straight line through a transparent medium of a single kind, then its corresponding science is '*De visu*' that studies direct vision. If light is reflected, then the science that investigates it is '*De speculis*', which studies the reflection of light, as is the case with catoptrics. As for the passage of light across several transparent media of different kinds, at the junctions of these the visual rays are refracted to form angles, '*primam partem couplet scientia nominata de visu; secundam illa, quae vocatur de speculis. Tertia pars apud nos intacta et incognita usque ad tempus hoc permansit*' (*De iride* 73). This phenomenon corresponds with the third type of science that interested Grosseteste, but he did not assign it a name, even though in classical terms it refers to the study of the principles and instruments of the refraction of light as the optical science of dioptrics. Grosseteste was impressed in this regard by Aristotelian inquiries (Hero of Alexandria 1900; Boyer 1945–1946) and the manner they describe how distant objects appear close, or how they let nearby objects appear significantly small, or how a small object placed at a distance can appear as large as one desires. To explain such applied optical aspects, Grosseteste states that the visual ray penetrating through several transparent substances of diverse natures is refracted at their junctions in angular transmissions, '*radius visualis penetrans per plura diaphana diversarum naturarum*' (*De Iride* 74) is refracted at their junction in angular transmissions, '*...in illorum contiguitate angulariter coniunguntur*' (Ibid). He holds that if an object is placed in a vessel, and the observer is stationed at a position from which the object cannot be seen, the object will become visible when water is poured in, and hence that light traverses across transparent media that do not possess a homogeneous nature. A visual ray is interrupted and changes direction at the interface between two transparent media of different material kinds and of varying levels of purity in their diaphanous properties. The deflection happens as a mean between continuity and discontinuity in the propagation of light, but not in a rectilinear fashion, rather at an angle. Grosseteste shows that the amount of divergence from rectilinear rays that are joined at an angle can be explained as follows: imagine a ray from the eye incident through air on a second transparent medium, and extended continuously and rectilinearly into the second medium, and imagine another line that is perpendicular to the interface drawn into the depth of the medium from the point at which the ray is incident on the second transparent medium. Grosseteste holds that it would then be the case that the path of the ray in the second transparent medium is along the line bisecting the angle enclosed by the ray, which we have imagined to be extended continuously and rectilinearly, and the perpendicular line drawn into the depth of the second transparent medium from the point of incidence of the ray on its surface. Accordingly one would say that the angle of refraction equals half the angle of incidence; hence that they are divided into equal angles, '*dividentis per aequalia angulum*' (*De iride* 74; Lejeune 1957). The refracted ray of light upon entering a denser transparent medium from a subtler one bisects the angle between the projection of the incident ray and the perpendicular to the interface. The size of the angle in the refraction of a ray may be determined in this way, and this is similar to those who discovered that the reflection of a ray upon a mirror takes place at an angle equal to the angle of incidence (*De iride* 74–75). It seems however doubtful that Grosseteste would have formulated his half-angle law had he known Ptolemy's *Optica*. Grosseteste's half-angle law of refraction was determined not through measurement but rationally, on grounds of symmetry and brevity of action (Eastwood 1967).

In *De lineis angulis et figuris*, Grosseteste notes that refraction is twofold (*dupliciter*), 'when the second medium is denser than the first (*Quoniam si illud corpus secundam est densius primo*), the ray is refracted between the prolongation of the direction of incidence and the perpendicular drawn from the point of incidence in the second medium (*tunc radius frangitur ad dexteram et vadit inter incessum rectum et perpendicularem ducendam a loco fractionis super illud corpus secundum*). When the second medium is subtler, then the ray is refracted by way of receding from the perpendicular beyond the prolongation of the incident ray, '*Si vero sit corpus subtilius, tunc frangitur versus sinistrum recedendo a perpendiculari ultra incessum rectum*' (*De lineis angulis et figuris* 63).

Grosseteste observed that an object seen through several transparent media does not appear as it really is but appears to be situated at the intersection of the ray emitted by the eye, as it extends in a continuous straight line, and the line drawn perpendicularly from the visible object to the surface of the second transparent medium that is nearer to the eye. These are taken by him as being the principles of the 'third science in the division of perspective', namely as what is conventionally named 'dioptrics', which studies the refractive laws of light. The study of the rainbow is subordinate to this science of perspective, '*et huic tertiae parti Perspectivae subalternata est scientia de iride*' (*De iride* 75), namely as a science of refraction, or dioptrics.

Grosseteste argues that the rainbow could not be formed by means of solar rays falling in a straight line into the concavity (*concavitatem*) of a cloud,[6] for they

[6] The Aristotelian view is expressed in: Aristotle, *Meteorologica*, *op. cit.*, III.2-5. On the history of the theory of the rainbow (see Boyer 1959), and refer also to division 63.2 from the *De iride et radialibus impressionibus* (*On the rainbow and the impressions of irradiances*) of Theodoric of Freiburg (Theodoricus 1914).

would produce a continuous illumination of that cloud, and that the rainbow cannot also be formed by the reflection of solar rays from the convexity (*convexitatem*) of the mist descending from a cloud as from a convex mirror, in such a way that the concavity of the cloud would receive the reflected rays and the rainbow would thus appear, '*ut concavitas nubis recipiat radios reflexos et sic appareat iris*' (*De iride* 74–75). He notes that if this were the case, then the rainbow would not always be in the figure of an arc (*arcualis figurae*), and as the sun rises, the rainbow would become proportionately larger and higher, or as the sun sets, the rainbow would become smaller, and the senses testify in all cases to the contrary. Therefore, the rainbow must be formed by the reflection of solar rays in a misty convex cloud, with its inside as being concave, whilst being less than a hemisphere, even though it may appear to be hemispheric '*appareat in visu semisphaera*' (*De iride* 76). Given that the mist descends from the concavity of the cloud, it also must be a conical convex figure at the top, descending to earth, and therefore, it would be more condensed nearer to the earth than in its higher part (Lindberg 1966).

Grosseteste goes on by stating that there are four transparent media through which a sunray penetrates: firstly is the pure air that contains the cloud; secondly the cloud itself; thirdly the higher subtle mist coming from it; fourthly the lower dense part of that same mist. In accordance with the refraction of rays, and the size of the refractive angle at the interface between two transparent media, solar rays must be refracted first at the interface between the air and the cloud, and then at the interface between the cloud and the mist. By these refractions the rays converge in the density of the mist, as if from the vertex of a pyramid, and they spread out into a cone that expands in the opposite direction from the sun. The rainbow assumes the shape of an arc, with a vertex of its cone near the earth and expanding away from the sun. Around sunrise or sunset the rainbow appears semi-circular and larger, while when the sun is in other positions, the rainbow appears as a part of a semicircle; the higher the sun, the smaller would be the visible part of the rainbow. According to Grosseteste's take on Aristotle, the apparent size of the rainbow is due to variations in luminosity through varying multitudes of vapours.

2.3 Colour

When it comes to colours, Grosseteste holds that they consist of light intermixed with a transparent medium, '*cum autem color sit lumen admixtum cum diaphano*' (*De iride* 77), whereby the latter is diversified according to purity and impurity, and light is divided in a fourfold manner according to: brightness and darkness, and to multitude and paucity, '*claritatem et obscuritatem ... multitudinem et paucitatem*' (*De iride* 77). All colours are generated from the combinations of these six differential aspects, and the variety of colours in different parts of one and the same rainbow occurs due to the multitude and paucity of solar rays. Where there is a greater multiplication of rays, the colour appears clearer and more luminous, and where there is a smaller multiplication of rays, the colour appears more bluish and obscure. Moreover, the difference in colour between one rainbow and another arises from the purity and impurity of the recipient transparent medium, and from the brightness and darkness of the light impressed on the medium, '*diversitas vero unius iridis ad aliam in coloribus suis tum accidit ex puritate et impuritate diaphani recipientis*', and from the brightness and the darkness of the light impressed on the medium, '*tum ex claritate et obscuritate luminis imprimentis*' (*De iride* 77–78). If the transparent medium is pure and the light is bright, then the colour of the rainbow will be whitish and similar to light. But if the recipient transparent medium contains a mixture of smoky vapours, and the light is not very bright, as it is nearer the rising or the setting of the sun, then the colour of the rainbow will be less brilliant and darker. Similarly, the production of all the variations in colour of the variegated rainbow is due to other combinations of brightness or darkness of light, copiousness or scarcity of lighting, and purity or impurity of the receiving transparent medium.

In *De colore*, (Grosseteste 1912f) he also emphasized that colour is light incorporated in a transparent medium, '*color est lux incorporata perspicuo*' (*De colore* 78).[7] This statement is also noted in the *Hexaemeron* by defining colour as 'light incorporated in a humid diaphanous medium', '*lux namque incorporata in perspicuo humido color est*' (Grosseteste 1982).[8] The essence of whiteness consists of three aspects: the intensity qua abundance of light, its clarity, and the purity of the diaphanous medium (*cum enim albedinis essentiam tria constituant, scilicet lucis multitudo, eiusdemque claritas et perspicui puritas—De colore* 78–79). Grosseteste also hints at his method, or that which is expected in this line of inquiry, by stating that one who knows deeply and inwardly the principles of natural science and optics does so not only by reasoning but also by experience, '*verum etiam experimento manifestum est his, qui scientiae naturalis et perspectivae profundius et interius noverunt principia*' (*De colore* 79). Grosseteste offers herein an account of the essence of colouration as being a phenomenon of light that interacts with the nature of transparent media, and hence colours are not seen as being ontologically distinct from light like it was for instance the case with Ibn al-Haytham's speculations about this matter, which were later reformed in the research of fourteenth-century opticians who were based on critical revisions of his science of meteorological optics, like it was the case with Kamāl al-Dīn al-Fārisī and Theodoric (Dietrich) of Freiburg. Both modelled the raindrop experimentally in a *camera obscura*, whereby they represented it in an enlarged scale in the form of a transparent spherical glass vessel filled with water and subjected to controlled rays of light that refracted through it and reflected within it.

It is possible that Grosseteste conducted inquiries associated with his *De natura locorum*—the nature of places (Grosseteste 1912g)[9]—on the burning sphere by studying the comportment of the rays of light as they get refracted through a spherical glass vessel ('*vitreum plenum rotundi corporis*'—*De natura locorum* 71).

[7] See also Dinkova-Bruun et al. (2013).

[8] See Dinkova-Bruun et al. (2013).

[9] This treatise is also supported by studies in Crombie (1961) and Eastwood (1968).

A geometric study of the burning sphere is set in a manuscript diagram that is attributed to Grosseteste, or possibly originating from Roger Bacon's interpretation of Grosseteste's work. It is unclear whether this geometric figure is originally drawn in Bacon's *Opus maius* or *De multiplicatione specierum*, or whether it originated from a manuscript of Grosseteste's *De natura locorum*. The attribution of this figure to Bacon is based on archiving this thirteenth-century manuscript from England at the British Library, as classified under the BL shelf-mark: 'Royal 7 F. VIII'. The diagram appears on folio number: f.25 of the manuscript that is entitled: *De multiplicatione specierum*, and catalogued under the name of Roger Bacon.[10] Crombie described the diagram as an illustration of Grosseteste's theory in *De natura locorum* as it focuses the sun's rays using a spherical lens (Crombie 1953).[11] Another attribution of this diagram to Roger Bacon makes reference to the *Opus maius*, iv. ii. 2, MS Roy. 7. F. viii, f. 25v.

2.4 The Tradition of Ibn al-Haytham

Having considered the principal aspects of Grosseteste's meteorological optics we shall now aim at a comparative elucidation of some of the leitmotifs of the tradition of Ibn al-Haytham and its reception within the Arabic scientific milieu.

The principal adaptive recension of Ibn al-Haytham's *Kitāb al-manāẓir—Book of Optics*; *De aspectibus*; *Perspectiva* (Ibn al-Haytham 1983, 1989, 2002)[12]—is embodied in the *Tanqīḥ al-manāẓir*—Revision of the *Optics* (Al-Fārisī 1928–1929)—by Kamāl al-Dīn al-Fārisī (d. 1320 CE) who completed his studies at the Maragha observatory under the tutorship of the astronomer Quṭb al-Dīn al-Shirāzī (d. 1311 CE). Al-Fārisī expanded Ibn al-Haytham's findings in the *Optics* in terms of advancing a novel theory that explicated the phenomenon of the colouration of the rainbow (*qaws quzaḥ*) based on analysing the geometric structure of the double refractions and reflections of light rays as they passed through spherical rain droplets. This natural phenomenon was simulated by al-Fārisī in an experimental model that consisted of a large transparent spherical glass vessel filled with water, to represent a single rain droplet, which was subjected to controlled light beams passing through selected apertures, while being situated within a *camera obscura*. Al-Fārisī analysed the comportment of light and its trajectories as it passed through this spherical vessel, while registering the images of the colour spectrum that emerged from it. He was able to offer a new understanding of the ontological nature of colour that rectified Ibn al-Haytham's speculations about the existential separation between illumination and colouration. Al-Fārisī removed Ibn

[10] Grosseteste studied the properties of the sphere in '*De sphaera* (Grosseteste 1912h).' Supporting studies are set in: Panti (2001, 2003, 2013) and Pacchi (1965).

[11] C. A. Ronan confirms Crombie's thesis in Ronan (1983).

[12] See also El-Bizri (2005) and Rashed (2002).

al-Haytham's doubts by showing that the colour spectrum is none other than a form of light that appears due to specific refraction and reflection conditions. He came very close in this to explaining what was later systematically explicated through Newton's theory about the decomposition and composition of white light. However, al-Fārisī was not the only one to offer this detailed explanation of the colouration of the rainbow in an experimental context, given that his contemporary, the Dominican friar, Theodoric (Dietrich, Teutonicus) of Freiburg, also arrived independently at similar conclusions shortly after him.[13] Each of these fourteenth-century opticians worked on separate manuscripts of Ibn al-Haytham's *Optics*, respectively in Arabic (with al-Fārisī) and Latin (with Theodoric), without being in touch with one another. The precision in their inquiries and proofs can be also credited to the accuracy in Ibn al-Haytham's description of the particulars of his measurable experimental procedures and observational data, which also facilitated the reforming of his theory of colour (El-Bizri 2005, 2009).

Ibn al-Haytham aimed at completing the work initiated by Ptolemy in the science of optics through novel methods of scientific inquiry that combined geometry with physics with isomorphism in the context of controlled experimental testing. He reworked the explications offered by the antique Aristotelian physicists (*al-ṭabīʿiyyūn*) who advocated intromission theories of vision according to which the form of a visible object is introduced into the eye as abstracted from its matter when the transparent medium between the eyes of the observer and the objects of vision is actualised by physical illumination. He also adapted the geometric models of the Greek mathematicians (*aṣḥāb al-taʿālīm*), mainly of Euclidean and Ptolemaic orientations, who upheld an emission theory of light, and according to which visible objects are seen by way of light rays that consist of non-consuming gentle irradiating fires, which are subtly emitted from the eyes in the shape of a cone or a pyramid of vision. Critically reassessing the unresolved disputes over these incommensurable ancient theories, Ibn al-Haytham rejected the mathematicians' claim that vision occurs by way of the emission of a light ray from the eye, while at the same time revealing the geometric principles that underpin the intromission theory, which was obscured by the conceptual ambiguities of the physicists cum natural philosophers. Ibn al-Haytham demonstrated that sight resulted from the introduction into the eyes of light rays that are emitted from the lit and visible surfaces of the opaque and coloured objects of vision, while taking into account that these light rays propagate rectilinearly across a given single homogeneous transparent medium (such as air), and that they enter the eye in the shape of a virtual geometric cone of vision (*makhrūṭ al-shuʿāʿ*), with its vertex at the centre of the eye and its base intersecting with the lit and visible surfaces of the seen objects (Ibn al-Haytham 1989 [*Optics* I.2, I.6, II.3–4]; El-Bizri 2005, 2009).

[13] This is mainly set in the Theodoric of Freiburg's *De iride et radialibus impressionibus*, and also associated with his treatises *De luce et eius origine* and *De coloribus* (see Theodoricus 1914; Calma 2010; Flasch 2007).

Ibn al-Haytham made a distinction between the conditions of sight from those of light. He showed that the rays of light irradiate spherically in every direction and from every point on the lit visible surfaces of the objects of vision. He also demonstrated that they travel across a given homogeneous transparent medium in straight lines, if unobstructed by opaque bodies, or if they did not pass through distinct transparent media that differed in their index of refraction.

Ibn al-Haytham separated pure sensation (*mujarrad al-ḥiss*), which only perceives light qua light and colour qua colour, from the psychological workings of vision in terms of recognition (*maʿrifa*), judging discernment (*tamyīz*), and comparative inferential measure (*qiyās*), as they get aided by imagination (*takhayyul*), memory (*dhikr*), and at times by prior knowledge. Ibn al-Haytham argued that sensation in connection with vision was ultimately effected by what he referred to as 'the last sentient' (*al-ḥāss al-akhīr*; *sentiens ultimum*), which, according to his analysis, was located in the anterior part of the brain (*muqaddam al-dimāgh*). He noted that when sight perceives individuals of the same species repeatedly and continually, a *forma universalis* (*ṣūra kulliya*) of that species takes shape in the imagination and gets recollected by recognition, while consequently assisting in the grasping of the quiddity (*māhiyya*) of the corresponding visible object and its inspected seen properties (Ibn al-Haytham 1989 [*Optics* II.3–4]).[14]

Ibn al-Haytham examined the essence of light and inquired about its propagation through variegated transparent media. He explored catoptrics, using finely polished surfaces, including parabolic, cylindrical and spherical mirrors, as well as studied the effects of visual magnification. His studies in these domains also rested on his critical analysis and extension of the findings of his predecessors, and most notably, the dioptrical research of Abū al-ʿAlāʾ ibn Sahl (tenth century CE) who discovered a principle akin to what is more commonly known since the seventeenth century as 'Snell's law of refraction' (namely a principle attributed to Willebrord Snellius that determines the refractive index of a transparent medium in connection with a given geometric shape, which can act as a basis for designing lenses).[15]

Ibn al-Haytham's studies in catoptrics and dioptrics, as the respective sciences of the reflection and refraction of light with their optical instruments, were associated with meteorological explorations of the halo and the rainbow (*Maqāla fī al-hāla wa qaws quzaḥ*). These inquiries were connected with his speculative take on the possibility that colour was ontologically distinct from light even though that it was only actualized by illumination in the sense of being propagated always alongside accidental secondary lights that are emitted from the lit surfaces of the objects of vision as they are illuminated by substantial primary lights such as those irradiating from fire or the sun. Ibn al-Haytham held that every opaque body (*jism*

[14] I examined these aspects in detail elsewhere in El-Bizri (2005).

[15] Regarding Ibn Sahl's anaclastic research in dioptrics see Rashed (1990, 2005). If n_1 is the refractive index of air and n_2 is the refractive index of water, then $n_2 > n_1$, and the incidence angle Θ_1 would be larger than the refraction angle Θ_2 (both measured against the normal to the surface at the point of incidence and refraction), then: $n_1 \sin \Theta_1 = n_2 \sin \Theta_2$.

kathīf) is coloured (*Optics* I.2 [12]),[16] and its colour (*lawn*) would be a visible property that is intermixed with the secondary accidental light that is emitted from that object and propagates with it. The form (*ṣūra*; *eidos*) of colour is akin to that of light, and it is a form in the lit and visible opaque coloured object (*Optics* I.3 [113–116]). Colours always accompany light and will never appear without illumination (*Optics* I.3 [114–166]; I.4 [129–131]). In the *Discourse on Light* (*Qawl fī al-ḍaw'*), Ibn al-Haytham distinguishes between accidental light that is secondary (*'araḍī thānī*) and emitted from lit surfaces of opaque objects, and substantial light that is primary (*jawharī awwal*) and irradiated from luminous bodies. Both types of light follow the same principles of emission, rectilinear propagation, reflection and refraction.

The distinction between substantial and accidental lights is inconsequential when considering the laws governing the comportment of light, however, the epistemic significance of such differentiation in the kinds of lights is evident in the case of investigating the essence of colours and their kind of being. This becomes manifest in the manner colours are closely associated with secondary accidental lights to the point that their separation from them is almost redundant. Although Ibn al-Haytham distinguished colour from the accidental light that it always accompanies, he was still unsure whether they were ontological distinct or whether they amounted to being one and the same. After all, observation and experience do not offer a proof in phenomenological terms whether light and colour are the same or whether they were distinct in kind and being despite the evidence that they obey the same laws (El-Bizri 2009).

The existential independence of colour from light was a matter of speculation for Ibn al-Haytham, and subsequent research on the rainbow disclosed the reality of colouration as a phenomenon of illumination. The eyes receive colours due to the transmission that is secured via accidental secondary lights that are emitted from the lit and visible surfaces of an object of vision as it is illuminated by substantial primary lights that are irradiated from self-luminous bodies. Ibn al-Haytham does not associate the essence of colours and light with the nature and properties of the transparent body in which they travel. The transparent media affect the propagation of light whether they spread rectilinearly, get reflected or refracted. He thus rejects the Aristotelian doctrine of the *diaphanous* wherein light actualizes the potencies of the transparent medium (*De anima* 418a 31-32, 418b 9-11).

Ibn al-Haytham explicates the rainbow in terms of his studies in catoptrics. He takes the rainbow to be the result of light reflected off a humid and dense air or cloud that has the properties of an arc segment of a concave spherical mirror. In consequence, the rainbow can be explicated geometrically by studying the reflective properties of a concave surface of a spherical mirror (Ibn al-Haytham's *Optics* Book V).[17]

[16] The Arabic critical edition of Ibn al-Haytham's *Qawl fī al-ḍaw'* is contained in Ibn al-Haytham (1938–1939). The annotated French translation of this treatise is contained in Rashed (1992) (see also El-Bizri 2009).

[17] Refer also to Rashed (1970).

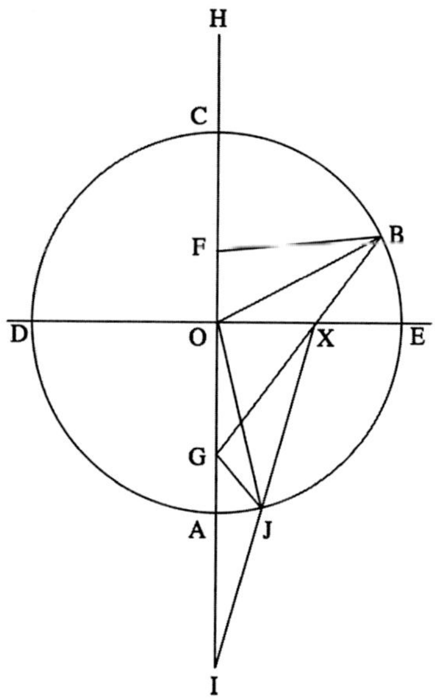

Fig. 2.1 Spherical concave mirror

To account for the geometric modelling of this meteorological phenomenon in terms of studying the reflective properties of a concave spherical mirror that has the observer placed within its envelop, and at a common horizontal plane with its centre, let us consider the following demonstration by Ibn al-Haytham (Fig. 2.1).

Let there be a circle that contains a semicircle *ABC* as a cross-section of a segment of a concave spherical mirror (as a model in catoptrics of a cloud in meteorology), and let its diameter extend vertically as *AC* while passing by the centre of the circle *O* (*qua* the centre of the spherical mirror). Extend *AC* to a point *H* in the direction beyond *A* and to a point *I* in the direction beyond *C*. Extend from *O* a line perpendicular to *AC* that cuts the circle at *E*. Let the point *X* demarcate the position of the eye of the observer (*al-baṣar*) within the circle on line *OE*. Join point *I* to *X* in such a way that *IX* cuts the circle at point *J*. Connect *J* to *O* such as the angle *GJO* is equal to the angle *OJX*. The point *I* is herein an *image* (*khayāl*; lit. qua 'shadow') of *G*, and every point on *AI* is an *image* of a corresponding point on *AO*. If we connect *G* to *X* and extend *GX* to cut the circle at point *B*, then we connect *B* to *O* and let the angle *XBO* be equal to the angle *OBF*, whereby *F* is the point of the perpendicular falling from *B* unto *OC*, then *G* is the *image* of *F*, and every point on *AO* is the *image* of a corresponding point on *OC*. If we extend *XO* through *O* it cuts the circle in *D*, and every point on *CH* if extended infinitely as a line would have its form (*ṣūra*) reflected into the eye of the observer at *X* through the mirroring arc *AD*, with its *images* falling on *AO*, since any point on *CH* that is linked to any point on the arc *AD* would cut the line *OD*. If any of these points is joined to *O* and an angle is generated that is equal to the angle that occurs at that point, the second line would cut *AO*, and if it is extended it would reach the eye at point *X*, and the same applies to every point on *OC*. Therefore, every point on the line *OAI* can be an *image* of the visible points, and the same with lines *AI* and *CH* if extended to infinity then every point on the diameter *AC* when extended in either direction to infinity would then be an *image* of one of the observed points. Every point that is visible by the observer through this spherical concave mirror would have a range of 4 images at a maximum and 1 image at a minimum.[18]

In contrast with Ibn al-Haytham's explication of the rainbow, Avicenna (Ibn Sīnā) held that a cloud could not act as a mirror, and that the rainbow results from a reflection of light on the totality of the raindrops in a light shower (*rashsh*) and as dispersed in the atmosphere, and without there being behind the cloud an opaque coloured body. Rather taking the example of a crystal, it does not act as a mirror if it is not masqued from behind by an opaque coloured object, and if a transparent expanse is behind it, then it does not act as a mirror (Ibn Sīnā 1965). A similar view is also encountered in the *Epistles of the Brethren of Purity* (*Rasā'il Ikhwān al-Ṣafā'*) a few decades before the times of Avicenna (c. 960 s CE) in the context of commenting on Aristotle's *Meteorology* (III.2, 4, 5). The Brethren of Purity (Ikhwān al-Ṣafā') noted that the cause of the occurrence of the rainbow is due to the irradiation of sunlight on a humid and dense vapour that fills the atmosphere (Ikhwān al-Ṣafā' 2013). However, there is no evidence that Avicenna benefited directly from their meteorological explications despite the wide dissemination of their encyclopaedic epistolary compendium in his epoch.

Al-Fārisī moves away from Ibn al-Haytham's and Avicenna's focus on the reflection of light, but takes from the discussion the idea that a geometrical study can be conducted on a single rain drop, to be modelled experimentally in the way it reacts to light projected in a controlled manner on it, with a focus on the study of spherical concavity in dioptrics and catoptrics. Al-Fārisī geometrically treated the experimental model of the single raindrop as a burning sphere in optics (*Tanqīḥ*, II, 259, 263-264, 285, 331, 337). The geometrical deductions and demonstrations would have significance within physics, and physical phenomena would be subject to geometry in the context of experimentation. A natural phenomenon that is otherwise inaccessible in direct testing would be modelled analogically in an experimental context that is informed by an isomorphic composition of geometry with physics. This scientific experimental method was itself developed by Ibn al-Haytham (El-Bizri 2005, 2009).

Al-Fārisī disclosed that colours were phenomena of light that are due to refractions and reflections, which are associated with the trajectories of light when traversing media of differing transparencies. Colours are lights and do not have an ontic reality in themselves (*Tanqīḥ*, I, 48–49). Light generates colours due to its angles of refraction and to its reflections as it passes from one transparent medium into another that differed in refractive properties (*Tanqīḥ*, II, 258–409). Two arcs of the rainbow appear with a band of greyness in between them as the sunlight rays are refracted and reflected in raindrops.

[18] See Ibn al-Haytham (2002), Book V in catoptrics, Figure [*Shakl*] 5/28, Part I, pp. 318–9, Part II, p. 198—*English translation is author's*.

The internal and primary arc of the rainbow, which is nearer to earth, has the colourations from below going upwards as: red, yellow, green, blue, violet. This results from two refractions of the decomposed white light and of one reflection in each rain droplet (as in Fig. 2.2). The sunlight ray that is incident at point *A* refracts as *AB* and *AC*, and in-between *AB* and *AC* is the spectrum colours of the decomposed white light. *AB* and *AC* reflect respectively as *BE* and *CD*, and then at points *D* and *E* the reflected lights are refracted at their exiting of the rain droplet as a spectrum of colour that constitutes the colouration of the primary rainbow with the totality of rain droplets that have these refractions and reflections within them. The refracted exiting ray at *D* is violet and at *E* is red (the middle is green and in the direction of red is the yellow, while in the direction of violet is the blue).

The external and secondary arc of the rainbow, which occurs higher in the sky and above the grey zone separating it from the internal arc, has the colouration sequence from below then upwards as: violet, blue, green, yellow, red. This results from two refractions of the decomposed white light, and of corresponding two reflections in each rain droplet (as in Fig. 2.3).

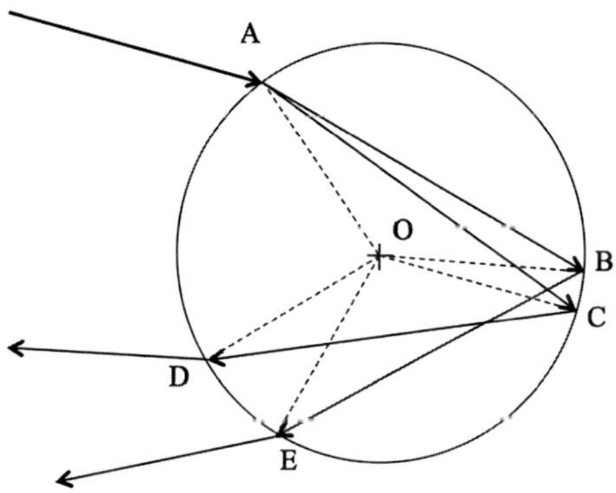

Fig. 2.2 Double refractions and single reflection in a raindrop

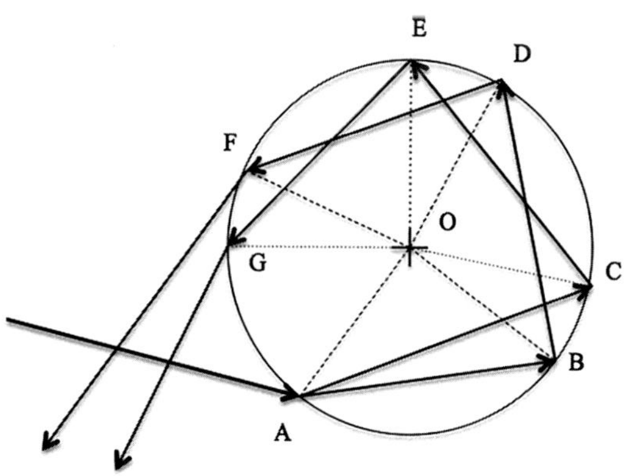

Fig. 2.3 Double refractions and double reflections in a raindrop

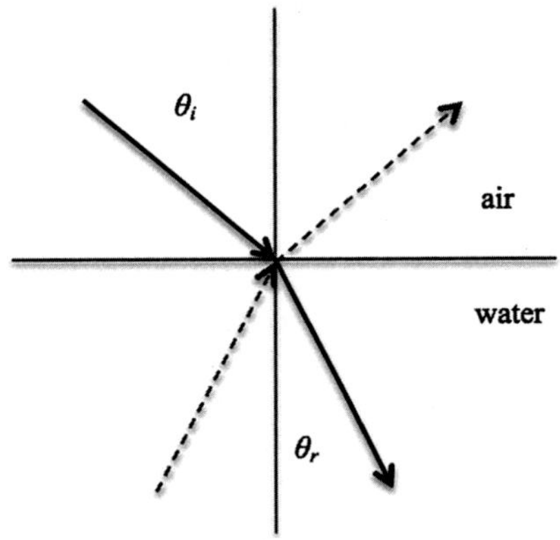

Fig. 2.4 Refraction towards the normal (air to water). Refraction away from the normal (water to air)

To sum up, al-Fārisī conducted his controlled testing experiments (*i'tibār*) in procedures that are guided by geometric models and using a spherical glass vessel filled with water to act as an enlarged model of a rain droplet, and placed within a *camera obscura* to observe the behaviour of controlled natural light as it passes through it (*Tanqīḥ* II, 340–342). Natural phenomena that cannot be studied directly are therefore investigated within the context of a physical experiment that is structured in a geometric model. Al-Fārisī relied in this inquiry on Ibn al-Haytham's experimental procedures in testing, on Ibn Sahl's and on his own studies on the burning sphere (*al-kura al-muḥriqa*), and on Ibn Sīnā's meteorology (the latter believed that colour is generated when sunlight rays are reflected on the totality of rain droplets dispersed in the atmosphere [*Tanqīḥ*, II, 337], Ibn Sīnā 1965; Rashed 1970).[19] Al-Fārisī also benefited from the observations, data and rules that were advanced by Ibn al-Haytham in connection with the refraction of light. The principle that governs this phenomenon is that ultimately, a transparent body resists the movement of light, and the denser it is, the greater its resistance would be, whereby it acts in the direction of the component parallel to the surface.[20]

The general refraction rules arrived at by Ibn al-Haytham can be summarized and annotated as follows (Fig. 2.4):

Let i_1, i_2 be two angles of incidence, and d_1, d_2, with r_1 and r_2 as their respective angles of deviation and refraction, and let $i_1 > i_2$; thence:

$$d_2 > d_1;$$
$$d_2 - d_1 < i_2 - i_1;$$

[19] Refer to the Aristotelian model as explicated in Sayili (1939).

[20] See Nazif (1942–1943). Lindberg observed in this regard that Ibn al-Haytham's interpretation is 'suspiciously Cartesian' in a hint that Descartes' *Dioptrique* may have benefited from an adapted Latinized transmission of Ibn al-Haytham's thesis (see Lindberg 1983; Pavlos 2008; Bellosta 2002).

$$d_2/i_2 > d_1/i_1;$$
$$r_2 > r_1;$$

In subtle to dense refraction: $d < 1 = 2i$;
In dense to subtle refraction: $d < ½\,(i + d)$ $[d > ½r \text{ if } r > i]$;
A denser transparent medium refracts light toward the normal;
A subtler transparent medium refracts light away from the normal.[21]

References

Al-Fārisī, K.-D. (1928–1929). *Kitāb Tanqīḥ al-manāẓir li dhawī al-baṣar wa-al-baṣā'ir* (Vol. 2). Hyderabad: Osmania Oriental Publications Bureau.

Aristotle. (1910). *De generatione animalium (On the Generation of Animals)* (A. Platt, Trans.). Oxford: Clarendon Press.

Aristotle. (1936). *Physics* (W. D. Ross, Ed.). Oxford: Clarendon Press.

Aristotle. (1952). *Meteorologica* (H. D. P. Lee, Ed. & Trans.). Loeb Classical Library. Cambridge, MA: Harvard University Press.

Aristotle. (1961). *De anima* (W. D. Ross, Ed.). Oxford: Clarendon Press.

Bellosta, H. (2002). Burning instruments: From Diocles to Ibn Sahl. *Arabic Sciences and Philosophy, 23*, 285–303.

Biard, J., et al. (2009). *Blaise de Parme, Questiones super pespectiva communi*. Paris: Vrin.

Boyer, C. B. (1945–1946). Aristotelian references to the law of reflection. *Isis, 36*, 92–95.

Boyer, C. B. (1954). Robert Grosseteste on the rainbow. *Osiris, 11*, 247–258.

Boyer, C. B. (1958). The theory of the rainbow: Medieval triumph and failure. *Isis, 49*(4), 378–390.

Boyer, C. B. (1959). *The rainbow: From myth to mathematics*. New York: Yoseloff.

Calma, D. (2010). *Le poids de la citation. Etude sur les sources arabes et grecques dans l'œuvre de Dietrich de Freiberg*. Fribourg: Academic Press.

Crombie, A. C. (1953). *Robert Grosseteste and the origins of experimental science, 1100-1700*. Oxford: Clarendon Press.

Crombie, A. C. (1961). Quantification in medieval physics. *Isis, 52*(2), 143–160.

Dinkova-Bruun, G., Gasper, G. E. M., Huxtable, M., McLeish, T. C. B., Panti, C., & Smithson, H. (2013). *The dimensions of colour: Robert Grosseteste's De colore*. Toronto: Pontifical Institute of Medieval Studies.

Eastwood, B. S. (1966). Robert Grosseteste's theory of the rainbow. *Archives internationales d'histoire des sciences, 19*, 313–322.

Eastwood, B. S. (1967). Grosseteste's quantitative law of refraction: A chapter in the history of non-experimental science. *Journal of the History of Ideas, 28*, 403–414.

Eastwood, B. S. (1968). Medieval empiricism: The case of Grosseteste's optics. *Speculum, 43*, 306–321.

El-Bizri, N. (2005). A philosophical perspective on Alhazen's optics. *Arabic Sciences and Philosophy, 15*, 189–218.

El-Bizri, N. (2009). Ibn al-Haytham et le problème de la couleur. *Oriens-Occidens: Sciences, mathématiques et philosophie de l'antiquité à l'Âge classique, 7*, 201–226.

[21] Refer to Omar (1977).

El-Bizri, N. (2010). Classical optics and the *perspectiva* traditions leading to the renaissance. In C. Carman & J. Hendrix (Eds.), *Renaissance theories of vision* (pp. 11–30). Aldershot: Ashgate.

Flasch, K. (2007). *Dietrich von Freiberg. Philosophie, Theologie, Naturforschung um 1300.* Frankfurt am Main: Klostermann.

Grant, E. (Ed.). (1974). *A source book in medieval science.* Cambridge, MA: Harvard University Press.

Grosseteste. (1912a). De potentia et actu. In *Die Philosophischen Werke des Robert Grosseteste, Bischofs von Lincoln.* Münster: Aschendorff.

Grosseteste. (1912b). De lineis angulis et figuris. In *Die Philosophischen Werke des Robert Grosseteste, Bischofs von Lincoln.* Münster: Aschendorff.

Grosseteste. (1912c). De motu corporali et luce. In *Die Philosophischen Werke des Robert Grosseteste, Bischofs von Lincoln.* Münster: Aschendorff.

Grosseteste. (1912d). De iride. In *Die Philosophischen Werke des Robert Grosseteste, Bischofs von Lincoln.* Münster: Aschendorff.

Grosseteste. (1912e). De luce. In *Die Philosophischen Werke des Robert Grosseteste, Bischofs von Lincoln.* Münster: Aschendorff.

Grosseteste. (1912f). De colore. In *Die Philosophischen Werke des Robert Grosseteste, Bischofs von Lincoln.* Münster: Aschendorff.

Grosseteste, R. (1912g). De natura locorum. In *Die Philosophischen Werke des Robert Grosseteste, Bischofs von Lincoln.* Münster: Aschendorff.

Grosseteste. (1912h). De sphaera. In *Die Philosophischen Werke des Robert Grosseteste, Bischofs von Lincoln.* Münster: Aschendorff.

Grosseteste. (1942). *On light (De luce)* (C. C. Riedl, Trans.). Milwaukee, WI: Marquette University Press.

Grosseteste. (1982). *Hexaemeron* (R. C. Dales & S. Gieben, Eds.). Oxford: Oxford University Press.

Grosseteste. (2011). La luce. In C. Panti (Ed.), *Introduzione, edizione, traduzione e commento a cura di.* Pisa: PLUS, Pisa University.

Hero of Alexandria. (1900). De speculis. In L. Nix & W. Schmidt (Eds.), *Heronis Alexandrini Opera quae supersunt omnia* (Vol. 2). Leipzig: B. G. Teubneri.

Ibn al-Haytham. (1938–1939). *Qawl fī al-ḍaw.* In *Majmū' al-Rasā'il* (epistolary collection). Hyderabad: Osmania Oriental Publications Bureau.

Ibn al-Haytham. (1983). *Kitāb al-manāẓir I–III* (A. I. Sabra, Ed.). Kuwait: National Council for Culture, Arts and Letters.

Ibn al-Haytham. (1989). *The optics of Ibn al-Haytham, Books I–III, On Direct Vision* (A. I. Sabra, Trans.). London: Warburg Institute.

Ibn al-Haytham. (2002). *Kitāb al-manāẓir IV–V* (A. I. Sabra, Ed.). Kuwait: National Council for Culture, Arts and Letters.

Ibn Sīnā. (1965). Metals and meteors. In I. Madkour, et al. (Eds.), *Kitāb al-Shifā'* (Book of Healing) (Part V). Cairo: *al-Hay'a al-miṣriyya al-'āmma li'l-ṭibā'a wa'l-nashr.*

Ikhwān al-Ṣafā'. (2013). *The Epistles of the Brethren of Purity* (*Rasā'il Ikhwān al-Ṣafā'*). *On the Natural Sciences. An Arabic Critical Edition and Translation of EPISTLES 15-21* (C. Baffioni, Ed. & Trans. With a Foreword by Nader El-Bizri). Oxford: Oxford University Press.

Lejeune, A. (1956). *L'Optique de Claude Ptolémée dans la version latine d'après l'arabe de l'émir Eugene de Sicille.* Louvain: Bibliothèque de l'université, bureaux du receuil.

Lejeune, A. (1957). *Recherches sur la catoptrique grecque.* Bruxelles: Palais des academies.

Lindberg, D. C. (1966). Roger Bacon's theory of the rainbow: Progress or regress? *Isis, 57*, 235–248.

Lindberg, D. C. (1983). *Studies in the history of medieval optics.* Chicago, IL: University of Chicago Press.

Little, A. G. (1926). The Franciscan School at Oxford in the thirteenth century. *Archivum Franciscorum Historicum, 19*, 803–874.

Longeway, J. L. (2007). *Demonstration and scientific knowledge in William of Ockham*. Notre Dame, IN: University of Notre Dame Press.
Lucretius. (1975). *De rerum natura* [*On the Nature of Things*]. Loeb Classical Library 181. Cambridge, MA: Harvard University Press.
Miccoli, L. (2001). Two thirteenth-century theories of light: Robert Grosseteste and St. Bonaventure. *Semiotica, 136*(1-4), 69–84.
Nazif, M. (1942–1943). *al-Hasan bin al Haytham: Buhuthuhu wa kushufuhu al-basariyya*. Cairo: *al-Hay'a al-miṣriyya al-'āmma li'l-ṭibā'a wa'l-nashr*.
Omar, S. B. (1977). *Ibn Al-Haytham's optics, a study of the origins of experimental science*. Minneapolis: Bibliotheca Islamica.
Pacchi, A. (1965). Ruggero Bacone e Roberto Grossetesta in un inedito hobbesiano del 1634. *Rivista critica di storia della filosofia, 20*, 499–502.
Panti, C. (2001). *Moti, virtu e motori celesti nella cosmologia di Roberto Grossatesta: studio ed edizione dei trattati De sphera, De cometis, De motu supercelestium*. Florence: SISMEL: Edizioni del Galluzzo.
Panti, C. (2003). Robert Grosseteste's early cosmology. In E. A. Mackie & J. Goering (Eds.), *Editing Robert Grosseteste: papers given at the thirty-sixth annual Conference on Editorial Problems. University of Toronto, 3-4 November 2000* (pp. 135–165). Toronto: University of Toronto Press.
Panti, C. (2013). Robert Grosseteste's *De luce*. A critical edition. In J. Flood, J. Ginther, & J. Goering (Eds.), *Robert Grosseteste and his intellectual milieu* (pp. 193–238). Toronto: Pontifical Institute of Medieval Studies.
Pavlos, M. (2008). Developing ideas of refraction, lenses and rainbow through the use of historical resources. *Science & Education, 17*, 751–777.
Plato. (1929). *Timaeus. Critias. Cleitophon. Menexenus. Epistles* (R. G. Bury, Ed. & Trans.). Loeb Classical Library 234. Cambridge, MA: Harvard University Press.
Rashed, R. (1970). Le modèle de la sphère transparente et l'explication de l'arc-en-ciel: Ibn al-Haytham et al-Fārisī. *Revue d'histoire des sciences, 23*, 109–140.
Rashed, R. (1990). A pioneer in Anaclastics: Ibn Sahl on burning mirrors and lenses. *Isis, 81*, 464–491.
Rashed, R. (1992). *Optique et mathématiques: Recherches sur l'histoire de la pensée scientifique en arabe*. Aldershot: Variorum.
Rashed, R. (2002). *Les mathématiques infinitésimales du IXe au XIe siècle* (Vol. 4). *Ibn al-Haytham, méthodes géométriques, transformations ponctuelles et philosophie des mathé matiques*. Wimbledon: al-Furqān Islamic Heritage Foundation.
Rashed, R. (2005). *Geometry and dioptrics in classical Islam*. Wimbledon: al-Furqān Islamic Heritage Foundation.
Ronan, C. A. (1983). *The Cambridge illustrated history of the world's science*. Cambridge: Cambridge University Press.
Sayili, A. M. (1939). The Aristotelian explanation of the rainbow. *Isis, 30*, 65–83.
Theodoricus Teutonicus de Vriberg. (1914). *De iride et radialibus impressionibus* (J. Würschmidt, Ed.). Münster: Aschendorff.
Turbayne, C. M. (1959). Grosseteste and an ancient optical principle. *Isis, 50*, 467–472.
Ver Eecke, P. (1959). *Euclide L'Optique et la catoptrique*. Paris: Blanchard.
Zemplén, G. Á. (2005). *The history of vision, colour, and light theories: Introductions, texts, problems*. Bern: Bern Studies in the History and Philosophy of Science.

Manuscripts

Ibn al-Haytham. (undated). *Kitāb al-manāẓir*: Ms. 3215 Fatih, Istanbul, fol. 267; Ms. 2448 Aya Sophia, Istanbul, fol. 465.

Chapter 3
Robert Grosseteste and the Pursuit of Learning in the Thirteenth Century

Jack P. Cunningham

3.1 Avicebron, the *Fons vitae* and Robert Grosseteste

Solomon Ibn Gabirol (c. 1020–1052), known most commonly by his Latin name Avicebron was a Spanish Jew and one of the first teachers of Neoplatonism in Europe. Avicebron's chief philosophical work, the *Fons vitae*, has been called the 'climax of the Neoplatonic tradition in medieval Jewish philosophy' (Zwi Werblowsky and Wigoder 1997). Written in Arabic, but translated into Latin in Toledo by the Jewish convert, Ibn Daud (John of Spain) the *Fons vitae* is presented as a dialogue between a student and master as the latter sets out to elucidate the fundamental question of how a material universe can have its source in a purely spiritual being. In response to his pupil's enquiries the master presents an ontology which posits a Neoplatonic 'First Essence,' which we might call 'God' or 'first maker.' God exists above all things and is infinite and eternal; we may know only of His existence but not His essence. From this God there emanated a 'Divine Will' and subsequently from the Divine Will there emanated substances that are composed of matter and form. The philosophical notion that all things (including spiritual beings, the soul and the intellect) are composed of matter and form is known as Universal Hylomorphism. According to Avicebron all the material and spiritual world (once again including the soul and the intellect) are created in an emanation from the divine which is the 'Fountain of Life.' He calls this emanation the substance of universal intelligence and its dissemination is likened, not only to a fountain, but crucially also to a diffusion of light.

> . . .like the light of the sun that is diffused in the air, penetrates it and yet does not appear visible on account of the subtlety of air, until it meets a solid body, like the earth: then the light becomes sensible because it cannot penetrate the parts of this body. . .In a similar

J.P. Cunningham (✉)
Bishop Grosseteste University, Lincoln, UK
e-mail: j.p.cunningham@bishopg.ac.uk

J.P. Cunningham, M. Hocknull (eds.), *Robert Grosseteste and the pursuit of Religious and Scientific Learning in the Middle Ages*, Studies in the History of Philosophy of Mind 18, DOI 10.1007/978-3-319-33468-4_3

> manner the lights of the simple substances penetrate and flow through each other without the perception of the senses, on account of the subtlety and simplicity of these substances (Avicebron 2008).

From this the author upholds the Neoplatonic principle that an inferior being emanating from a superior being contains something of its source. *Ispo facto*, God is present in all things, '...the power of the holy penetrates all things, exists in all things, and acts in all things beyond time' (Ibid.).

Although feted by the intelligentsia of the Latin West, Avicebron's work was poorly received among the Jewish communities for some easily discernible reasons. In the *Fons vitae* we have a purely rationalistic attempt to trace the divine origins of the universe; there is not a single reference to either the Old Testament or Talmud. Attempts to unravel the mysteries of life without reference to God's word or the sacred tradition were unthinkable in medieval Judaism. However the Christian West was more receptive, in particular these ideas had an influence on the Franciscans in the thirteenth century and there is strong evidence to suggest that they made a profound impression on their teacher at Oxford, Robert Grosseteste (Miccoli 2001). The *Fons vitae* has none of the corporeity of the first principle that is contained in *De luce*; yet other similarities are striking. We have the flowing out from the single source in which all things self-propagate in the same fashion as light. Elsewhere, in Avicebron's most famous poem, *Keter Malkhut* (A Kingdom's Crown) there is a nothingness that awaits the form to bring it forth into existence, 'To bring out the stream of existence from Nothing, like light flowing from sight's extension,' which resonates with Grosseteste's light giving form, and thus dimension, to matter (Avicebron 2001). When we consider that Avicebron's 'Divine Will' came directly from God and was therefore partly Divine, and that Grosseteste's light has the role Christians would have naturally associated with the *Logos* or Christ, the parallels draw even closer. When we go on to consider the infusion of all created things with the Divine Will we must understand that the author of *De luce* was agreeing with the Jewish scholar that something that 'proceeds from the Father,' to quote the Creed, is in all living things.

It is here in these assertions that both a theological problem and perhaps the key to dating the Bishop of Lincoln's work lies. When these authors suggest, or the logical implications of their formulas imply, that God is somehow or in some way, present in all created beings they are getting close to heretical notions of pantheism. We must be clear that this is not the same as saying that Grosseteste was a pantheist, any more than Avicebron. In fact both authors might best be identified as panentheists. This term was coined by Karl Freidrich Krause in the nineteenth century but what it describes may be traced back to the Neoplatonists of the Middle-ages with their emphasis on the emanation of material from the immaterial (Krause 1829). It is no coincidence that a good number of the doctrine's twentieth-century adherents wrote on the subject of Neoplatonism.[1] Panentheism maintains that God

[1] Philip Clayton studied Nicholas of Cusa and Norman Pittenger commentated on Erigena. For an examples of Panentheism (see Clayton 1998; Pittenger 1950).

is *in* the cosmos, as opposed to the pantheistic notion that God *is* the Cosmos, and nothing more. The panentheistic God might be present in the created order but there is more to him than this: he exceeds it, he existed before it, and he will exist after it.

It also should be made clear that Avicebron is seemingly aware that he is on doctrinal thin ice and there are in the *Fons vitae* overt attempts to divert accusations of pantheism. The first act of God is the creation of something from nothing. He is also insistent that at the moment of creation resemblance, similitude, union and harmony between Creator and Creation are removed (Avicebron 2008). Such qualifications did little to appease contemporaries and he was accused of pantheism in his lifetime; and it is possible to conjecture that this perceived tendency was an additional reason why he was not taken up by his coreligionists. The fact that these anxieties about pantheism did not ultimately confine themselves to the Jewish community might provide us with a reason why the connections that exist between Grosseteste's early works and Avicebron's emanation theory are not apparent in his later works. In the twelfth century Christianity was also encountering doctrines and methods that it found just as vexatious. Certain individuals connected with the School of Chartres were as inclined as Avicebron towards rationalism and a concomitant pantheism, and it was an inclination that was not going unnoticed.

3.2 The School of Chartres and the Problem of Pantheism

Without doubt the biggest perceived threat to established orthodoxy in the thirteenth century came from the dualism of the Cathars. Their doctrine posited a universe in many ways the diametrical opposite of pantheism. Their universe had nothing of God in it, at least nothing of the right god since it had been created by an evil demiurge. The history of the Cathar threat to the established Church need not detain us here but it is important to note that it was largely responsible for goading the Church into a frenzy of counter-heretical activity.

Western Europe had first witnessed glimpses of potential pantheism in the twelfth century in the neighbourhood of Paris at Chartres. Aristotle's physical works made their first appearance in the Latin West in the School there which set itself the task of reconciling Platonic and Aristotelian philosophies. It can also be credited with introducing the Latin world to Aristotle's hylomorphism. Building on this Bernard of Chartres (d. 1124) proposed that forms (*forma nativae*) were copies of the ideas of God. All this was perfectly congruent with established thought but other intellectuals of the School such as Thierry of Chartres (c. 1100–c.1155) or Clarembald of Arras (c.1110–c.1180) made statements that often strained the boundaries of orthodoxy. Thierry provides us with his explanation of Creation in his *De sex dierum operibus* which is a commentary on the first chapter of Genesis. In his introduction he explains that his explanation will be focused on the physical science and on the letter, *secundum physicam et ad litteram*. In other words he will not concern himself with the allegorical or moral interpretations which have been adequately expounded upon by the 'holy expositors.' Giulio D'Onofrio has taken

Thierry at his word and argued that his intension to explain the origins of the universe in purely physical terms was based on an assumption that the Patriarchs of Christendom had done such a complete job of elucidating the theology of Genesis that it would be otiose to pursue that line of enquiry further.

> According to Theodoric [Thierry] it is therefore permissible and even indispensable to offer an exclusively physical, that is historical and literal reading of the text now that its allegorical, spiritual and moral depths have been sufficiently studied and explained by the Fathers of the Church (D'Onofrio 2008).

If this is true then Thierry's methodology is based on a type of conservatism that assumes the canon of the Church does not need to be challenged. However there is a problem with this analysis in that his approach is highly distinctive and what he has to say is too radical to take the notion of acute deference seriously. Édouard Jeauneau has argued that Thierry and others of the School 'cannot-nay must not', content themselves with Genesis in order to explain the physical world (2009). The implication being that their *modus operandi* was not so much motivated by a profound respect for what had already been written, in spite of what Thierry claimed, but more by a strong will to add to the sum of knowledge by finding truth in unapproved sources. Writing an account of Creation with little reference to the theological was an exercise in itself that might well be accused of heresy. It was a lesson that another writer associated with Chartres, William of Conches learnt well when his *De philosophia mundi* was attacked precisely for attempting to tackle the Christian mysteries armed only with the tools of science. His persecutor was that scourge of the doctrinally suspect, William of Thierry who accused the author of being a 'Mono physicus'. The author of the *De sex dierum operibus* was perfectly aware that his solely scientific approach would have required an explanation and it was a stroke of expert disingenuity to claim that it was out of respect for the theologians. What they did in Chartres was a bold step and has more fittingly been described as '*audace rationaliste*' (Duhem 1954) and by the great Étienne Gilson as an 'experimental justification of Genesis' (1928).

According to *De sex dierum operibus* the Divinity is the 'cause of all existence (*forma essendi*). From this the author concludes that since all things derive their existence from God, the Divinity can rightly be said to be everywhere entirely and essentially, '*Si Deus forma essendi est, Deus ubique est totus et in omnibus essentialiter est*' (Häring 1955). Thierry goes on to tell us that every being that exists does so because it is one. A statement that in and of itself is a central tenet of pantheism and Nikolaus Häring has pointed out that Plato makes a similar claim in the *Paramenides* where he states that if one exists, the one is in all things. Häring conjectures that if Thierry knew of this text then it might well have been the source of the pantheistic tendency which Chartres was accused of (1955).

Clarembald of Arras was a pupil of Thierry and he imbibed the teachings of his master well. In his commentary on the *De Trinitate* of Boethius he states his agreement that God is the *forma essendi* of all things and since this is so He must be present in all things, God is essentially present everywhere (Janssen 1926). The same text sets out a form of Monism that is also drawing on his teacher. Here he maintains that there is but one and the same humanity in all people. Differences in individuals can be accounted for by a simple matter of '*propter accidentium varietatem*' (Janssen 1926).

Such perceptions of the universe often meant that these authors were naturally disposed to be receptive of the concept of the World-Soul. In William of Conches' *De philosophia mundi* we find a brave assertion that the *anima mundi* is one and the same as the Holy Spirit: brave because it was a concept that had already been condemned at the Council of Soissons in 1121 where Peter Abelard's *Theologia Sumnii Boni* had been ordered to be burnt. Taking up the theme Thierry describes the Spirit of the Lord as the *artifex* that gives form and order to matter. Following Conches he identifies it clearly with pagan writers.

> The philosophers call this power by different names. Mercurius calls it, 'spirit' in his *Trismegistus*. Plato calls it the 'world soul' in his *Timaeus*. Virgil refers to the 'spirit' in a poem [*Aeneid*]...Moses and Solomon speak of 'the spirit of the Lord' while David [Psalms] calls this power the 'word of the Lord'. The Christians call it the Holy Spirit (Häring 1955).

A good deal of ink has been expended on the question of whether the thinking that emerged from Chartres in the twelfth century can rightly be classified as pantheism. The great nineteenth-century medievalist Barthélemy Hauréau was clearly attracted to 'free-thinkers' in general and celebrated Thierry of Chartres methodology, describing it thus '*Son système est un panthéisme avoué*' (Hauréau 1872). Commenting on this assessment Éduard Jeauneau, otherwise a great admirer of his predecessor, has written, 'On this point, Hauréau fell victim to a figment of his imagination, for there is not an ounce of pantheism in the thought of Thierry of Chartres' (Jeauneau 2009).

The great Catholic historian of philosophy Fredrick Copleston was more circumspect and though he seemed anxious to exonerate the School he did point out dangers inherent in their system.

> The doctrine that natural objects are composed of matter and form, the form being a copy of the exemplar, the Idea in God, clearly makes a distinction between God and creatures and is non-pantheistic in character; but certain members of this School used terminology which, if taken literally and without qualification, would naturally be understood to imply pantheism (Copleston 1966).

The question rests on a crucial issue: if divinity is synonymous with reality and it is the intrinsic principle of all things, then thinkers such as Thierry of Chartres might be described as out-and-out pantheists. However another great historian of philosophy from the nineteenth century, Clemens Baeumker asked us to consider an additional aspect of their thought which distinguishes between the individual essence, which is unique to each individual thing, and the formal essence which is the divine. Because of this distinction we cannot rightly charge the School with pantheism (1890).

It is certainly true that immediately Thierry has told us that the divine form is the form of all things he is telling us that this is only by virtue of it being the perfection and integrity of all things (Janssen 1926). With this qualifying statement he has steered his doctrine onto the doctrinally solid ground of exemplarism. The divine cannot be the individual essence of a man, a horse or a stone. In a similar manoeuvre Clarembald restores himself by telling us that forms of things are 'images' and in so doing he reinstates himself as conventional. Häring for one was convinced by Thierry's claim that the One is transcendent not immanent, it therefore surpasses all things.

> If this is pantheism, i.e., the doctrine that God is everything and everything is God, the term must be used very loosely by those who accuse the School of Chartres of such an error (Häring 1955).

There is little doubt that if the question is: Was the School of Chartres made up of unabashed pantheists? Then the answer must be a resounding, No! However if we ask ourselves whether certain writers associated with the School betrayed pantheistic traits then the answer must be affirmative. In addition we would do well to consider a third interpretation of their seeming ambiguity which is that they were at times engaged in self-censorship. They knew well the parameters of orthodoxy but once they had breached these limits they also knew well how to retrieve themselves. This is something we have witnessed with Avicebron above and we may be already aware that members of the School were perfectly capable of doing this if we chart the career of William of Conches. He was attacked, as we know, for the heterodoxy of certain aspects of *De philosophia mundi* for which the Abbot of Saint-Thierry attempted to have him condemned with the vehement declaration, 'From out of this serpent's root has come forth an adder' (*Corpus Christianoum* 89A.61). After this condemnation we can clearly trace a change in what he subsequently claims or is prepared to say about the World-Soul. The confident identification with the Holy Spirit seems to become decidedly more guarded until we get to the *Dragmaticon* in which we find a deafening silence on the subject. Dorothy Elford maintained that this development was explained by a growth in confidence. There was a principal motivating impulse in Conches to 'discover what underlies the world and what holds it together' (1992). In spite of the initial attraction of the World-Soul for providing a link between the World and its source as a concept it ultimately proved too imprecise. As his career developed Conches seems to have grown in his faith that the properties of matter, as given by God, would provide a sufficient explanation for physical processes (Ibid.). However, a more likely explanation is that Conches' development reflects a growing uneasiness rather than any assuredness. When challenged by William of Thierry, Conches riposte was that he was a Christian and not a member of the Academy (Copleston 1966). Both he and Abelard were on the sharp end of a great deal of criticism and it is likely that Conches knew that the weight of ecclesiastical authority was categorically not on his side. In these circumstances one was well advised to do what Thierry of Chartres and Clarembald did, which was to qualify their most controversial statements, and when this does not work you omit them altogether.

If Grosseteste was influenced by this type of rationalism, if he was subsequently prone to the type of quasi-pantheism it produced, he may have also been influenced by the idea of the World-Soul. As a metaphysical concept the World-Soul has a long tradition and it can be encountered in the pages of Seneca, Augustine of Hippo and Macrobius. However, as James McEvoy has pointed out it gained few adherents in the Latin world until the reading of Plato's *Timaeus* c. 1120. McEvoy tells us that Peter Abelard was the first to be 'seduced' by the temptation to harmonize biblical faith and Neoplatonic spirituality. Abelard identified the *anima mundi* with the Holy Spirit which indwells in all Creation (McEvoy 1982). He was duly condemned at Soissons, but the idea was resilient up until the early decades of the thirteenth century. McEvoy went on to note eight references in Grosseteste to the World-Soul, but even more interestingly he has noted what he called 'an evolution of considerable moment in his thought' (Ibid.).

In *De Sphera* (1215–1220)[2] Grosseteste quite clearly identifies the efficient cause of the diurnal motions of the heavens as the *anima mundi*. In *De motu corporali* and *De motu supercaelesti* (c.1230) Plato's one soul becomes multiplied when Grosseteste writes that the heavenly bodies are the efficient cause of celestial motion, they have a single faculty of knowledge and desire. In other words there is a plurality of celestial souls. When we get to the texts written between 1230 and 1240 he is less committed to the idea, or at least less frank about it. In the *De operationibus solis*, for example, he says that 'certain philosophers' claim that there is a living principle of heavenly motion- it may or perhaps may not be the soul of the heavens. By the time we get to the *Hexaemeron*, McEvoy tells us that Grosseteste is feeling the weight of Patristic scholarship confuting ideas of mundial or celestial souls; here he quotes St John Damascene saying that the heavens are both inanimate and insensate. Grosseteste's final words on the subject come in his commentary on the *Celestial Hierarchy* where he tells us categorically that the celestial movers of the spheres are not conjoined to them. 'This conclusion brings us to the end of a long process of development in which Grosseteste personally made the transition from twelfth-century Platonism to thirteenth-century cosmology' (Ibid.). It might also be added that it marked the development of Grosseteste into a theologian who was first and foremost a Churchman.

3.3 David of Dinant

In the Paris of the next century other commentators were markedly less inclined to shepherd their theories away from heresy than their predecessors and the first case of full-blown pantheism we encounter is that of David of Dinant (c.1160–c.1270) who was a physician, philosopher and cosmologist from Belgium.

Once again the identification of our subject as a pantheist is not without controversy. G. Théry writing in the 1920s insisted that David of Dinant was a

[2] Except for the *Hexaemeron*, the following dating has been taken from McEvoy (1983). The *Hexaemeron* is dated by S. Harrison Thomson to no earlier than 1240 (1940).

heretic but advised that his 'pantheistic realism' was better placed alongside the pre-Socratic Eleatic tradition which argued against the Physicalists that there is a universal unity of being that lies behind the existence of all things. Théry takes Dinant to task for having a limited knowledge of Aristotle and a unilateral and simplistic mind that was unequipped to grasp the great philosopher since he did not understand the theory of analogy (Théry 1923). Enzon Maccagnolo seems to suggest than Dinant did little more than introduce Paris to the naturalistic writings of Aristotle, for which he was condemned, through an association or 'fantastic marriage' with contemporary heresies, 'by those to whom David's translations from the Greek…were unknown.' Which effectively meant that he was tarnished with the same heretical brush (Maccagnolo 1988).

However since exerts from Dinant's *Quaternuli* (little notebook) were rediscovered in the 1930s and published in the 1960s we now know a good deal more about the author. It has been established that he travelled in Greece and encountered first-hand the scientific works of Aristotle which he translated and commented on in the *Quaternuli* (Kurdzialek 1963). Tristan Dagron, in his more recent assessment of Dinant, has described him as having a 'vast knowledge' of Aristotle. He is rightly impressed that Dinant was able to read the texts in their original declaring, 'This is a remarkable feat at a time when a lot of translations tend to be based on Arabic versions…' (Dagron 2003/2004). In addition Dagron was able to rehabilitate Albertus Magnus' assessment of Dinant as a reader of 'the commentator' Alexander of Aphrodisias, when he described the heresy they had in common:

> Claiming that every creature is God, this is a heresy of Alexander who said that the primary substance, God, and the *nous*, that is to say, the substantial intellect, are a single substance, Alexander was later followed by a certain David of Dinant' (Théry 1923).

We are probably only in the process of gaining an appreciation of Dinant's contribution to learning in the West at this time and David Luscombe, for one, suspects that he is still, 'rather an underestimated transmitter and interpreter of Aristotelian natural philosophy' (2011). Dinant is described as 'Magister', a title which qualified him to teach and was almost certainly derived from the University of Paris where he was perhaps a master in the Arts faculty. Rather surprisingly, given his heretical legacy, he seems to have been close to Pope Innocent III who described him in familiar terms as '*capellanus noster*' (our chaplain) in a letter addressed to the Abbot and Chapter of the church at Dinant (*Patrologiae Latina*, CCXV). The relationship was noted with disapprobation in some quarters, the anonymous chronicler from Laon grumbled that, 'Master David another heretic, of Dinaunt [sic] and the inventor of this novelty, was frequently in the company of Pope Innocent because the Pope was passionately dedicated to subtle questions. Because David was more subtle than was appropriate' (*Anonymi Laudunensis Canonici* 1822).

Also from Albertus Magnus, in his *Compilatio de novo spiritu* we learn that David was the author of a text called *De tomis, id est de divisionibus* (On the Divisions) a work which is probably the same text as the *Quaternuli*. What startled the ecclesiastical authorities in these writings was not only the author's naturalistic explanations of certain biblical miracles but also a philosophy that amounted to materialistic pantheism. Here he tells us that reality is divided into three 'indivisibles', that is: bodies, minds and eternal substances. However the first two indivisibles, *hyle* and *nous*, are in fact one and therefore all things have one essence and that essence is God (Théry 1923).

> It is therefore manifest that there is only one single substance, not only of all bodies, but also of all souls, and this substance is none other than God himself (Dagron 2003/2004).

He seems to have arrived at these conclusions from his study of Aristotle and when Aquinas went to the trouble of personally castigating him as 'David of Dinant who most absurdly taught that God was primary matter', it was probably with a view to rescuing his beloved Greek philosopher from association with heresy (*Summa Theologica*, I, iii. 8). The works of Dinant were condemned at the Council of Sens in 1210 (see below) and the Synod decreed.

> David of Dinant's notebooks are to be handed in before Christmas to the Bishop of Paris and burnt; and no lectures are to be held in Paris either publicly or privately using Aristotle's books on natural philosophy or the commentaries and we forbid all this under pain of excommunication. If, from the birthday of our Lord onwards, anyone is found to be in possession of Master David's *Quaternuli*, he shall thereafter be considered a heretic (*Chartularium Parisiensis* I).

Dinant's absence from Paris during the storm that proceeded the dissemination of his ideas has led to conjecture that he fled the city, and though he does seem to disappear from the annals, the more likely explanation is that he ended his days in the services of the Curia at sometime around 1214 (Maccagnolo 1988). His absence from Paris did not mean however the end to the city's conflict with pantheism.

3.4 The Amalricians

If God was, for David of Dinant, all matter, for the Amalricians he was all form.[3] This group was born out of the teachings of a brilliant, if maverick, professor of logic and theology at the University of Paris by the name of Amaury (Amalric) of Bêne. The Laonese historian tells us that Amaury 'absorbed the errors' of David of Dinant but this is a chronological error since he died before these writings had arrived in Paris (*Cronica, recueil*). William the Breton tells us that the real source of his errors was Aristotelian metaphysics.

> During those days certain short writings [libelli], said to be by Aristotle and teaching metaphysics were being read in Paris, having recently been brought from Constantinople and translated from Greek into Latin. These writings provided an opportunity not only for

[3] For a full account of this sect see Dickson (1989).

the subtle doctrines of the Amalrician heresy but also for other doctrines which had not yet been invented (Maccagnolo 1988).[4]

It is also worth noting that Amaury's place of birth was in the diocese of Chartres where he studied and no doubt first discovered a taste for Greek cosmology. His passionate nature and his challenge to orthodoxy proved an attractive combination to students and he built up a devoted group of followers. His colleagues were clearly much less enamored and when they initiated a case against his teachings he felt obliged to appeal to the Curia in order to be examined. Here he received a fair hearing from Innocent III who nevertheless unsurprisingly judged in favour of the University scholars and consequently Amaury was compelled to recant (Ibid.). In the end the Parisian professor was a dutiful servant of the Church and he submitted to its reproaches; though it was reported that the doctrine he repudiated with his lips he continued to hold in his heart. He died in 1206 a broken man, though one in full communion with the Church.[5]

If Amuary wrote anything we no longer have it but we can have an idea of what he was perceived to have said from the writings of more conservative contemporaries. Aquinas tells us that Amaury (or at least the sect that followed him) claimed that 'God is the formal principle of all things' (*Summa*, I, 8). Henry of Suse (Ostiensis) tells us that Amaury had taught that God was in all things, *'dixit quod Deus erat Omnia.*' The master's disciples blended his pantheism with the popular idea that men could be justified by the spirit within. Garnier of Rochefort has left us a tract entitled, *Contra Amaurianos* which provides a valuable source with its descriptions of their beliefs. This text tells us that the God of the Amalricians is *ubique* (everywhere) since he is everywhere he must be in all things; in all people, in the stones beneath our feet, in an ordinary piece of bread as much as the Eucharist. If God is everywhere then it follows that Heaven is inside us and it is incumbent on the believer to recognize this and rejoice (Davenport 1997). One alarming conclusion followed another as they proclaimed a new age of the Holy Spirit in which man might aspire to be a purely spiritual being. This was based on perhaps their most enduring idea that the Holy Spirit was one in the same as the *intellectus agens* (activating intellect) which acted in all men. From this they concluded that knowledge made them spiritual and therefore neither the sacraments nor the Church were necessary to their salvation. Indeed, even a Jew with knowledge of the truth need not be baptized (*Contra Amaurianos*).[6]

In the autumn of 1210 we are told that a large crowd of Parisians gathered in the market-place outside the gates of St Honoré and watched as ten followers of Amaury were burnt at the stake (Thijssen 1996). Of the condemned at least six were priests, two were deacons and three were sub-deacons (*Caesarii Heisterbachensis*). Six days previous to their execution they had been laicized at

[4] The scant information we have regarding the life of Amaury is contained in the (*Gesta Philippi II Augusti*).

[5] See also (*Ditionnied Histoire et de Géographie Ecclésiastiques*).

[6] A translation can be found in Russell (1971).

the Church of St Honoré. Defiant to the last one of their number, a laicized sub-deacon, had declared at the trial, '...all is one, since all that which is, is God; consequently, insofar as I am, I cannot be burned, nor executed, since insofar as I am I am God' (Cohn 1957). The sentence on the unfortunate heretics had been passed by the Council of Sens hastily convened by Archbishop Peter Corbiel. As mentioned above, at the same council the works of David of Dinant were declared anathema and orders were given that they, along with the works of Amaury of Bêne, were to be burnt. A decree was also promulgated that Amaury's body was to be exhumed and cast into unconsecrated ground (Thijssen 1996). If the Parisian philosopher had not been condemned in his lifetime by his own words he was most certainly condemned in death by those of his disciples.

Compared with the threat posed by the Cathars we might be tempted to regard the Amalricians as little more than an annoying thorn in the side of Mother Church, but we should be wary of allowing their limited numbers to blind us to the significance of their movement. Firstly, far from being a demotic and ill-defined movement of enthusiasts they were a highly educated group. Among their number in Paris were four masters and seven others had studied theology, some for a considerable amount of time. Gary Dickson has gone as far as to ask us, '...within the context of their time...would it be an exaggeration to consider them an elite group of clerics?' (Dickson 1989). The University moved decisively to quash the heresy at its source and the moratorium of 1210 on the writings of Amaury (under pain of excommunication) was repeated in 1215 by the papal legate, Robert Courçon. It was unfortunate for these emerging movements that their activities could not have been more untimely.

3.5 Backlash and Accommodation

The nature and tone of intellectual enquiry in the West changed dramatically in the first half of the thirteenth century as the centers of learning gradually migrated from the monasteries to the newly emerging universities. The new academic institutions were never going to entirely replicate the intellectual endeavours of their forebears and one of the clearest indications that certain quarters of these scholarly communities were keen to plough a new furrow was in the attempt to emancipate philosophy from its servitude to philosophy. As Philipp W. Rosemann has pointed out, for the first time in the history of the Church this movement considered philosophy as no longer the handmaiden of theology tasked to assist it with its interpretation of Scripture; it was in and of itself a legitimate tool for approaching the quest for knowledge.

> For the new philosophy, language was not rooted in the divine Word as it had revealed itself in Scripture and the Incarnation; rather language was to be analyzed as an autonomous phenomenon, by means of the tools of logic and semantics (Rosemann 2013).

The process was not, of course, without its problems and one of them was, as Hastings Rashdall pointed out more than a century ago, 'an outburst of pantheistic thought' which at times bordered on 'pure materialism' (Rashdall 1895). The inevitable conservative backlash was nowhere more apparent than in the University of Paris. As Jaques Verger demonstrated, here the first generations of theologians were the most directly confronted by the various heretical movements and at the same time there were among them scholars who were the first to incorporate Aristotelian natural philosophy into the discipline of theology (Verger 2005). For the Old School their difficulty was not just with heresy, it was also about preserving the purity and superiority of their discipline. Theology was the 'celestial philosophy' and any attempt to combine the discipline would result in polluting the sacred with the mundane. In addition, since man was a fallen creature efforts to provide a purely rational (human) explanation of the universe was hubristic and destined only to offend God. Christianity was not alone in responding in this way, all three of the main religions in the medieval world accused their philosophers who were influenced by the Greeks of heresy (Caldwell Ames 2015). The first official moves to return philosophy to it appropriate station in the hierarchy of learning came in the Council of Sens which, aside from executing the Amalricians, had also threatened excommunication for anyone reading the natural philosophy of Aristotle. Efforts to counter rationalism and independent thinking continued, in 1215 the papal legate, Robert de Courçon renewed the bans on the metaphysical and physical works of Aristotle, including summaries of them, in a body of statutes for the masters of the University. All candidates for the License of Arts were called upon to swear an oath not to read the works of David of Dinant or Amaury of Bêne. In 1225 when Pope Gregory IX wrote an alarmed letter to the Theology faculty, whose certain members in 'a spirit of vanity' were failing to subjugate philosophy and in doing so they were transgressing the boundaries that had been clearly established by the Fathers. He admonished them to 'teach pure theology unfermented by worldly learning and cease adulterating the Word of God with the fictions of philosophers' (*Chartularium Parisiensis*).

At a higher and more universal level the thirteenth century represented a period in the Church's history which was marked by a growing awareness and a commensurate intolerance of heterodox groups. As Spencer Young has indicated the cause of this inclination may not be entirely certain but the resulting outcomes of an 'institutionalization of theology' are entirely clear.

> Whether or not the cause was as actual proliferation of heretical activity or a diminished tolerance on the part of the institutional Church for deviance, there was undoubtedly a heightened anxiety over the definition and protection of doctrinal orthodoxy (Young 2014).

This century was also a time which saw the papacy reach the peak of efficiency toward which it been moving for a century (Watt 2015). At an international level the fight against heresy was waged on a grand scale by Innocent III via the most important, most ambitious and best attended council of the medieval period. A recent history of Christianity has claimed of Innocent that, 'Few Christian leaders have had such a transforming effect on their world' (MacCulloch 2009). Perhaps for the first time the Church found itself led by someone who made the fight against heresy, and what one commentator called 'intellectual frontiersmen', one of his primary occupations (Mundy 2000). Innocent's reign was followed by three popes who were cut from a similar cloth as Honorius III, Gregory IX and Innocent IV kept up the momentum of their predecessor's reforming zeal. In 1215 Innocent III called the Lateran Council IV as a huge show of strength by a Church that was determined to flex its muscles. The opening lines of the Pope's summons, *Veneam Domini*, amounted to something akin to a distress signal, 'Beast of many kinds are attempting to destroy the vineyard of the Lord of Saboath, and their onset has so far succeeded against it that over no small area thorns have sprung up instead of vines' (*Selected Letters* 1953). For their part the servants of the Church responded accordingly, in all seventy-one patriarchs and metropolitan bishops, 412 bishops, 900 abbots, as well as representative of several monarchies were eventually in attendance. Having got down to business the Council asserted emphatically that, 'there is no similarity between the Creator and His creatures that is not subsumed by a discernible dissimilarity' (*Corpus iuris canonici* Liber extra 1 and 2) The second canon ended with a specific denunciation.

> We also reprobate and condemn the perverse teaching of the impious Amaury (Almaricus, Amalricus) de Bêne, whose mind the father of lies has so darkened that his teaching is to be regarded not so much heretic as insane (Schroeder 1937).[7]

Then, remarkably, in 1231 we begin to witness the first signs that there might be a softening in attitudes to the new learning. When Gregory IX issued the bull *Parens scientiarum* which confirmed the University's mandate and put an end to the Great Dispersion, he renewed his previous prohibition on certain readings but crucially he added a condition which was pregnant with implications. That is, that the *libri naturales* of Aristotle were to remain off limits 'until they have been examined and purged from all heresy' (Bulaeus 1665). Ten days after the promulgation of the bull the Pope brought together a commission of scholars in order to prepare the *libri naturales* for study (Young 2014). Two decades later in 1255 almost the entire corpus of Aristotle's writings were prescribed text for the masters in the Arts faculty at Paris. This is an extraordinary development through which Aristotle's works pass from proscription to prescription in a matter of decades and it is one that requires some explanation. It will be argued here that by examining Grosseteste's scholarly development, together with his personal contribution to ecclesiastical polity, we may be provided with insights that go some way to shedding light on these extraordinary developments.

[7] Fr Gabriel Théry has noted that David of Dinant is not mentioned in the canons and suggests that this was because of 'a certain liking for David on behalf of the Pope (1923).

3.6 Robert Grosseteste and the Pursuit of Learning in the Thirteenth Century

In terms of Grosseteste intellectual development in general, and his cosmology in particular, I have argued elsewhere that there is a significant development that reflected the changes in the intellectual and ecclesiastical environment during the turbulent decades when he was writing (Cunningham 2014). In 1225 Pope Honorius wrote a letter to the episcopacy in France in which he condemned the *Periphyseon* of Erigena because of its perceived pantheism. Probably at this time Grosseteste wrote his famous letter which would become known as *De unica forma omnium*. In this communication, even in the face of the papal opposition, Grosseteste defines God as 'the form of all things.' In other early works such as *De luce, De operationibus solis* and the commentary on Aristotle's *Physics* we find the concept that light is corporeity, it is an idea that is conspicuously absent in later works such as *De statu causarum* and the *Hexaemeron* (Panti 2012). A comparison of two of these works on Creation, *De luce* and the *Hexaemeron* is also indicative of a general trend in which the early rationalism, the corporeity of light and the non-Christian sources give way to a Deocentric account with it reference points in Scripture and Patristics. It is argued here that Grosseteste the early scholar is a writer that is open to the new ideas that were emerging in and around Paris. The milieu he emerged from as a scholar was one that had a proximity and a debt to the scholars of Chartres. It was a milieu that was open to the newly discovered Aristotelian *Physics* and *Metaphysics*, it was seeking rational explanations of the universe of the type that we find in *De sex dierum operibus*. He was a scholar like Thierry of Chartres who took the radical step of searching for the key to the cosmos outside the pages of Scripture.

However Grosseteste appears to have developed as the world around him developed. He may well have been in Paris in 1210 when the Amalricians were being burnt and David of Dinant was condemned. Post Lateran IV he is a Churchman and one of the most impressive stalwarts of institutionalized theology. For this Grosseteste, Scripture is the preeminent source for any scholar intent on the pursuit of knowledge. Writing to the regent masters at Oxford in 1246 he tells them this in no uncertain terms.

> So the foundation stones of the building of which you are the master builders are the books of the prophets…and also the books of the apostles and the Gospels. There are no others anyone can find or place in the buildings foundation (Mantello and Goering 2010).

This is not to say by any means that Grosseteste relinquished the philosophers—his translations alone represent an enormous contribution to the introduction of Greek learning to the West, but later as a teacher, and ultimately as a Bishop, a characteristic feature of his approach to learning was to draw up a clear line of demarcation between theology and philosophy. What Grosseteste did was use Greek learning in conformity with the spirit of Lateran IV. As Gordon Leff has pointed out Grosseteste crucially did not try to harmonize the pagan philosophies with orthodox theology, Aristotle was not a Christian and treating him as one would only lead to heresy. At Oxford what he did do was take and utilize the Aristotelian methodology of demonstration and scientific method which he applied to the Neoplatonic proclivity for Mathematics, light and intelligible truth.

> That is not to say that Grosseteste rejected Aristotle; on the contrary, he did for Aristotelian scientific method what the theologians of the first part of the thirteenth century did for Aristotelian natural philosophy, in Neoplatonizing it. But with this difference: that Grosseteste brought to his work a mathematical insight and originality that all his contemporaries lacked in their speculative theology (Leff 1968).

However, in terms of the reception of Aristotle it was what Lincolniensis does with the Greek philosopher's epistemology that has the most profound impact. If Aristotle tells the world that new knowledge can be obtained through the senses then he is contradicting a generally held Christian assumption that sensory perception was disabled and consequently untrustworthy since the Fall of man. One of Grosseteste's finest achievements was to reassure the West that knowledge does not come 'from' the senses, but rather 'via' them. The true source of all wisdom is the Divine Intelligence. The senses, the lowest of all the human faculties, are in the service of the mind. 'The senses now perform the function of rousing the mind from its somatically induced slumbers, and indeed it is repeated sensory experiences that prosecute this task best of all' (Harrsion 2009). Aristotle's *scientia*, however useful for jolting the mind out of its slumber, is inferior to the *sapientia* that comes to us by illumination from God. This is a highly important development since once it recognizes the inferiority of *scientia* the Western world could, paradoxically, also recognize its profound, even sacred, utility.

Grosseteste was a developing scholar in the highly charged atmosphere of the thirteenth century and one of his greatest intellectual contributions is that by subsuming Greek philosophy to Scripture and Patristics he was maintaining their continued use in the only way possible. However we must not think of this intellectual acclimatization as a one way process in which the Church molded the methodology of Grosseteste. In continuing to employ these sources the Bishop of Lincoln, along with others, was demonstrating to the Curia, and to Roman Church in general, that it was possible to do so in a benign manner which did not threaten the strict orthodoxy that was being so rigorously applied. Rega Wood has maintained that a primary cause behind the softening of Gregory IX attitude toward the *libri naturales* and the ultimate full approbation that came mid-century, was that scholars celebrated for their piety were using these texts. She singles out Robert Grosseteste and William of Auvergne as principal examples (Wood 1995). The banning and then the endorsing of Aristotle within astonishingly few decades at the University of Paris was testament to the successes of Grosseteste and others. Grosseteste's great gift to the Church, and to Western thought, at this period in time was that he was not merely an outstanding student, but by his example he was also a highly important teacher.

References

Avicebron. (2001). *Selected poems of Solomon Ibn Gabirol* (P. Cole, Trans.). Princeton, NJ: Princeton University Press.

Avicebron. (2008). *The fountain of life* (H. E. Wedeck, Trans.). Charleston, SC: Bibliobazaar.

Baemker, C. (1890). *Das Problem de Materie in der Griechischen Philosophie*. Munster: Aschendorff.

Bulaeus, C. E. (1665). *Historia Universitatis Parisiensis*. Paris.

Caldwell Ames, C. (2015). *Medieval heresies: Christianity, Judaism and Islam*. Cambridge: Cambridge University Press.

Clayton, P. (1998). The case for Christian Panentheism. *Dialog, 37*, 201–208.

Cohn, N. (1957). *Pursuit of the Millennium*. London: Secker & Warburg.

Copleston, F. (1966). *A history of philosophy: Augustine to Scotus*. London: Burns and Oates.

Cunningham, J. (2014). *Lumen de Lumine*: Light, God and creation in the thought of Robert Grosseteste. In N. Temple, J. S. Hendrix, & C. Frost (Eds.), *Bishop Robert Grosseteste and Lincoln Cathedral* (pp. 81–98). Surrey: Ashgate.

D'Onofrio, G. (2008). *History of theology: The Middle ages*. Collegeville, MN: Liturgical.

Dagron, T. (2003/2004). David of Dinant—on the Quaternuli Fragment. *Revue de Métaphysique et de morale, 40*, 419–436.

Davenport, A. A. (1997). The Catholics, the Cathars and the concept of infinity in the thirteenth century. *Isis, 88*(2), 263–295.

Dickson, G. (1989). The burning of Amalricians. *Journal of Ecclesiastical History, 40*(3), 347–369.

Duhem, P. (1954). *Le Systéme du Monde III*. Paris: Hermann.

Elford, D. (1992). William of Conches. In P. Dronke (Ed.), *A history of twelfth-century philosophy* (pp. 308–327). Cambridge: Cambridge University Press.

Häring, N. (1955). The creation and creator of the world according to Thierry of Chartres and Clarenbaldus of Arras. *Archives d'histoire doctrinale et littéraire du moyen âge, 22*, 137–216.

Harrison Thomson, S. (1940). *The writings of Robert Grosseteste*. Cambridge: Cambridge University Press.

Harrsion, P. (2009). *The fall of man and the foundations of science*. Cambridge: Cambridge University Press.

Hauréau, B. (1872). *Histories de la philosophie scolastique* (3 Vols.). Paris.

Janssen, W. (1926). *Der Kommentar des Clarembaldus von Arras zu Boethius de Trinitate*. Breslau: Müller & Seiffert.

Jeauneau, É. (2009). *Rethinking the school of Chartres*. Toronto: University of Toronto Press.

Krause, K. F. (1829). *Vorlesungen über die Grundwahrheiten der Wissenschaft*. Göttingen: In Commission der Dietrich'schen Buchandlung.

Kurdzialek, M. (1963). *Davidis de Dinanto Quaternulorum* (Studia Mediewistyczne III). Warsaw: Państwowe Wydawnictwo Naukow.

Leff, G. (1968). *Paris and Oxford Universities in the thirteenth and fourteenth centuries*. New York: Wiley & Son.

Luscombe, D. (2011). Crossing philosophical boundaries c. 1150-c.1250. In S. E. Young (Ed.), *Crossing the boundaries of medieval Universities*. Leiden: Brill.

Maccagnolo, E. (1988). David of Dinant and the beginnings of Aristotelianism in Paris. In P. Dronke (Ed.), *A history of twelfth-century western philosophy* (pp. 429–442). Cambridge: Cambridge University Press.

MacCulloch, D. (2009). *A history of Christianity*. London: Allen Lane.

Mantello, F. A. C., & Goering, J. (2010). *The letters of Robert Grosseteste, Bishop of Lincoln*. Toronto: Toronto University Press.

McEvoy, J. (1982). *The philosophy of Robert Grosseteste*. Oxford: Clarendon.

McEvoy, J. (1983). The chronology of Robert Grosseteste's writings on nature and natural philosophy. *Speculum, 58*(3), 614–655.

Miccoli, L. (2001). Two thirteenth-century theories of light: Robert Grosseteste and St Bonaventure. *Semiotica, Journal of the International Association for Semiotic Studies, 136*, 69–84.
Mundy, J. H. (2000). *Europe in the high Middle-ages, 1150-1300*. London: Pearson Education Ltd.
Panti, C. (2012). The evolution of the idea of corporeity in Robert Grosseteste's writing. In J. P. Cunningham (Ed.), *Robert Grosseteste, his thought and its impact* (pp. 111–139). Surrey: Ashgate.
Pittenger, N. (1950). *Historic faith in a changing world*. New York: Oxford University Press.
Rashdall, H. (1895). The Universities of Europe in the Middle ages (F. M. Powicke & A. B. Emden, Eds., 1936 ed.). Oxford: Oxford University Press.
Rosemann, P. W. (2013). Philosophy and theology in the Universities. In C. Lansing & E. D. English (Eds.), *A companion to the medieval world* (pp. 544–560). Oxford: Blackwell.
Russell, J. B. (1971). *Religious dissent in the Middle ages*. New York: John Wiley & Son.
Schroeder, H. J. (1937). *Disciplinary decrees of the general councils: Text, translations and commentary*. St Louis, MO: Herder Books Co.
Théry, G. (1923). *Autour de Décret de 1210: I David de Dinant*. Le Saulchoir: Kain.
Thijssen, J. M. M. (1996). Master Amalric and the Amalricians: Inquisitorial procedure and the suppression of Heresy at the University of Paris. *Speculum, 71*(1), 43–64.
Verger, J. (2005). Conclusion. In F. Morenzoni & J. Tilliette (Eds.), *Autour de Guillaume d'Auvergne (d. 1249)*. Turnout: Brepols.
Watt, J. A. (2015). The Papacy. In D. Abulafia (Ed.), *The New Cambridge medieval history, c. 1198-c. 1300* (pp. 107–163). Cambridge: Cambridge University Press.
Wood, R. (1995). Richard Rufus' *Speculum anime*: Epistemology and the introduction of Aristotle in the West. In A. Speer (Ed.), *Die Bibliotheca Amploniana: Ihre Bedeutung im Spannungsfeld von Aritotelismus, Nominalismus und Humanismus* (pp. 86–109). Berlin: De Gruyter.
Young, S. E. (2014). *The scholarly community at the early University of Paris: Theologians, education and society, 1215-1248*. Cambridge: Cambridge University Press.
Zwi Werblowsky, R. J., & Wigoder, G. (Eds.). (1997). *Oxford dictionary of the Jewish religion*. Oxford: Oxford University Press.

Edited Primary Sources

Anonymi Laudunensis Canonici. (1822). *Cronica Recueil des historiens des Gaules et de la France XVIII*. Paris.
Caesarii Heisterbachensis Monachi Dialogus Miraculorum. (1851). *Ed Joseph Strange*. Cologne: H. Lempertz & Co.
Cheney, C. R., & Semple, W. M. (Eds.). (1953). *Selected letters of Pope Innocent III concerning England, 1198–1216*. London: Nelson.
Contra Amaurianos nach der Handschrift zur Troyes. (1926). In C. I. Baeumker (Ed.), *Beiträge zur Geschichte der Philosophie des Mittelatters XXIV*. Munster: Aschendorff.
Corpus Christianorum: Continuatio medievalis. (1966). Turnhout: Brepols.
Corpus iuris canonici. (1879–1881). 2 vols. Leipzig: Ex Officina Bernhardi Tauchnitz.
Denifle, H., & Chatelain, E. (Eds.). (1889–1897). *Chartularium Universitatis Parisiensis*, 4 vols. Paris.
Ditionnied Histoire et de Géographie Ecclésiastiques. (1914). Paris.
Gesta Philippi II Augusti. (1882). In H. F. Delaborde (Ed.), *Oeuvres de Rigord et de Guillaume le Briton, historiens de Philippe-Auguste*, 1. Paris.
Patrologiae cursus completes: Series Latina. (1844–1990). Paris: Migne & Garnier.

Chapter 4
All the Colours of the Rainbow: Robert Grosseteste's Three-Dimensional Colour Space

Hannah E. Smithson

4.1 Our Task: To Understand Colour

Our sensory experience is enriched by colour. Some of the most eye-catching displays in nature are coloured, and for centuries mankind has sought to understand colour and to discover how to produce, at will, all possible colours. A full understanding of colour demands answers to multifaceted questions, spanning physical, biological and psychological domains. It would encompass the origins and generation of colour, the regularities of colour perception, and the manipulation of colour experience. In this chapter we consider two accounts of colour—one from the thirteenth century and one from the twenty-first century. Although separated by nearly 800 years, comparison of these accounts is made possible through their shared reference to the colours of the rainbow, a persistent natural phenomenon that allows us to reach into the past with an objective standard.

Some colours are perceptually similar—red and orange; green and blue—while others are perceptually dissimilar—red and green; blue and orange. Such observations have driven the search for a colour-ordering system that captures these relationships (Kuehni and Schwarz 2007). There are indications that some colours have a special status, which should also be captured. Black corresponds to the absence of light. White is one of the so-called unique hues, appearing phenomenologically unmixed (although see Saunders and van Brakel 1997 for critical discussion). There is evidence that some colours are preferentially represented across languages and cultures (Berlin and Kay 1969). Does their special status arise from some characteristic of our biological make-up or from the physical environment around us? How can the relationships between colours be represented and would a

H.E. Smithson (✉)
Department of Experimental Psychology, University of Oxford, Oxford, UK
e-mail: hannah.smithson@psy.ox.ac.uk

J.P. Cunningham, M. Hocknull (eds.), *Robert Grosseteste and the pursuit of Religious and Scientific Learning in the Middle Ages*, Studies in the History of Philosophy of Mind 18, DOI 10.1007/978-3-319-33468-4_4

suitable representation allow us to make sense of the link between characteristics of visual stimulation and the perceptual experiences they evoke? Modern colour science grapples with these issues, but they are not new.

4.2 A Thirteenth Century Introduction to Colour

4.2.1 Grosseteste on Colour

We base our historical discussion on two treatises composed in the early thirteenth century by Robert Grosseteste: the *De colore* (On colour) and the *De iride* (On the rainbow). The *De colore* is a dense text of fewer than 400 words but one that presents a number of deep puzzles and challenges. It has received relatively little study and yet forms an important element within Grosseteste's 'scientific' canon, dating probably from the mid-1220s (Dinkova-Brunn et al. 2013). The *De iride* is among the last of the scientific works by Robert Grosseteste, dating to the period 1228–1232, a period of his life in which it is easier to establish his career (Panti 2013). The treatise is a sophisticated investigation into the phenomenon of the rainbow. The final section deals with colour variation in rainbows, and it is this section that we consider in detail here.

In the *De colore*, Grosseteste conceives and deploys terminology in a way that assumes very tight definitions, but he avoids linking those definitions to explicit colour terms. Instead, he uses his precise terminology to construct an abstract geometric space within which he enumerates exact mathematical combinatorics of colour. In the *De iride*, Grosseteste makes substantial use of the theory of colour and light expounded in the *De colore*, but now he links his terminology to specific properties of rainbows. In what follows, we start by summarising the proposals in the *De colore* and the *De iride*, and then summarize modern proposals, before making links between the two, with direct reference to the colour properties of rainbows. Finally, we evaluate the successes and failures of both the thirteenth century and twenty-first century accounts and their correspondences.

The *De colore* and the *De iride* are technical texts, and some terminology drawn from modern colour science proves very useful in discussing the issues that Grosseteste confronts in his writing. To do so is not to admit any form of anachronistic projection of a modern scientific framework onto the achievements of a thirteenth-century mind; rather it is to employ all the tools at our disposal to think ourselves back to what Grosseteste perceived his task to be, and what he accounted as progress toward its completion. The tools of modern colour science help us to form a sharper understanding of the perceptual phenomena that Grosseteste was analyzing.

4.2.2 *The De coloribus to the De colore*

The prevailing view of colour at the time was the one put forward by Aristotle. He identified seven 'species' of colour existing naturally in the world that could be arranged in a line from white to black, and proposed that all other colours were derived by mixture of the seven fundamental species (Aristotle trans. Beare 1984). A later anonymous work from the Peripatetic School, the *De coloribus*, was long-believed to have been written by Aristotle himself, since it systematically describes colours according to Aristotle's elemental admixture theory. The *De coloribus* identifies three things that generate colours: 'the light, the medium through which the light is seen, such as air and water, and thirdly the colours forming the ground, from which the light happens to be reflected' (Anon trans. Hett 1936). Any link between the seven fundamental species of colour and the three things that generate colours remains obscure in these texts.

In the *De colore*, Grosseteste introduces the phenomenon of colour as a property of light and matter. The text opens with his definition of colour: 'Colour is light incorporated in a diaphanous material' (Dinkova-Brunn et al. 2013). He then identifies three bipolar qualities of colour, using three pairs of Latin words. Two qualities belong to the description of light independent of the medium carrying it, and one arises only by reference to properties of the medium through which the light is passing. The light can be either *multa* or *pauca*, and either *clara* or *obscura*. The material can be *purum* or *impurum*. The three pairings are used consistently and without variation throughout the text. In the following analysis, we use Grosseteste's Latin terms without translation. For, as we shall see, the appropriate translation is not immediately obvious (Smithson et al. 2012).

It is common today to specify colour by three independent properties. For example, red, green and blue (RGB) on a computer display, or hue, saturation and value or brightness (HSV, or HSB) in other contexts. The use of three independent properties in these schemes is no coincidence, reflecting instead a fundamental constraint imposed by the biology of the human visual system, as we explain below. So, Grosseteste's use of three bipolar qualities is tantalizing. However, if the text included only a list of three bipolar qualities the structure of Grosseteste's colour space would remain ambiguous. The next part of the text is critical.

Grosseteste explicitly sets-up a combinatorial space of colours, based on his three bipolar qualities. He starts specifying colours in terms of their relative positions along the bipolar dimensions. Whiteness, for example, is specified by the triplet [*multa*, *clara*, *purum*], and blackness by the triplet [*pauca*, *obscura*, *impurum*]. By identifying whiteness with the positive poles of his three qualities and blackness with the poles of absence, Grosseteste makes a conceptual link with an Aristotelian one-dimensional scale. However, his next requirement cannot be accommodated in a one-dimensional scheme. He identifies a set of three colours that share two elements with whiteness, and that are diminished in the third element; a further set of three colours that share only one element with whiteness, and that are diminished in the other two elements; and a final colour that is diminished in all three elements at once. For Grossteste, the 'seven colours close to whiteness' (Dinkova-Brunn et al. 2013) result not from a close sequence of descent, but from counting the different combinations of presence and absence of three bipolar qualities. Borrowing efficient place-value notation from binary mathematics, we can count the possible combinations as 000, 001, 010, 011, 100, 101, 110 and 111, where the three places represent the three qualities, and 1 indicates presence and 0 indicates absence. If whiteness is already associated with 111, there are seven remaining combinations. The combinations are categorically different, rather than differing by degree, giving an unambiguous reason for specifically seven colours, 'no more, no fewer' (Dinkova-Brunn et al. 2013).

By linking Grosseteste's three bipolar linguistic qualities to the three axes of Cartesian space, width, depth, and height, it is possible to visualize Grosseteste's combinatoric account of colour space. In doing so, we make no claim about whether Grosseteste himself imagined his scheme geometrically, we simply use a representation that is familiar to us. In this visualization, the three spatial dimensions represent a change from *pauca* to *multa*, from *obscura* to *clara*, and from *impurum* to *purum* (Fig. 4.1a). So, the corners of a cube in this *space* represent combinations of these three bipolar qualities. Two corners, one representing [*multa*, *clara*, *purum*] and the other representing [*pauca*, *obscura*, *impurum*] have been identified as blackness and whiteness respectively (Fig. 4.1b).

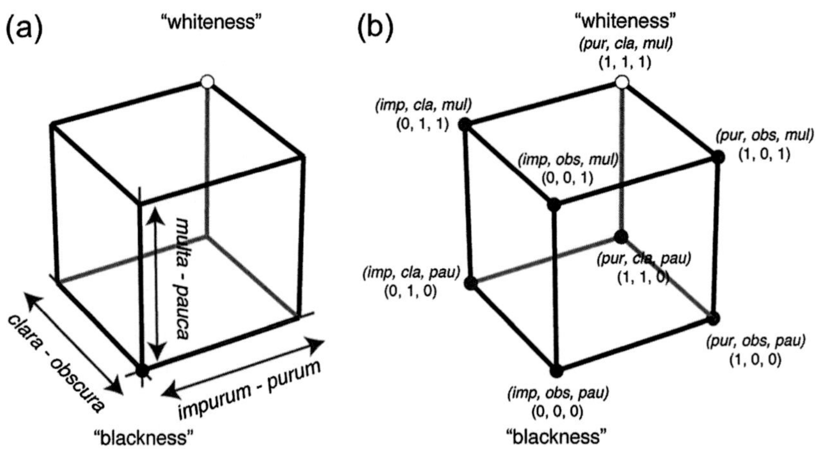

Fig. 4.1 Visualising Grosseteste's bipolar linguistic qualities in a three-dimensional Cartesian space. (**a**) The three spatial dimensions represent a change from *pauca* to *multa*, from *obscura* to *clara*, and from *impurum* to *purum*. (**b**) The corners of a cube in this space represent combinations of these three bipolar qualities. Here we use 0 and 1 to represent the two extremes, along each of three dimensions. Two corners, one representing [*multa*, *clara*, *purum*] (1, 1, 1) and the other representing [*pauca*, *obscura*, *impurum*] (0, 0, 0) have been identified as *blackness* and *whiteness* respectively. The eight possible combinations are represented in the figure, using abbreviations *pau* (*pauca*), *mul* (*multa*), *obs* (*obscura*), *cla* (*clara*), *imp* (*impurum*) and *pur* (*purum*)

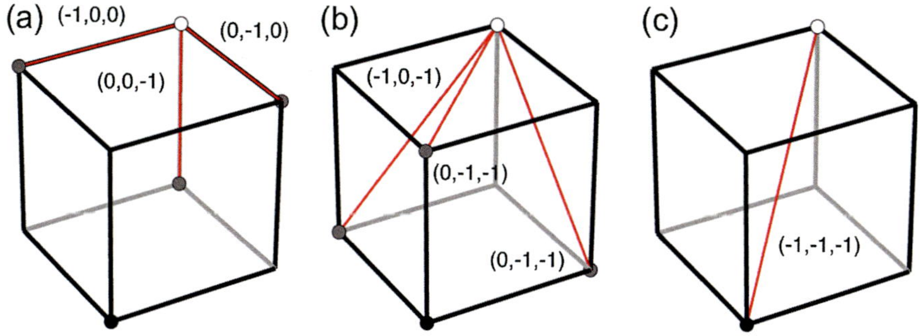

Fig. 4.2 Cartesian visualisation of the seven directions of descent from whiteness. From the whiteness corner, located at the coordinate (1, 1, 1), there are seven discrete directions that lead to the seven remaining corners. (**a**) The first set of three, defined by the Cartesian vectors [−1, 0, 0], [0, −1, 0], [0, 0, −1], correspond to decreasing exactly one quality while keeping the others fixed, and move along edges of the cube. (**b**) The second set of three, notationally [−1, − 1, 0], [−1, 0, −1], [0, −1, −1], correspond to decreasing two of the qualities while keeping one fixed, and describe the diagonals of faces of the cube. (**c**) The vector [−1, −1, −1] corresponds to the decrease of all three qualities at once and corresponds to the main diagonal of the cube

In identifying seven additional colours descending from whiteness, Grosseteste specifies seven directions, represented here by vectors joining the whiteness corner to the other corners. Following Grosseteste's categorisation of these displacements, the first set of three, defined by the Cartesian vectors [−1, 0, 0], [0, −1, 0], [0, 0, −1], correspond to decreasing exactly one quality while keeping the others fixed, and move along edges of the cube (Fig. 4.2a). The second set of three, notationally [−1, − 1, 0], [−1, 0, −1], [0, −1, −1], correspond to decreasing two of the qualities while keeping one fixed, and describe the diagonals of faces of the cube (Fig. 4.2b). The vector [−1, −1, −1] corresponds to the decrease of all three qualities at once and corresponds to the main diagonal of the cube (Fig. 4.2c), directly connecting whiteness to blackness. The Cartesian geometric construction presented here is exactly equivalent to Grosseteste's logic. Once the geometric space is recognized, a structure of stark clarity is perceived in the apparently dense language.

In the subsequent section, Grosseteste extends the discrete combinatorial space of his first enumeration into a fully continuous three-dimensional space. He generates a matching yet distinct set of seven colours by the symmetrical working from black in ascent (Fig. 4.3), and refers to colours as continuously distributed along these two sets of seven directions, so that they meet in a 'middle space'. This meeting is described by Grosseteste using the Latin word *concurrentes* (commonly signifying 'convergence'). Importantly it is used in this text not to imply a meeting of all the colours at a single point. Instead, it is a meeting of pairs of colour directions from whiteness and blackness, each pair meeting at its own point, rather like the fingers of two hands spread out and resting against each other, fingertip to fingertip. Grosseteste's ablative '*in medio*,' together with the geometric implications of his dual sets of colours emerging from the poles '*in idem*' towards it, indicates his contemplation of an extensive 'middle space' of colour that is in some

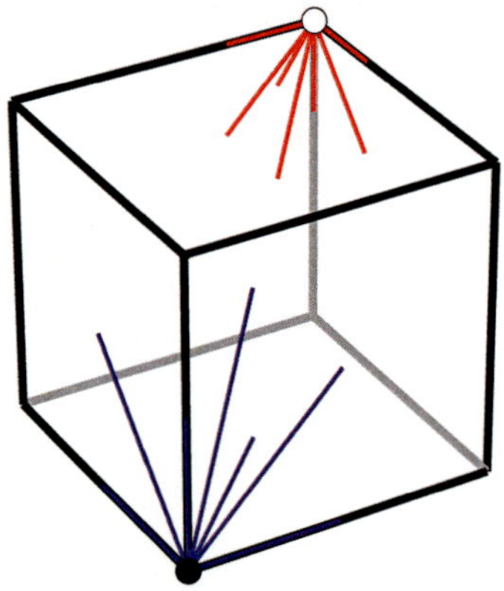

Fig. 4.3 Cartesian visualisation of the seven colours of descent from whiteness and a further seven of ascent from blackness. The cubic frame represents Grosseteste's three axes of colour variation. One corner (at the minimum extreme of each axis) is associated with blackness and the opposite corner (at the maximum extreme of each axis) is associated with whiteness. The *red lines* identify seven directions of descent from whiteness, associated with diminishing one, two or three qualities once. The *blue lines* identify seven directions of ascent from blackness, associated with increasing one, two or three qualities at once. The *red* and *blue lines* meet in a middle space, and not at a single point

sense midway between white and black. Critically, this is a space and not a single point (Dinkova-Brunn et al. 2013). The Cartesian geometric construction presented here provides a way to appreciate and understand the conceptual leap made in the *De colore* away from the Aristotelian linear series of degrees from whiteness to blackness.

4.3 Functional Interpretation of the *De colore*

4.3.1 How Significant Is the Three-Dimensional Scheme?

It is striking that in the *De colore*, Grosseteste uses no colour terms (apart from whiteness and blackness). How precise were Grosseteste's observations? Does he reach a unique conclusion about the phenomenon of colour, or simply present a neat account based on a synthesis of limited observations and mathematical or doctrinal convenience? Here we explore a range of interpretations of the treatise that differ in the weight they attach to the importance of the three-dimensional scheme. We question the extent to which the scheme captures a fundamental principle that is germane to human colour perception specifically, and we seek the appropriate interpretation of his key terms, used without variation to refer to the three dimensions.

Classification of phenomena by the presence and absence of particular properties or elements follows the style of Aristotelian physics, in which, for example, the four elements are classified by the presence and absence of the two qualities of heat and dryness. In general, enumeration of the presence or absence of two elements produces four things; enumeration of the presence or absence of three elements produces eight things; and enumeration of presence or absence of four elements produces 16 things. Human colour perception is to some extent categorical—we see many distinct colours in the world, and for a few of them we have specific colour names. This property is preserved across languages and cultures, though the number (and possibly the referents) of the terms may differ (Berlin and Kay 1969). Grosseteste in his corpus uses at least the following: *rubeus*, *croceus*, *viridis*, *ceruleus*, *hyacintinus*, *violacaeus*, *purpureus*. The categorical nature of colour perception, the number of commonly used colour terms, and the Aristotelian account of seven colours, calls for a classification based on the presence or absence of three qualities; two qualities produces too few colours ($2^2 = 4$), and four produces too many ($2^4 = 16$). In one interpretation of *De colore*, in which Grosseteste's account reflects the straightforward mathematical convenience of relating eight colours to combinations of three bipolar qualities, there is no scheme by which to identify particular patterns of presence and absence to particular colours, except in the case of blackness and whiteness.

It is clear however from the text, that Grosseteste means his scheme to be used to capture all possible variation in colour. He identifies discrete combinations of the three bipolar qualities, but also allows degrees of intensification and diminution to produce all possible colours. This moves away from a simple counting scheme to one that describes continuous gradations of colour. The relationships he describes between colour directions that differ in one, two or three qualities from white, and the extension of this to include degrees of difference, can be neatly visualised in a three-dimensional Cartesian space. It is not clear from the text whether Grosseteste had an explicit geometric representation in mind, but the three-dimensionality of Grosseteste's colour space is beyond question (Smithson et al. 2012).

Furthermore, its interpretation as a conceptual theatre in which specific manipulations of colour can be played out is heavily implied by the closing paragraph of the treatise:

> What is understood in this way about the essence of colours and their multiplication, becomes apparent not only by reason but also by experience to those who thoroughly understand the depth of the principles of natural science and optics. And this is because they know how to make the diaphanous medium either pure [*purum*] or impure [*impurum*], so that in it they can receive bright [*clarum*] light, or dim [*obscurum*] if they prefer, and through the shape formed in the diaphanous medium itself they can make scarce [*paucum*] light, or increase that same light at will; and so through skilful manipulation they can show visibly, as they wish, all kinds of colours (Dinkova-Brunn et al. 2013, pp. 19).

4.3.2 How Should We Interpret Each of the Dimensions?

Translation and interpretation of Grosseteste's bipolar qualities is more problematic. Grosseteste provides no explicit definition these qualities, except in the case of *multa-pauca*, where he elaborates, writing:

> I do not say that copious [*multa*] light is light diffused through a large expanse; rather, I say that copious [*multa*] light is gathered as if in a point when a concave mirror is positioned facing the sun and light falling over the whole surface of the mirror is reflected towards the centre of the sphere of the mirror. Thus, by the collection of this light in the very centre combustible material is very quickly set on fire. (Dinkova-Brunn et al. 2013, pp. 17).

One interpretation of the referential ambiguity of the *De colore* is that Grosseteste had in mind a general perceptual framework for colour, which is adaptable to different circumstances of materials and illumination, rather than a definite scheme. It is notable from a modern perspective that Grosseteste assigns two qualities to the description of light independent of the medium carrying it, and one arises only by reference to properties of the medium through which the light is passing. It is possible that, on examination, Grosseteste is not extending the dimensionality of Aristotle's colour-line at all, but merely incorporating the ideas of 'cloudiness' or 'transparency' (or a combination of these) within the concept of 'colour.' Perhaps the departure from Aristotle really consists of a move from discussing the appearance of coloured *lights* to discussing the appearance of coloured *materials*. This corresponds to quite a different dimensional extension of a colour line.

Alternatively, the failure to identify specific colour terms is equally to be expected if Grosseteste were describing a three-dimensional perceptual space in which the bipolarities identified subjective axes of colour variation, like hue, saturation and brightness. Although the *De colore* presents several strong constraints on the nature of the bipolar axes—specifying for example that whiteness is located where all three axes of variation are at their positive extremes—generating a mapping from the terminology in the *De colore* to modern colour terminology is possible only by forcing interpretations on the text. From the *De colore* interpretation of the key terms describing Grosseteste's axes of colour variation remains ambiguous.

4.4 The *De colore* to the *De iride*

Although the *De colore* presents the modern reader with an unresolvable puzzle, Grosseteste, in the *De iride*, provides us with a clue: the variation of colour in rainbows. In the *De iride*, the section on colour starts with a recapitulation of the framework that was laid down in the *De colore*. Again colour is inherently associated with the interaction of light and materials: '… colour is luminosity mixed with a diaphanous medium' (Smithson et al. 2014, pp. A342). Variation in colour results from variation in the qualities of the light and the medium: the 'diaphanous medium is differentiated according to *puritatem* and *impuritatem*, the luminosity is divided four ways; that is, according to *claritatem* and *obscuritatem*, and then according to *multitudinem* and *paucitatem*, and the generation and the diversity of all colours occurs according to the combinations of these six distinguishing characteristics' (Ibid.).

In the *De iride*, Grosseteste goes beyond this abstract conceptualization of colour to link these axes of variation to properties of rainbows. He writes, 'The variety of colour in the different parts of one and the same rainbow occurs chiefly because of the *multitudinem* and *paucitatem* of the solar rays. For where there is a greater multiplication of rays, the colour appears clearer and more luminous; and where there is a smaller multiplication of rays, the colour appears dim and close to purple' (Ibid.). And later, 'In fact, the difference in the colours between one rainbow and another arises sometimes from the *puritate* and *impuritate* of the recipient diaphanous medium, sometimes from the *claritate* and *obscuritate* of the luminosity impressed on it. For if the diaphanous medium is *purum* and the luminosity is *clarum*, the colour of the rainbow will be more similar to white and light. But if the recipient diaphanous medium should contain a mixture of smoky vapors and the *claritas* of the luminosity is scarce, as occurs around sunrise and sunset, the colour of the rainbow will be less brilliant and more obfuscated.'

This passage in the *De iride* therefore provides the potential link from Grosseteste's terminology to physically repeatable phenomena. One of his bipolar axes is assigned to different parts of the rainbow [*multitudinem et paucitatem*], another to the quality of the diaphanous medium giving rise to different rainbows [*puritate et impuritate*], and a third to the luminosity of the incipient light [*claritate et obscuritate*]. It is therefore possible to test the hypothesis that the colour variations exhibited by rainbows span perceptual colour space in a way that is consistent with the abstract description in the *De colore*.

One way to formalize the variation of colour within and between rainbows, would be to obtain a range of calibrated photographs, and to analyse the colour variations therein. An alternative approach is to undertake detailed physical modelling of the colours, or more correctly the light spectra, produced in different types of rainbow. Here we describe modelling (Ibid.) that was directly inspired from the observations and comparisons set out in the *De iride*.

4.5 Modelling the Colours in a Rainbow

A rainbow is formed by refraction and reflection of light within individual raindrops. According to the basic scheme, described by Descartes (trans. Olscamp 2001) and Boyer (1959), sunlight enters a raindrop and is reflected internally one or more times before finally exiting (Fig. 4.4a). A single internal reflection gives rise to the primary bow, and two internal reflections generate the secondary bow. The refractive index of the droplet determines the angular deviation of the ray at

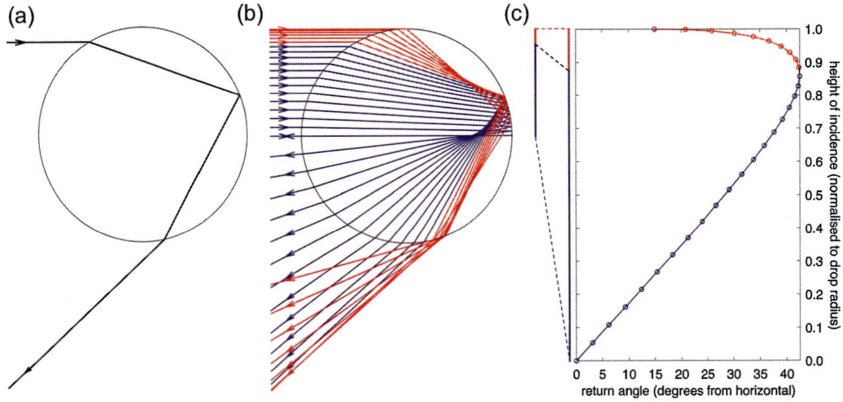

Fig. 4.4 Light refraction and reflection by a spherical raindrop. (**a**) Light enters a raindrop, and is refracted at the air-water boundary, before being reflected internally, and refracted again on exiting the raindrop. (**b**) A parallel bundle of rays from a distant source (such as the sun) take different paths through the raindrop, depending on their height and hence the angle at which they are incident on the air-water boundary. A ray that enters in line with the centre of the raindrop (height $= 0$) is reflected back horizontally. Rays that enter higher than the centre of the raindrop (height > 0) are reflected with some angular deviation from horizontal. As the height increases, the angular deviation increases (rays depicted in *blue*), until a critical height is reached, at which the angular deviation begins to decrease (rays depicted in *red*). The critical height depends on the refractive index of the raindrop. For a refractive index $(n) = 1.33$ it corresponds to an angle of incidence of 59.6° and produces a deviation of 42.5°. The exiting rays are concentrated along this so-called caustic angle. (**c**) A plot showing the relationship between incident height of rays on the raindrop, and the angular deviation of the exiting ray. The colour coding indicates the transition from increasing deviation with increasing height (*blue*) to decreasing deviation (*red*). The caustic angle corresponds to the turning point in this function, where incident power is concentrated over a narrow range of exit-angles

each air-water interface. A bundle of parallel rays from the distant sun that enter the droplet and undergo one internal reflection will be dispersed when exiting the raindrop, but remain largely concentrated along a particular path, known as the caustic ray (Fig. 4.4b, c).

Descartes was able to use this account to predict the location of the rainbow, but was missing an explanation of rainbow colours, which relied on Newton's observation (Mollon 2003) that different wavelengths of light, associated with different colours, have different refractive indices in a given medium. Refractive index increases through the spectrum: It is low for short wavelengths (which appear blue) and high for long wavelengths (which appear red). Hence, for each wavelength there is a different caustic angle (Fig. 4.5). Since sunlight contains energy at many wavelengths, the spectral content of the exiting light varies as a function of angle. Internal reflection within a raindrop and refractions at the air-water boundaries concentrate light at different caustic angles for each wavelength, and the superposition of wavelengths produces the familiar colours of the rainbow.

A naïve interpretation of the sequence of colours within a rainbow is that it corresponds only to a variation in hue, captured by the hue-terms red, orange,

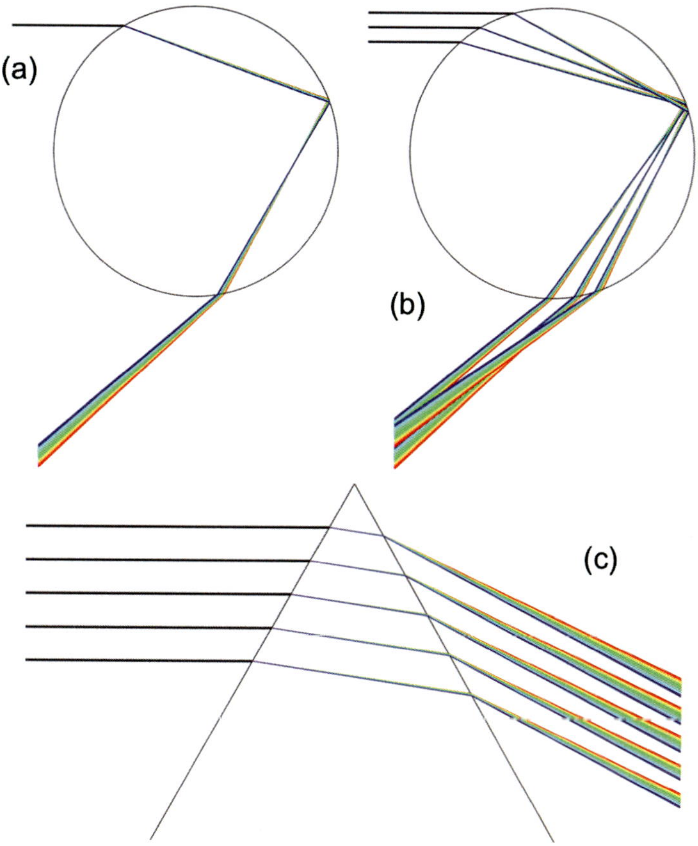

Fig. 4.5 Wavelength dependence of refraction in a raindrop and a prism. (**a**) White light containing energy at many wavelengths enters the raindrop. Each wavelength of light has a different refractive index in water, so each wavelength is refracted through a different angle, and exits the raindrop at different angles. In this image, different wavelengths are colour coded (*blue*, *cyan*, *green*, *yellow*, *orange*, *red* for the progression from short to long wavelengths). The diagram is generated with a realistic range of refractive indices for visible wavelengths of light in water, which is just sufficient to show the wavelength-dependent dispersion. (**b**) A bundle of rays entering the raindrop at different heights is each dispersed. The exiting light is a superposition of dispersed wavelengths from each of these incident rays. This superposition produces the familiar colours of the rainbow. The light exiting the raindrop at a given angle is not confined to a single wavelength, but is instead an additive mixture of several wavelengths. (**c**) Alternatively, when the refracting medium is a triangular prism, the parallel bundle of light rays entering the prism meets the refracting surface at a fixed angle, and the light exiting the prism at a given angle corresponds only to a single wavelength. This geometry was critical in Newton's experiments for it allowed him to isolate single wavelengths of light

yellow, green, blue, indigo and violet. Similarly, it is common in nontechnical accounts of the rainbow to link rainbow colours to spectral colours. However, as is clear from Fig. 4.5b, the light exiting the raindrop at a given angle is not confined to a single wavelength, but is instead a complex additive mixture of several wavelengths. To further illustrate the differences between spectral colours and rainbow colours, it is worth considering the effects of refraction by a prism. In this case, due to the straight sides of the prism, parallel rays of white light meet the refracting boundary at a constant angle, irrespective of the location along the side of the prism, which in turn means that the light exiting the prism at a given angle corresponds only to a single wavelength (Fig. 4.5c). The shape of the refracting object profoundly influences the way in which colour varies with scattering angle. The straight-sided prism allowed Newton to isolate single wavelengths of light, and it is this simplicity that supported further inferences he was able to make about colour and wavelength.

Predictions from geometric optics do not capture two further important physical characteristics of rainbows: the existence of supernumerary arcs, and the dependence of rainbow colours on droplet size. Using wave theory, Airy developed an excellent approximation of the primary rainbow (Airy 1838). A rigorous model of all of the scattering processes caused by a spherical droplet of water, such as external reflection, multiple internal reflections, surface waves, and diffraction is provided by Mie theory. A reformulation of Mie theory, known as the Debye series (Hovenac and Lock 1992), also provides an exact solution, and that is what we use in our modelling.

4.6 Linking the Model to Grosseteste's Account in *De iride*

We first consider two of Grosseteste's dimensions of colour variation, one assigned to different parts of the rainbow [*multa–pauca*], and the second to the quality of the diaphanous medium that gives rise to different rainbows [*purum–impurum*]. To investigate Grosseteste's colour space, we must operationalize these descriptions by linking them to physical parameters in the model. The spatial separation of different colours in a rainbow arises because of the variation in light spectrum with scattering angle. The wavelengths that reach the eye from a particular spatial location are those that exit the raindrops at the appropriate angle (Fig. 4.6). The [*multa–pauca*] dimension is therefore directly associated with scattering angle.

Although for a given wavelength most light exits at a particular angle (the caustic angle), there is some dispersion that depends on the position at which the light enters the raindrop (which in turn determines the angle between the incoming ray and the surface of the droplet) and some additional constructive interference at other angles that is captured by Mie theory. The intensity of light at each wavelength is therefore maximal at one scattering angle, with subsidiary peaks at other angles (Fig. 4.7). The light reaching the eye from each angle is a supposition of multiple wavelengths in different proportions.

The colours produced in rainbows depend heavily on the nature of the rain or mist that produces them. A fine mist or fog produces desaturated, pastel bows, whereas large droplets of rain produce highly colourful bows. We have assumed

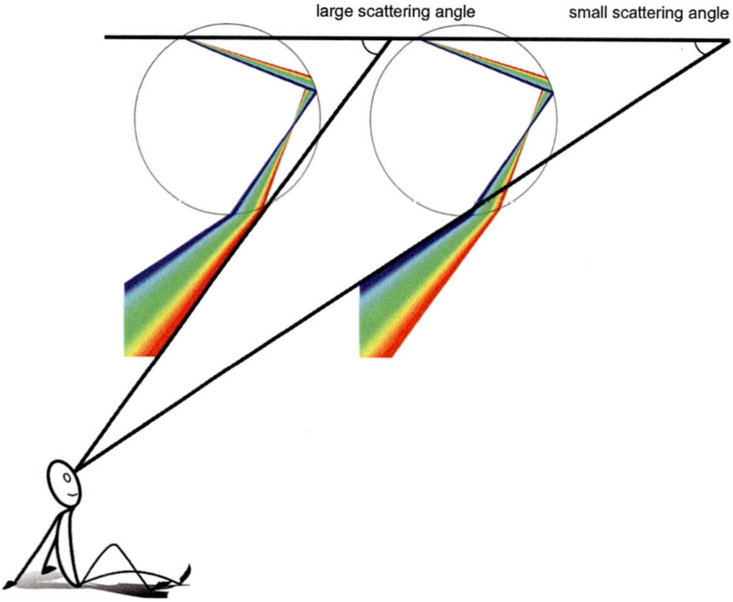

Fig. 4.6 Spatial separation of colours in a rainbow depends on scattering angle. The spectral dispersion of light through different scattering angles in raindrops means that the light reaching the observer's eye has a spectral content that varies with elevation. At low elevations, the stimulus is dominated by short-wavelength light; at high elevations, the stimulus is dominated by long-wavelength light

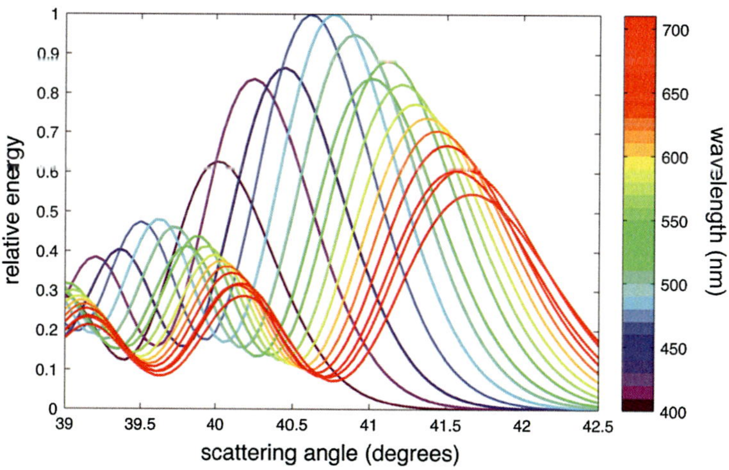

Fig. 4.7 Energy at each scattering angle for wavelengths from 400 to 700 nm in steps of 20 nm producing 16 curves. For each wavelength, most light exits the raindrop at a particular scattering angle (the caustic angle), generating the primary peak in each curve. As wavelength increases this primary peak moves to higher scattering angles. Scattering within the raindrop generates subsidiary peaks, which give rise to supernumerary arcs

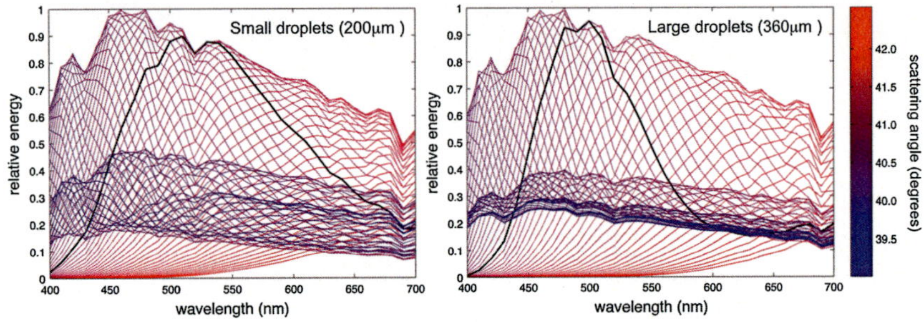

Fig. 4.8 Example spectra for a range of scattering angles and for two droplet sizes. Each plot shows spectra (energy as a function of wavelength) for a range of scattering angles, depicted here in different colours as indicated in the *colour bar*. One spectrum is plotted in *black* to facilitate the comparison in spectral shape between the *left*- and *right*-hand plots. The plot on the *left* is for small droplet sizes (200 μm) the plot on the *right* is for larger droplet sizes (360 μm). The main effects of changing scattering angle, shown by the different curves within a plot, are primarily to change the peak wavelength in the spectrum and secondarily to flatten the spectrum. The main effects of changing droplet size are primarily in spectral shape, and secondarily in the location of the spectral peak. For small droplet sizes, the spectra are broader than for large droplet sizes. This is most easily seen for the single spectra plotted as *black lines*, but generalises across all scattering angles as the intersecting curves on the *left* create horizontally stretched diamonds, whereas those on the *right* create vertically stretched diamonds

that the most significant source of the difference in colours between rainbows is the size of the raindrops. Indeed, Grosseteste identifies cases where 'the diaphanous medium should contain a mixture of smoky vapors,' which is consistent with the appearance of mist or fog in which the droplet size is small. We therefore choose to associate the [*purum–impurum*] dimension directly with droplet radius.

The output of our model is a specification of the light spectrum (i.e. the relative energies at each wavelength) as a function of scattering angle and droplet size. This is a description of the physical stimulus. Figure 4.8 shows example spectra obtained for a range of scattering angles and droplet sizes. Scattering phenomena can produce complex variations in spectral energy distributions. Later we shall analyse these variations and their consequences for colour perception, but for now we note only that there are systematic effects of the two physical parameters we manipulate. Firstly, the effect of scattering angle is largely to change the wavelengths most strongly present in the spectrum. Secondly, the effect of decreasing droplet size is to flatten the spectra, reducing variation between wavelengths. But these manipulations interact in interesting ways.

The third dimension of Grosseteste's colour space, characterized by *clara-obscura* in the *De colore*, is also referenced in the *De iride*. He notes cases where 'the *claritas* of the luminosity is scarce, as occurs around sunrise and sunset.' The spectrum of sunlight that impinges on water droplets to produce a rainbow will depend on the amount of atmosphere the light has encountered, which can be

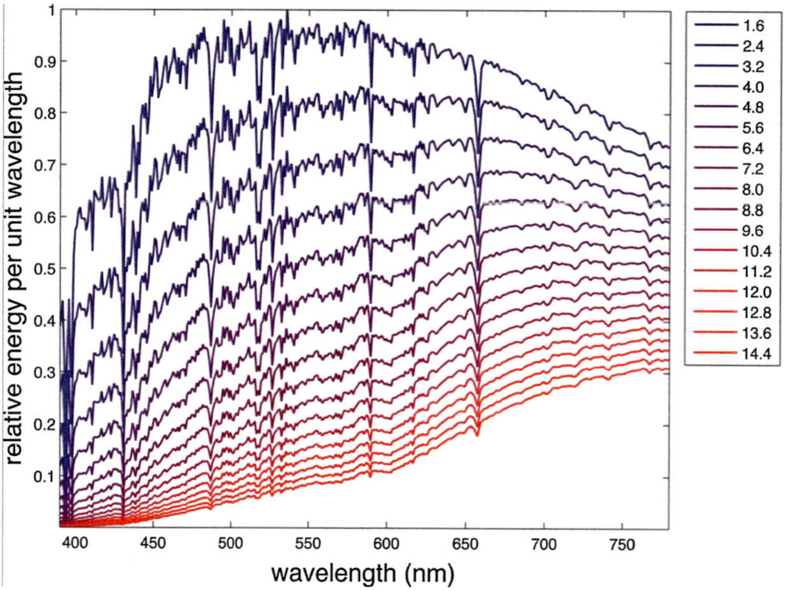

Fig. 4.9 The effect of solar elevation angle on the solar spectrum. The family of spectra (energy per unit wavelength) obtained with air mass values from 1.6 to 14.4 in steps of 0.8 [corresponding to solar elevation angles from 38.6° to 3.2° (Kasten and Young 1989)] and ozone and aerosol factors of 1, based on the extinction model using molecular and aerosol scattering (Allen 1973) and ozone absorption (Bogumil et al. 2001)

conveniently parameterized by the total air mass along the solar ray, which has a one-to-one relationship to solar elevation angle under particular atmospheric conditions (Kasten and Young 1989). Modelling the spectral effects of solar elevation angle allows us to consider changes imposed on the sunlight spectrum by atmospheric factors. Different incident spectra re-weight the relative wavelength composition of the rainbow. Figure 4.9 shows the modelled changes in spectrum. The effect of an increase in air mass is to reduce overall intensity and to drag the distribution to longer wavelengths.

The language of modern colour science is strict in its separation of terms that describe *physical* stimuli and terms that describe human *perceptual* experience. Grosseteste's primary method of measurement will have been to rely on his own sensory systems. To interpret the physical model presented here in relation to Grosseteste's observations we need also to know the properties of the human visual system. The perceptual analysis of the modelled spectra is greatly enhanced by appealing to modern colour science. Formal correspondence between modelling the physical phenomena that Grosseteste describes and the perceptual colour space requires an introduction to the twenty-first century account of colour.

4.7 A Twenty-First Century Introduction to Colour

4.7.1 The Biological Basis of Trichromacy

There is no straightforward definition of colour for it depends both on properties of the physical world and on characteristics of the biological observer. Human colour vision begins with the absorption of photons by the light-sensitive cells—the cone photoreceptors—that tile the retinal surface at the back of the eye. The light that is captured by these cells carries with it—in its wavelength composition and in its spatial and temporal distributions—signatures of its origin (the light source) and of the materials with which it has interacted (through reflection, refraction, scatter and absorption) on its path to the eye. The perceptual apparatus of the observer has evolved to extract these signatures, thereby recovering information about the physical world.

In the human eye there are three classes of cone photoreceptor that are each sensitive to different but overlapping ranges of wavelengths. These are called the S-, M- and L-cones because their sensitivities peak in the short, middle and long wavelength regions of the visible spectrum (Fig. 4.10b). The absorption of photons initiates a chain of biochemical reactions that lead to a voltage difference between the inside and the outside of the cone cell. This is the process of sensory transduction in which the physical light stimulus is converted to an electrical neural signal. For each photoreceptor, the neural signal generated varies in only one dimension (the voltage difference can get larger or smaller) while the number of photons absorbed depends on two independent properties of the light (its intensity and its wavelength composition, Fig. 4.10a). A single cone photoreceptor is truly colour-blind as it confounds changes in wavelength and changes in intensity. To disentangle wavelength and intensity, the visual system must compare the outputs of different classes of cone (Fig. 4.10c). Yoked changes in the signals from the S-, M- and L-cones are likely to indicate changes in light intensity, whereas changes in the relative activations of the S-, M- and L-cones indicate a change in the spectral composition of the light.

4.7.2 Manipulating Colour Experience

One useful definition of colour is that it is the perceptual correlate of changes in the spectral composition of light. But the human visual system senses only the relative photon catches in the three classes of cone. It in no way measures the intensity of light at every wavelength: for that we need a spectroradiometer. Importantly, since the cones form the input stage of the human colour perception, the cone excitations are the only information available to the human observer about colour. So all

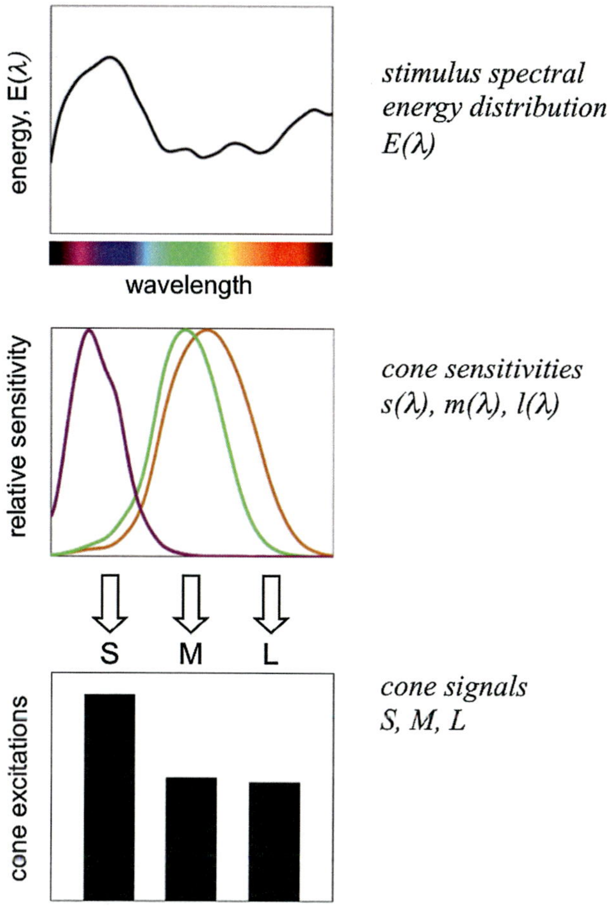

Fig. 4.10 The relationship between the spectral energy distribution of light and the cone signals. (**a**) A light can be characterised by its spectral energy distribution (energy as a function of wavelength). (**b**) There are three classes of human cone photoreceptor, the L-, M- and S-cones, labelled according to whether their spectral sensitivity functions peak in the Long, Middle or Short wavelength regions of the visible spectrum. The cone spectral sensitivity functions describe the relative probability that light of a particular wavelength will be absorbed. (**c**) The cone signal is determined by the summed absorption of light across all wavelengths. The signal in each cone class is determined by the amount of light available and the probabilities of absorption of each of the constituent wavelengths. After absorption in a single cone, independent information about wavelength and intensity is lost, and the signals transmitted to downstream neural processing are simply the relative excitations of the three cone classes—a trichromatic signal [L, M, S]

physical lights are reduced to just three values (Fig. 4.10c), and the space of all possible colours can be constructed by representing the S, M and L cone excitations as coordinates within a three-dimensional plot.

This is a very powerful result. Any two lights that produce the same cone excitations will be indistinguishable to the human observer, and in that situation, where the signals are matched at the input stage, there is nothing that downstream visual processing can do to undo the match, since all other information about the origin of the lights has been lost. Colour reproduction technology, from printing to television, depends on this fundamental limitation on human colour perception. The vibrant green of a newly unfurled leaf in springtime can be captured on television by choosing a particular balance of red, green and blue (RGB) primaries so as to elicit the same cone signals as produced by the complex natural spectrum of light reflected from the leaf. Three-colour printing (Le Blon 1725), a technology driven by the desire for cheap colour reproduction, preceded scientific understanding of physiological trichromacy (Young 1802). With strong echoes of the final paragraph of the *De colore*, the success of Le Blon's technique was widely held to stem from his selection of appropriate primary inks and his skill in determining through observation and mental experimentation the components of the colour to be reproduced (Mollon 2003).

A summary of human trichromacy implies that colour perception is fully understood. Certainly, it is possible to predict when two lights viewed in isolation will match, because they offer the same triplet of cone signals. However, predicting the appearance of the matching patches of light is very difficult, and even more so when the lights are viewed in context. Colour appearance is heavily influenced by surrounding colours, and by colours recently viewed. These spatial and temporal interactions hint at the significant neural processing the cone signals undergo to support our perceptual experiences. Trichromatic theory, though very powerful, cannot be said to fully explain colour experience.

The purpose of our perceptual systems is to provide information about the physical world in which we live and interact. Yet the information carried by the pattern of cone activations across the retina provides only an incomplete and indirect sampling of that environment. The surface of an opaque object will reflect some wavelengths of light and absorb others. Pigments are used in paint to modify which wavelengths are absorbed and, perceptually, we can say that different paints change the 'colour' of an object. But the eye senses this property only indirectly. Since it receives only the light reflected to the eye by the object, the spectral composition of that light depends both on the composition of the light illuminating the object, and on the reflectance properties of the object. Objects have additional perceptual qualities, such as glossiness and lustre that depend on surface properties and translucency that depends on subsurface, volumetric properties. These too are carried in the information provided by the pattern of cone activations across the retina. Only recently have these aspects of our perceptual world been systematically studied (Adelson 2001). It is not at all clear that through perceptual observation alone, the trichromatic nature of human colour perception could be unambiguously deduced.

4.8 Bridging Thirteenth and Twenty-First Century Accounts to Translate Grosseteste's Key Terms

4.8.1 *Quantifying the Variety of Colour in Natural Rainbows*

Although the plots in Fig. 4.8 provide a full account of the spectral variation of light within a rainbow, and between rainbows generated from droplets of different sizes, they do not provide an intuitive summary of the perceptual experiences produced by these spectra. We follow Lee (1998) in using a pseudo-colour diagram—known as a Lee diagram—to show how the appearance of rainbows varies with the size of the scattering water droplets. The right-hand panels in Fig. 4.11 are modified Lee diagrams showing the pseudocolour representation of the spectrum of light obtained at a range of scattering angles (between 42.5° and 39.0°) and a range of droplet radii (between 200 and 1000 μm). We link the variations in this diagram to the ones referenced by Grosseteste's terms [*multa–pauca*] and [*purum–impurum*]. Variation within a rainbow [*multa–pauca*] corresponds to moving parallel to the ordinate; variation between rainbows [*purum–impurum*] corresponds to moving parallel to the abscissa.

Lee diagrams provide a good qualitative description of the variety of colours produced, but for centuries colour scientists have attempted to describe quantitatively the relationships between colours. Trichromacy implies that a three-dimensional space should suffice to capture such relationships. Two different spectra that generate the same [L, M, S] triplet of signals in the cones should plot at the same point; two lights that offer only slightly different signals to the cones should plot at similar locations; and two lights that offer vastly different signals should be well separated. Cone-excitation spaces, in which the three axes represent the signals in the three classes of cone, successfully capture the different *physiological* signals produced by different *physical* spectra. However, largely because of additional stages of neural processing beyond the photoreceptors, cone-excitation spaces do not directly capture the *perceptual* differences between colours. There are a number of alternative colour spaces, produced by applying mathematical transformations on cone spaces, that are approximately perceptually uniform so that any two lights that exhibit a fixed magnitude of perceptual difference are separated by a fixed distance when plotted in the space. Here we choose to use the CIE 1976 L*, a*, b* (CIELAB) space. The L* axis represents perceived differences in light and dark; the a* and b* axes correspond to variation along red-green and blue-yellow perceptual dimensions respectively.

The left-hand panels in Fig. 4.11 are projections onto the equal-lightness plane of CIELAB space, spanned by the a* and b* axes. The coloured squares show examples of spectra derived from three vertical sections through the Lee diagram (upper panels) and eleven horizontal sections through the Lee diagram (lower panels). The grey mesh reproduced in both CIELAB plots depicts the loci of colours obtained with a range of scattering angles (between 42.5° and 39.0°) and droplet

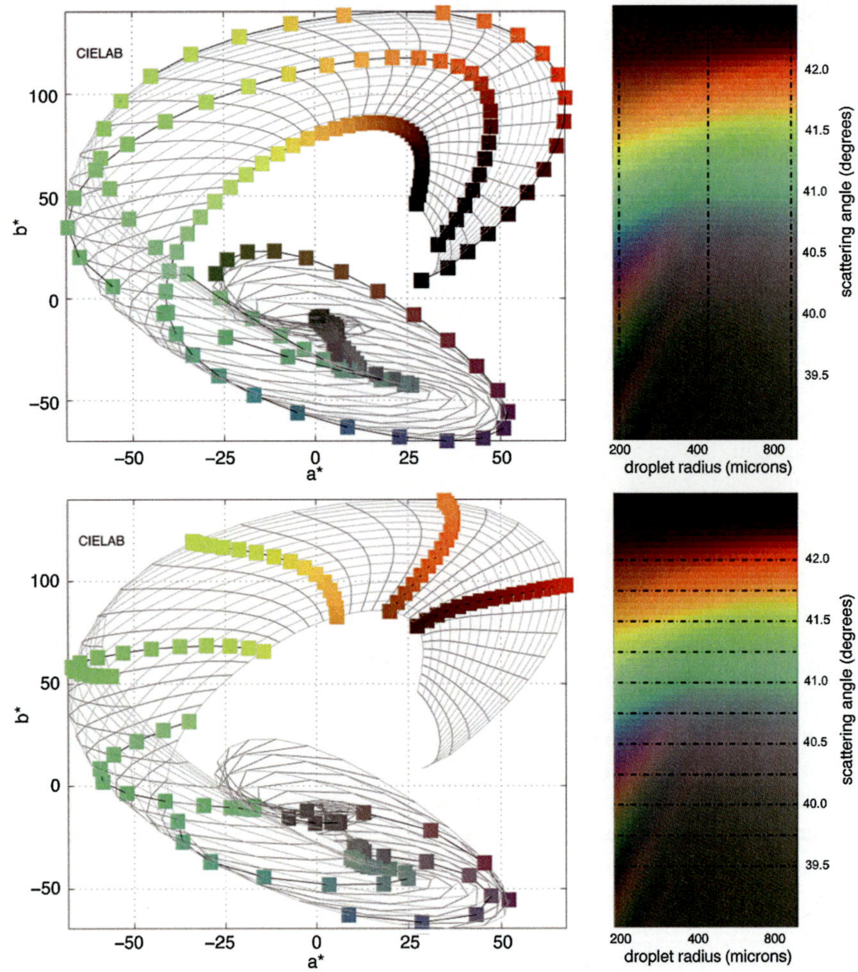

Fig. 4.11 CIELAB coordinates and pseudo-colour representation of the spectra produced within rainbows of different droplet sizes. The panels on the *right* are Lee plots. They provide a pseudo-colour representation of the spectrum produced at particular droplet radii (abscissa) and scattering angles (ordinate). The panels on the *left* show the chromatic plane of CIELAB colour space, spanned by the a* and b* axes. *Coloured square* symbols correspond to samples from the Lee plots. The *upper panel* shows three series that correspond to vertical sections through the Lee plot (showing the effect of scattering angle, at three levels of drop radius); the *lower panel* shows 11 series that correspond to horizontal sections through the Lee plot (showing the effect of droplet radius, at 11 levels of scattering angle). The sections through the Lee plots are indicated by the *dashed black lines*

radii (between 200 and 1000 μm). These plots reveal the structure in the complex changes seen in the spectral plots above (Fig. 4.8).

There are two important points to note. Firstly, variations in scattering angle and in droplet radius generate families of spectra that span a good range of possible variation in perceived colour: The chromatic plane is almost completely filled by the grey spiral mesh in these figures. Secondly, spectra produced within a rainbow of constant droplet size (upper panels) plot on spiral loci in this colour space, and spectra produced from different droplet sizes plot as a second set of interlocking spiral loci. These two dimensions of variation therefore provide a coordinate system for navigating colour variation, as specified by location in the chromatic plane of a trichromatic colour space.

The proposed coordinate grid projects onto the chromatic plane as shown, but the spectra differ systematically in their lightness. Plotted in the full three-dimensional CIELAB space, the coordinate grid lies on a spiral surface. Figure 4.12 shows this spiral surface viewed from two different directions. The effect of solar elevation angle on the solar spectrum (Fig. 4.9) imposes an additional dimension of variation, whose effect is to translate the spiral surface through colour space, sweeping out a three-dimensional volume. The grid lines in Fig. 4.12b are coloured depending on the physical dimension that varies along them. The blue coordinate characterises the differences in colour within a within a rainbow; the red coordinate characterises the differences in colour between rainbows created from differently sized droplets; and the black coordinate characterises the differences in colour between rainbows created with incident light from different sun elevations.

As Grosseteste states in the final paragraph of the *De colore*, a person who is skilled in manipulating the physical dimensions he identifies—selecting light that is either *multa* or *pauca,* and either *clara* or *obscura*, and a recipient medium that is either *purum* or *impurum*—can show visibly, as they wish, all kinds of colours. The rainbow inspired co-ordinate system provides a reasonably effective means of navigating the perceptual space of coloured lights, which we now understand to be constrained to three-dimensions by biological human trichromacy.

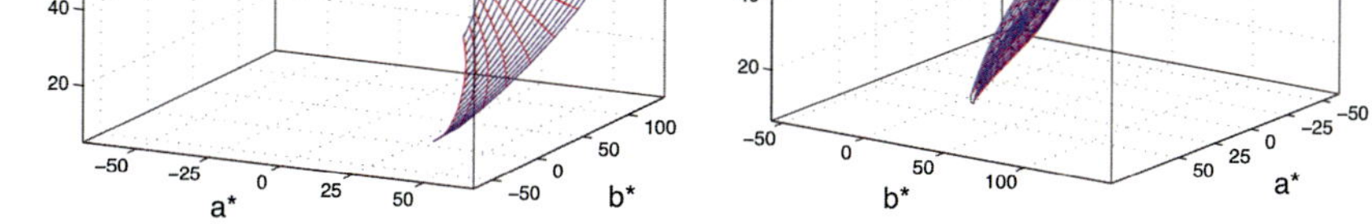

Fig. 4.12 A rainbow-inspired coordinate grid that spans human trichromatic colour space. Panels (**a**) and (**b**) show two views of the three-dimensional CIELAB colour space. (**a**) The spiral surface spanned by variation in scattering angle (*blue coordinate lines*) and droplet radius (*red coordinate lines*). (**b**) The effect of solar elevation on the locus of rainbow colours is to sweep the spiral surface through colour space (*black coordinate lines*)

4.8.2 Consistencies and Inconsistencies in the Scheme

The account of colour variation in the *De colore* has strong similarities to a modern abstract colour space, perhaps describing colour variation in terms of hue (the dimension that distinguishes red, yellow, green and blue), saturation (the dimension that separates red and pink), and brightness (the difference between dark and light), but this particular interpretation presents unresolvable internal conflicts. A primary difficulty with identifying one of Grosseteste's colour dimensions with hue is that he is very clear that all three dimensions must terminate at whiteness. Hue is associated with wavelength and only by mixing more than one wavelength is it possible to produce a light that appears white or achromatic. In modern colour spaces, this is accommodated by representing hue circumferentially, centred on white. In the hue-saturation-brightness cylindrical coordinate system of human color perception, lines of constant hue are radial while lines of constant saturation are concentric.

The mapping of the Lee diagrams onto the three dimensions of perceptual CIELAB space immediately suggests another approach to covering colour space with a coordinate system, using coordinates that are spiral in configuration. Confining discussion initially to a two-dimensional space (e.g., of a* and b* in the chromatic plane), a simple mathematical operation transforms the orthogonal Cartesian grid into twin intersecting sets of spiral coordinate grids (Fig. 4.13). A parameter in the transform 'tunes' the coordinate system so that one of the two sets of spirals is more or less radial, and the other more or less circumferential. A significant property of this set of coordinate systems is that the central point lies at one extremity of both coordinates. It is only in the limit of the purely radial-circumferential system that this property is lost. To take the illustrative example that the mapped space is the chromatic plane of hue and saturation, any set of twin intersecting sets of spiral coordinate grids permits the neutral (white) point to be the source of both coordinates. The modern radial and circumferential system of hue and saturation is the limit where this fails.

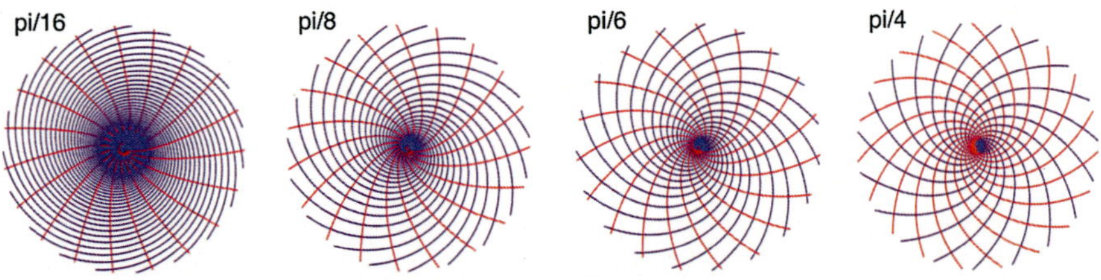

Fig. 4.13 Examples of 'logarithmic-polar' coordinate systems. A parameter in the transform 'tunes' the coordinate system so that one of the two sets of spirals is more or less radial, and the other more or less circumferential. A value of pi/4 generates symmetric sets; decreasing values result in one set becoming tighter, the other looser, until the purely radial-circumferential system emerges when the parameter is set to zero

In the *De iride*, Grosseteste gives us an essential clue to unlocking the meaning of his terms that remained obscure to us in the *De colore*. Comparing Figs. 4.12 and 4.13, we find that his association of two of the dimensions of colour onto variation within a rainbow and variation between rainbows produced in different media does indeed map a significant portion of the chromatic plane with a spiral coordinate system, albeit of generalized form. The rainbow-spiral coordinate system is not perfect—it is not orthogonal at every point like the ideal logarithmic-polar system. However, it does share the same topology, and, essentially for Grosseteste, the property of mapping a colour plane with independent coordinates, both of which originate from white.

In the *De iride*, Grosseteste describes the *pauca* extreme of the *multa-pauca* axis with the Latin adjective *hyazinthinus*, from the substantive *hyacinthus*, which we choose to translate as purple. The sources here are complex and are based on medieval references to gem stones and other color terminology. So, the identification with any particular color is blurred, but on balance we believe that violet or purple with some red is an appropriate interpretation. In the (perceptual) hue circle (but not on a wavelength scale) violet, purple and red are adjacent. However, the purple or violet end of the rainbow sequence spirals towards white, becoming desaturated by the superposition of several wavelengths. This leaves us with something of a puzzle since whiteness in the *De colore* is explicitly associated with the triplet [*multa*, *clara*, *purum*] and blackness by the triplet [*pauca*, *obscura*, *impurum*]. The *De colore* therefore links *multa* to whiteness and *pauca* to blackness whereas the *De iride* associates *pauca* with purple, which would leave the *multa* to *pauca* direction running towards white, moving vertically downwards in the Lee diagram and tracing the spirals in the upper part of Fig. 4.11 from outside (saturated red) to inside (desaturated violet).

Similarly, in the *De iride*, Grosseteste associates *impurum* with cases where 'the diaphanous medium should contain a mixture of smoky vapors'. If we interpret this smoky appearance as mist or fog in which the droplet size is small, the *purum* to *impurum* direction would run towards white, moving horizontally from right to left in the Lee diagram and tracing the spirals in the lower part of Fig. 4.11 from outside to inside. Again, this sits uncomfortably with the association of *purum* with whiteness and *impurum* with blackness in the *De colore*.

The three-dimensional representation of rainbow colours in CIELAB shows the strong variation in lightness that accompanies colour changes through the rainbow. The reddish hues at the top of the rainbow, associated with high scattering angles, are dim. This feature is associated with the phenomena known as Alexander's Dark Band—the dark region that appears outside the rainbow, beyond the red hues, and is bounded by the secondary bow if one is visible. The processes of reflection and refraction within a raindrop that concentrate light at the caustic angle effectively do so by removing light from other angles, producing a dark region. For this reason, we might associate high scattering angles and large droplet radius (the upper right-hand corner of the Lee diagram) with blackness. The green colours in the rainbow plot at particularly high values of L*, largely because the wavelength content of these lights aligns with the peak wavelength sensitivity of the eye. The violet colours at the bottom of the rainbow are desaturated, arising from the superposition of several wavelengths, and are also relatively high in L*. The concentration of light in the centre of the arc of a rainbow is sometimes very striking. Although the polarity of the progression from *multa* to *pauca* is unclear, the gathering of light by a rainbow, and the association of *multa-pauca* with scattering angle in the *De iride*, links back to the statement in the *De colore* that *multa* refers to the intensification of rays by a burning glass.

The third dimension, labelled *clara-obscura*, and linked to solar elevation, associates *obscura* with sunrise and sunset, reducing the intensity of the incident light and biasing the spectrum towards longer wavelengths (Fig. 4.9). This has the effect of sweeping the coordinate grid downwards (lower L*) and leftwards (higher a*) in CIELAB space (Fig. 4.12). *Obscura* is then comfortably associated with blackness. *Clara* is associated with whiteness, pointing to light and desaturated colours.

There is very good evidence for biological trichromacy. Metamerism—in which two lights with different spectral energy distributions are indiscriminable because they offer the same triplet of cone signals—implies that the three-dimensional space of cone signals is exhaustive in describing the gamut of colour experience. This is true under certain limited conditions of observation, for example when a small patch of light is seen in isolation against a black surround, as if through an aperture. However, if we consider regions of extended spatial extent, modern descriptions of colour perception become more complex. For extended spatial regions that are nonhomogeneous in colour and lightness, the dominant mode of perception is that of illuminated surfaces. The surface colours perceived under these conditions have additional qualitative dimensions: for example they can appear glossy or matte; rough or smooth; cloudy or transparent. These qualities are associated with particular signatures of colour and lightness variation across space. The correspondences between the physical and perceptual variables associated with these higher qualities remain relatively poorly understood (Adelson 2001). With advances in computer graphics, it has become possible to generate physically accurate renders of materials and their interaction with the light that illuminates them, thus allowing carefully controlled experiments on perception of surface-colours. It is striking that Grosseteste, in the *De colore*, places the interaction between light and the material within which it is incorporated at the centre of his definition of colour. Although modern colour science typically separates the colour perception of objects, and of the materials from which they are made, from the perception of isolated lights, or of surfaces viewed through an aperture, it is not clear that Grosseteste also made this distinction, but nor is it clear that such a distinction is appropriate for a full account of colour perception.

4.9 Conclusions

Robert Grosseteste in his short treatise the *De colore* creates an explicitly three-dimensional abstract space of color. He replaces Aristotle's linear arrangement of colors between white and black by a scheme in which colors are generated from whiteness (and symmetrically from blackness) by the independent adjustment of three qualities. He painstakingly leads the reader through an explicit combinatorial exercise, identifying $7 = 2^3 - 1$ unique directions from whiteness along which colour can vary. The space is continuous, with infinite degrees of intensification or diminution along each of the directions he identifies. He introduces this space through abstract reasoning, but in the closing paragraph he refers to technical expertise in manipulating light and materials to generate all possible colors. Whether Grosseteste actually carried out such manipulations, or just conceived of them, is a tantalizing question, coming as it does from an era that saw the first stirrings of ideas that later led to the experimental method.

The *De colore* introduces a conceptual theatre within which colour can dance, but at the end of the treatise, the nature of the dimensions that span the space is unclear. In the *De iride*, Grosseteste operationalizes these parameters, as the variation within a rainbow and variation between rainbows produced in different media, and with different phases of sunlight. The simulations presented here indicate that the three components of variation identified in the *De iride*—variation within a rainbow, parameterized by scattering angle, and variation between rainbows, parameterized by droplet size and by the effect of air mass on solar spectrum—can be used to navigate perceptual colour space reasonably effectively.

This analysis provides an example of how modern methods within the scientific fields descendent, in some verifiable measure, from thinking in the medieval period can illuminate the questions, assumptions, and goals of scientific writing then.

References

Adelson, E. H. (2001). On seeing stuff: The perception of materials by humans and machines. *Human Vision and Electronic Imaging VI, 4299*, 1–12.

Airy, G. B. (1838). On the intensity of light in the neighbourhood of a caustic. *Transactions of the Cambridge Philosophical Society, 6*(3), 397–403.

Allen, C. W. (1973). *Astrophysical quantities*. London: University of London/The Athlone Press.

Anon. (1936). *De Coloribus* (W. S. Hett, Trans.). Aristotle: Minor Works I. London: Heinemann: Loeb Classical Library.

Aristotle. (1984). *De Sensu et Sensato (Sense and Sensibilia)* (J. I. Beare, Trans.). The Complete Works of Aristotle: The Revised Oxford Translation. Princeton: Princeton University Press.

Berlin, B., & Kay, P. (1969). *Basic color terms: Their universality and evolution*. Berkeley: University of California Press.

Bogumil, K., Orphal, J., Burrows, J. P., & Flaud, J. M. (2001). Vibrational progressions in the visible and near-ultraviolet absorption spectrum of Ozone. *Chemical Physics Letters, 349*, 241–248.

Boyer, C. B. (1959). *The rainbow: From myth to mathematics*. New York: Yoseloff.

Descartes, R. (2001). *Discourse on method, optics, geometry and metereology* (P. J. Olscamp, Trans.). Indianapolis: Hackett.

Dinkova-Brunn, G., Gasper, G. E. M., Huxtable, M., McLeish, T. C. B., Panti, C., & Smithson, H. E. (2013). *The dimensions of colour: Robert Grosseteste's De colore*. Toronto: Pontifical Institute of Mediaeval Studies.

Hovenac, E. A., & Lock, J. A. (1992). Assessing the contributions of surface waves and complex rays to far-field Mie scattering by use of the Debye Series. *Journal of the Optical Society of America A, 9*(5), 781–795.

Kasten, F., & Young, A. T. (1989). Revised optical air-mass tables and approximation formula. *Applied Optics, 28*(22), 4735–4738.

Kuehni, R. G., & Schwarz, A. (2007). *Color ordered: A survey of color systems from antiquity to the present*. Oxford: Oxford University Press.

Le Blon, J. C. (1725). *Coloritto*. London.

Lee, R. L. (1998). Mie theory, Airy theory, and the natural rainbow. *Applied Optics, 37*(9):1506–1519.

Mollon, J. D. (2003). The origins of modern color science. In S. Shevell (Ed.), *Color science* (pp. 2–36). Washington, DC: Optical Society of America.

Panti, C. (2013). Robert Grosseteste and Adam of Exeter's physics of light, remarks on the transmission, authenticity and chronology of Grosseteste's scientific Opuscula. In J. Flood, J. Ginther, & J. Goering (Eds.), *Robert Grosseteste and his intellectual milieu* (pp. 165–190). Toronto: PIMS.

Saunders, B. A. C., & van Brakel, J. (1997). Are there nontrivial constraints on colour categorization? *Behavioral and Brain Sciences, 20*(2), 167–228.

Smithson, H. E., Anderson, P. S., Dinkova-Bruun, G., Fosbury, R. A., Gasper, G. E. M., Laven, P., et al. (2014). Color-coordinate system from a 13th-century account of rainbows. *Journal of the Optical Society of America A, 31*(4), A341–A349.

Smithson, H. E., Dinkova-Bruun, G., Gasper, G. E. M., Huxtable, M., McLeish, T. C. B., & Panti, C. (2012). A three-dimensional color space from the 13th century. *Journal of the Optical Society of America A, 29*(2), A346–A352.

Young, T. (1802). The Bakerian lecture. On the theory of light and colours. *Philosophical Transactions of the Royal Society of London, 92*, 12–48.

Part II
Purity: Physical and Spiritual

Chapter 5
Medicine for the Body and Soul: Healthy Living in the Age of Bishop Grosseteste c. 1100–1400

Christopher Bonfield

5.1 Secular Medicine and Divine Care

> As Christians, we know that there are two kinds of medicine, one of earthly things, the other of heavenly things. They differ in both origin and efficacy. Through long experience, earthly doctors learn the powers of herbs and the like, which alter the condition of human bodies. But there has never been a doctor so experienced in this art that he has not found some illnesses difficult to cure and others absolutely incurable. . . . The author of heavenly medicine, however, is Christ, who could heal the sick and raise the dead from the grave (Fulbert of Chartres 1865; Park 1992; Ell 1981).

This extract, from an eleventh-century hymn attributed to Fulbert of Chartres, who was bishop of the Cathedral of Chartres from 1006 till 1028, demonstrates the denigration by medieval theologians of secular medicine to divine care (Risse 1999). It was a point of view reinforced by Jacques de Vitry, an early thirteenth-century theologian, who complained of: '. . . the blindness of the sick, who call to their bedsides the physicians of the body rather than the physicians of the souls [priests], preferring mud to treasure, straw to grain, dregs to wine and the body to the soul' (Rawcliffe 1999). Indeed, the study of health and disease during the Middle Ages has historically been viewed through the lens, or hierarchy, of *two medicines*—the body below and the soul above (Rawcliffe 2002). On the one hand was Christ, Himself described as a physician (*Christus medicus*) who, according to theologians from the time of St Augustine (c. 340–430) onwards, had the ability to recapture the precious harmony of body and soul which proved so effective a medicine against the malignant effects of the Fall (Augustine of Hippo 1972). On the other was an older Classical Greek tradition of medical theory and practice, as mediated through Greek, Jewish and Muslim scholars, and made famous by such

C. Bonfield (✉)
University of Bath, Bath, UK
e-mail: c.a.bonfield@bath.ac.uk

J.P. Cunningham, M. Hocknull (eds.), *Robert Grosseteste and the pursuit of Religious and Scientific Learning in the Middle Ages*, Studies in the History of Philosophy of Mind 18, DOI 10.1007/978-3-319-33468-4_5

names as Hippocrates (d. c. 377 BCE), Galen (d. c. 200 CE) and Avicenna (d. 1037) (Rawcliffe 1995; Nutton 1995; Lindberg 1992). Most medical historians today, however, would suggest that rather than being diametrically opposed, the relationship between medicine for the soul (*Medicina sacramentalis*) and therapeutic medicine in the past was symbiotic (McCleery 2014). In other words, just as the body and soul were unified entities, so physical and spiritual health was essential for health and well-being.

This paper will flesh out these ideas and concepts in more detail, setting out how men and women in the age of Bishop Grosseteste would have explained health and disease. It will address three main points: firstly, how the Church defined good health, and what wider impact this had on institutions caring for the sick and poor. Second, to what extent 'religious' explanation of sickness and disease sat alongside so-called 'medical', or 'scientific', explanations of healthy living. Here, particular emphasis will be placed on the genre of self-help guides to health called the *regimen sanitatis* (regimen of health). This advice took many forms, including a wide and varied collection of manuscript and early printed copies of the *Regimen*, commonplace books, popular collections of medical recipes, regulations promulgated by urban magistrates, political propaganda and homiletic literature.[1] Having thus established the importance and interconnected nature of sin and sickness, the final part of the paper will examine the impact that medical explanations had on wider spiritual concepts of health and disease, focusing in particular on homiletic literature, such as the handbook for preachers, *Fasisculus morum*. Written about fifty years after Robert Grosseteste's death in 1253 by a Franciscan friar, it contains a selection of popular religious and moral stories that would have been heard from the pulpit throughout England (Wenzel 1989).

5.2 Medicine and the Church

Church authorities were never slow to point out that death and disease, and all man's other troubles, were seen to be a direct consequence of Original Sin. The fateful event incurred punishment for Adam and Eve and all their descendants ranging from the pains of childbirth and menstruation to illness, epidemics and even death itself. The consequences were felt by everyone, rich or poor, young or old. St Augustine maintained that the Fall had upset the harmonious relationship between body and soul which had hitherto preserved Adam and Eve in a perfect state of health (Ziegler 2001; Brown 1988). As St Thomas Aquinas (1215–1274) would later argue, this resulted in the balance of health being 'so utterly wrecked that life is destroyed; so as to cause sickness' (Aquinas 1982). The characteristics of Fallen man included 'mortality, death, misery, suffering, crimes, [and] the war of flesh

[1] For the *Regimen sanitatis* and its reception in late medieval England (see Gil Sotres 1998; Bonfield 2006).

against the spirit' (Deane 1963). Put simply, in the final analysis all imbalances could be traced to Original Sin. In fact, Robert Grosseteste, who is known to have been influenced by the works of the famous physician and scholar Avicenna, himself noted that rational thought and the ability to see clearly were impeded because of the corrupt nature of the body. He wrote that as a consequence of the Fall, mankind had to rely on the senses which, according to Richard Southern, was for Grosseteste like 'a blind man's use of a white cane' (1986; Harrison 2007).

Yet there was hope. As noted above, Christ had the ability to heal. Augustine urged that the sick man should drink from the 'bitter cup [of death] in order to become well'. Do not fear to drink from this cup, he reassured the reader, for:

> ...to dispel your fear the Physician [i. e. Christ] drank first, that is, the Lord drank first the bitterness of the passion. He had no sin, he had nothing to be cured; yet he drank. Drink until the bitterness of this ages passes away, until there comes a time when there will be no scandal, no anger, no wasting disease, no bitterness, no fever, no deceit, no enmities, no old age, no strife (Trans. Arbesmann 1954).

In addition to the heavy load of Original Sin, which was carried by everyone, came the burden of individual, personal wrongdoing, likely to bring down the wrath of God, either in this world or—even worse—in the next. It is perhaps no surprise, therefore, that in 1215, the Fourth Lateran Council ruled throughout the whole of Western Christendom that:

> As sickness of the body may sometimes be the result of sin ... so we by this present decree order and strictly command physicians of the body, when they are called to the sick, to warn and persuade them first of all to call in physicians of the soul so that after their spiritual health has been seen to they may respond better to medicines for their bodies; *for when the cause ceases so does the effect* (García 1981).[2]

These ideas endured and gained popularity; they also found concrete expression in institutions responsible for 'caring' and 'curing', such as hospitals (Horden 2007).[3] There were well over a thousand hospitals documented in medieval England, which between them ranged considerably in size and means, and dealt with a variety of diseases (Carlin 1989). Four main types can be identified: leper houses; hostels for pilgrims; institutions for the sick poor; and alms-houses. Generally, these were primarily religious institutions 'with liturgy at their heart', and in their wards patients received a combination of a therapeutics, such as clean bedding and a good diet, and *Medicina sacramentalis* (the medicine of the soul) (Rawcliffe 2008; Park and Henderson 1991). The latter has recently been termed the 'true' medicine of hospitals which, unlike drugs or invasive surgery, did not require the presence of physicians or surgeons yet could potentially affect the body. Indeed, the Mass and the Daily Offices were performed daily in hospitals, and many patrons invested heavily in liturgical items and vestments. For instance, at St Leonard's hospital (York), the liturgy was sufficiently important to justify an impressive complement of thirty secular choristers—then more than Exeter and Salisbury

---

[2] My emphasis.

[3] For the therapeutic regime (see Bonfield 2013).

cathedrals combined (Bowers 1975). Records from a number of larger institutions, including St Katharine's by the Tower (London), St Leonard's (York) and St Mary in the Newarke (Leicester), reveal also that regulations were in place to ensure high standards of performance, and, equally important, that prayers would be said at each of these times for the spiritual health of patrons or benefactors (Jamison 1952; Dugdale 1817–1830; Thompson 1937).

If we jump ahead to the end of the sixteenth century, we can even note the continued influence of these enduring ideas on hospital architecture. The Savoy, London, whose earliest surviving statutes, dated 1523, were based on the model established at Santa Maria Nuova, Florence, include detailed specification about the linen on the hospital's 100 beds; an overriding concern with cleanliness; and special provisions for the care of the sick, who were to be attended twice daily by a physician, an apothecary and a surgeon, each salaried by the hospital (British Library, MS Cotton Cleopatra C V, ff. 25^{r}–28^{v}.). Clearly, the physical health of patients was import; yet, even here, spiritual health was given (almost on an hourly basis) precedence over that of the body. Furthermore, the hospital's layout was, like its Florentine counterpart, based on a cruciform ground plan: the hospital literally embodied Christ's redemptive cross.

Evidently, religious explanations of health, sin and disease impacted upon both the theory and practice of medicine. However, although heavy emphasis was placed on the soul and spiritual medicine, there was also current a very different idea: that healing and medicine came from God, and that he wanted man to be fit and derive the most from life on earth. Medicinal plants and other cures came from God, and were to be exploited to the full. Indeed, the absorption of Greek, Arab and Jewish concepts of man and the natural world in the twelfth and thirteenth centuries encouraged this more positive outlook.[4] The force of this tradition dominated medical teaching in the West from Bishop Grosseteste up to the seventeenth century—the best part of 500 years—and dated back to the fourth century BCE. Indeed, Grosseteste obtained a considerable level of medical knowledge himself, and his theology in known to have been influenced by Galenic medicine (Murray 1991; Grosseteste 1861). This brings us to the powerful impact of non-Christian tradition on Western medicine.

5.3 'Medical' Ideas About Health

Ideas about the preservation of health have a long pedigree, for an impressive corpus of medical treatises circulated in Ancient times. The Western tradition of medicine can, indeed, be traced back to the Ancient Greeks, as:

> ...generations of doctors and surgeons have proclaimed their intellectual descent from Hippocrates of Cos (d. 377 BC) and their adherence to a practice of medicine based on ethical, rational, and independent judgement, sound experience and fine learning (Nutton 1995).

[4] For the translation and transmission process (see Lindberg 1992).

Even among his contemporaries Hippocrates, who is now known as the father of medicine, was so revered that authors were keen to attribute their works to his name; he was often invoked by medieval writers to authenticate their writings and appeal to a wider public. Certainly, although a medical practitioner called Hippocrates did live on the island of Cos over 400 years before the birth of Christ, the *corpus*, or body, of works attributed to him was compiled much later in Alexandria.[5] In fact, the legendary Hippocrates of medical times acquired such a lasting reputation because he was praised by both Aristotle (383–322 BCE) and Galen (d. 129 CE), who acquired even greater posthumous celebrity than he did. Indeed, it was the Greek physician and surgeon, Galen, who commanded the most respect in the medieval medical world. Trained at Pergamum he was a practitioner, scientist and prolific author (Touwaide 2014). He was also a physician to Marcus Aurelius in Rome. It was he who hailed Hippocrates as a great authority, and elaborated many works in the Hippocratic Corpus, which themselves were disseminated and transmitted to the West, along with works by Galen and later commentaries and compendiums, such as the *Canon of Medicine* by Avicenna (Gruner 1930).[6]

It is in Ancient Greece that we find a series of precepts on diet and hygiene meant to preserve health, and that these precepts were a step in the progressive discovery of a *regimen* of life (Gil-Sotres 1998).[7] In the earlier Hippocratic treatise *De natura hominis* (*On the Nature of Man*), it was argued that the body owed its existence and growth to an admixture of four humours: sanguine (hot and wet); choleric (hot and dry); phlegmatic (cold and wet); and melancholic (cold and dry) (Littré 1839–1861). As humoral imbalance appeared to be responsible for disease, maintaining a state of equilibrium was of vital importance. The all-pervasive Classical doctrine of health hinged upon the avoidance of *dyscrasia* or excessive imbalance, the favoured means of achieving which was through diet or a broader *regimen* of health.

By the time these ideas reached Galen, the greatest advocate of preventive medicine, *On the Nature of Man* had been joined together with another treatise called *Regimen in Health* which recommended that:

> ...he who aspires to treat correctly of human regimen must first acquire knowledge and discernment of the nature of man in general—knowledge of its primary constituents and discernment of the components by which it is controlled...These things therefore the author must know, and further the power possessed severally by all the foods and drinks of our regimen... Even when all this is known, the care of man is not yet complete, because eating alone will not keep a man well; he must also take exercise (Hippocrates 1931).

Galen went on to argue that physical and spiritual well-being relied on the existence of an ideal equilibrium between two extremes, warning that, in order to

[5] For an introduction (see Jouanna 2001).

[6] Robert Grosseteste, who took a keen interest in such medical writers, is known to have referred to the *Canon* (Crombie 1971).

[7] For an examination of medicine in the Graeco-Roman World (see Jouanna 2012).

enjoy good health, a man must follow strict rules of hygiene. He also advocated an approach based on moderating what would later become known in the West as the *Sex res non naturales* (six non-naturals) (Gil-Sotres 1998; García-Ballester 2002). These included such vital factors as diet (the first instrument of medicine), the elimination of bodily fluids (through activities ranging from sex to purgation and phlebotomy), a clean and bracing environment, exercise, rest and the psychological state of the individual.

From the time of Grosseteste onwards, and certainly well into the Tudor period, the management of this set of rules was elaborated upon in a therapeutic body of literature called *regimina* (Nicoud 2007). This medical advice gained increasing popularity through the medieval period, first in Latin and later in a number of vernacular translations and associate texts, such as the *Secreta secretorum* (Secret of Secrets) and its close relation, the *Regimen sanitatis Salerni* (Salernitan Regimen of Health) (Bonfield 2006; Hardingham 1985).

Collectively, such guides to health instructed people how to safeguard their own precarious mental and physical well-being, and effectively promoting a culture of medical self-sufficiency. There were also some of the most popular advice works circulating in England during the Middle Ages (Slack 1979). The seeds of their success had first been sown in the ninth century, when Arabic copies began to appear. Latin translations of the *Secreta* survive in no fewer than five hundred manuscripts, whilst English translations of the pseudo-Aristotelian text, as can be seen in Fig. 5.1, were made by John Lydgate (1370–1449), Thomas Hoccleve (1369–1426) and Sir William Forrest (1548), to name but a few. Part of their success was also the story told in the dedicatory preface, which recounted how Aristotle was summoned by King Alexander to join him on his expedition to Persia. The King, desperate to learn of the 'Poweer of planetys/And mevyng of al sterrys', knew that Aristotle understood these things and wanted him at his side. Aristotle, however, 'was [too] feble and Oold' to make the journey.[8] This put him in a difficult position: should he endanger his own health, or risk alienating a powerful patron? Fortunately there was a compromise; he would write a treatise entitled *De regimine principum*, which promised to teach the King the secrets of a healthy, happy and contented life.

This tale is, of course, a mere figment of the author's pen—a literary device designed to entertain the reader. Yet it also served another, more serious, function. Indeed, the fact that both the *Secreta* and the *Regimen sanitatis Salerni* were actually translations of the Arabic *Kitāb Sirr al-asrār* (The Book of the Secret of Secrets) did not matter to the reading public; what really concerned them was the authority bestowed by Aristotle's medical wisdom. However, the provenance of this Arabic text, which was translated into Latin, and eventually into the vernacular languages of English and French, is a fascinating one; and it is worth briefly repeating to underscore the translation and dissemination of medical advice from self-help guides during the time of Grosseteste and his contemporaries.

[8] For example, see Paynell (1528).

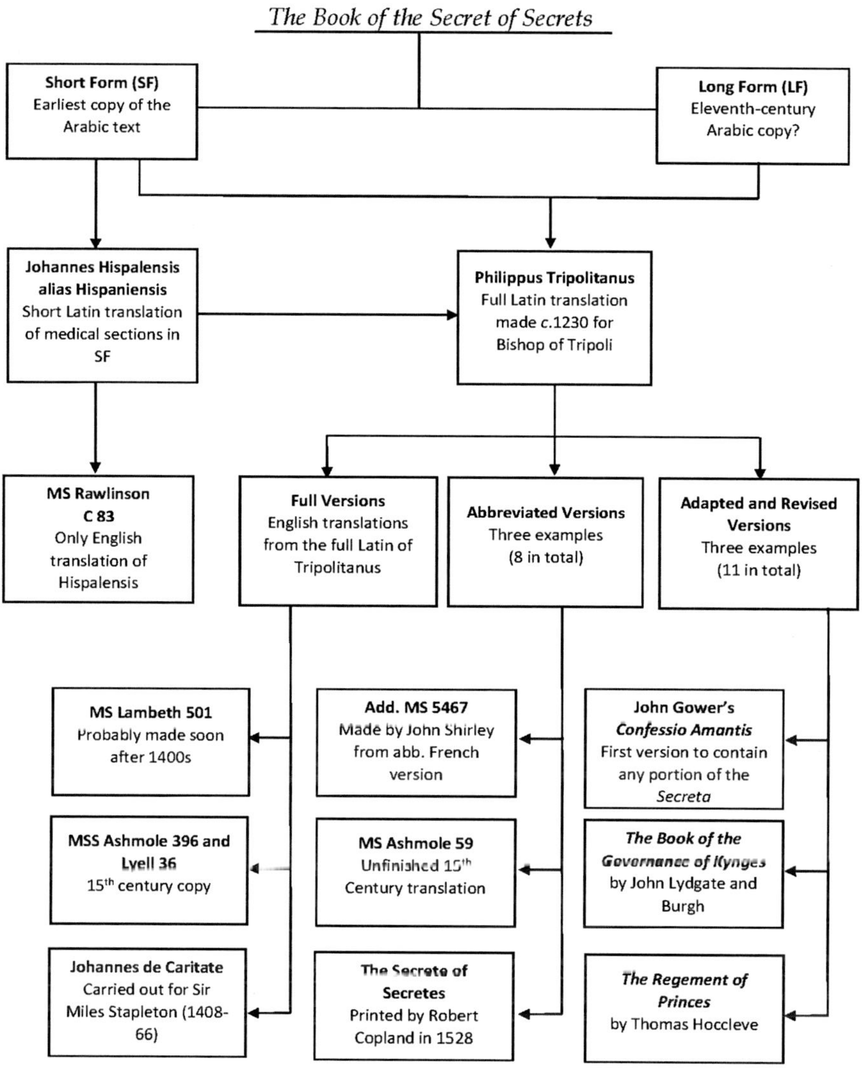

Fig. 5.1 Dissemination of *The Book of the Secret of Secrets*

5.4 Self-Help Guides

5.4.1 Arabic Texts

The *Secreta* was probably translated into Arabic by the ninth-century translator, Yahyä ibn ul-Bitrĭq (John, son of the Patrician). The proem claims that John translated the *Book* from Greek into Syriac and from Syriac into Arabic.[9] It was soon translated into various other languages as well, including Hebrew, Turkish, Latin, Russian and English. There are two surviving recensions of his text in England: one is known as the Short Form (British Library, Add. MS 2453); and the other as the Long Form (Bodleian Library, MS Laud A. 88). The Short Form (SF) appears to be the earliest portion of the work and was divided into seven or eight books, whilst the Long Form (LF), split into ten books, is later in date, having been assembled some time in the eleventh century (Grignaschi 1982).

5.4.2 Latin Texts

The next transformation of the *Secreta* was its translation into Latin, probably (but not definitely) at some point in the first half of the twelfth century, by Johannes Hispalensis.[10] The question of the translator's real identity, however, is problematic and the 'identity of Johannes has never been indisputably established' (Hardingham 1985). Indeed, Maureen Robinson suggests the following possible surnames which include, among others, Hispalensis, Hispaniensis, Hispanus and Hispano (Robinson 2000). A further difficulty is that Johnannes might also have been known as John of Seville and John of Toledo, yet he can hardly have been associated with both cities at once (Ibid.).

We *do* know that Johannes (whoever he might have been) was also the translator of Arabic texts on astrology, and that at least 150 manuscripts of his version of the *Secreta* survive, comprising the Latin dedication and a major part of the Arabic proem. Surprisingly, although Johannes' work was known in England, only one English translation of the *Secreta* was based on his text: Bodleian Library, MS Rawlinson C. 83. This translation only runs to seven pages; yet, despite its comparatively small size, it none the less manages to cover the essentials of healthy living. The 15 'doctrines' range from diet to the 'iiij ceasons of the Þe yere'. The focus of the work is hygiene, which is hardly surprising considering this is what Johannes had been asked to write about (Manzalaoui 1954).

The second Latin version of the *Secreta* (see Fig. 5.2) was made some time between 1000 and 1300 in verse form. However, just like the original text of the *Secreta*, the authorship of the *Regimen santiatis Salerni*, as the poem is now known, is shrouded in a mist of academic and popular speculation. It was originally believed to have been written for the benefit of Robert, Duke of Normandy, the eldest son of William the Conqueror (it is also said that Robert visited Salerno in

[9] According to the 'Ashmole Version' of the *Secreta* John translated the text from the Greek (which no longer survives), into Syriac and 'fro Þat into Arrabike': Mahmoud Manzalaoui (1977). *Secretum secretorum: Nine English Versions*, 29 and ix–xiv. EETS, 226. Oxford: OUP.

[10] It is assumed that the text was translated at some point between 1135 and 1150: Melitta W. Adamson (1995). *Medieval Dietetics: Food and Drink in Regimen Sanitatis Literature from 800 to 1400*, 51. New York: Peter Lang.

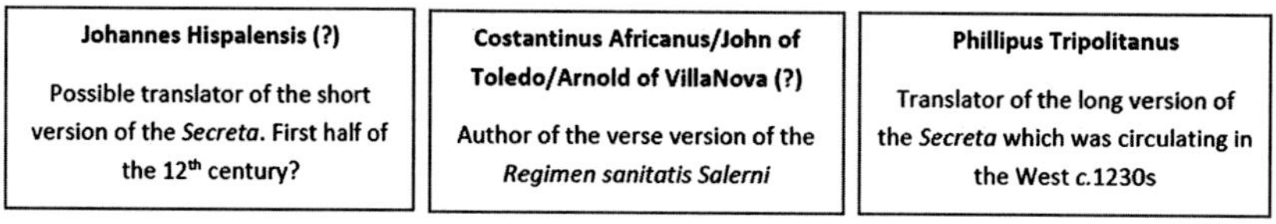

Fig. 5.2 Latin Versions of the *Secreta*

c. 1099) (Packard 1922). In all probability the work was not created for him, yet quite who wrote it and when is still a matter of historical debate.

The third and last translation of the Book into Latin, made by Philip of Tripoli for his patron, Guy de Vere, Bishop of Tripoli, was circulating in the West by the late 1230s or thereabouts (Lindberg 1992). Philip, who was made canon of Tripoli under Gregory IX (c. 1217–1241), was a well-educated man who also had a keen interest in medicine. This interest is evident throughout his translation, which is the longest and most complete version known to have been made. Indeed, Philip not only incorporated the medical parts of Johannes Hispalensis's translation, and added what he had left out, but is also the first translator who actually states that he worked directly from the Arab original. He found the text in Antioch, or so he claims, whilst he was accompanying his uncle Ranier, vice chancellor of the curia under Honorius III (1216–1227), to his new patriarchal see (Paravicini-Bagliani 2000). His work spread throughout Europe and provided the basis for virtually all of the vernacular translations now known to exist. Moreover, his work was also manipulated and adapted by successive generations, as it can be distinguished in two adaptations: the full version (Vulgate) and the abbreviated version (AbTrip).

To summarize, we can be reasonably certain that three Latin versions sprang from the *Secreta*: those of Johannes Hispalensis (short version), Philip of Tripoli (long version, which itself gave rise to the Vulgate and abbreviated versions) and the *Regimen sanitatis Salerni* (verse). Each of these texts shared a common ancestry, yet each was constantly being revised and refashioned during the age of Grosseteste in order to accommodate the many different religious, medical and cultural environments that it encountered. Furthermore, as the three originals were disseminated throughout the various countries of Europe, including England, they came into the hands of such eminent scientists as Albertus Magnus (1206–1280) and Roger Bacon who, in turn, imposed their own ideas upon these malleable texts (Getz 1998). None the less, one of their most radical transformations was still to come: translation from Latin into the vernacular. Twenty-two versions survive in the English language alone, with at least eight deriving at second hand from a French recension of Tripolitanus. All three followed in the footsteps of their progenitor as they were handbooks for princes, but the novelty lay in their wider readership (Bonfield 2006).

Together, these works counselled patients on what foods to eat and wines to drink. Indeed, it is no exaggeration to state that such texts offered their readers a complete guide to health, telling them when and how to get up in the morning and how much sleep to take, as well as advising on almost everything else in between, including whom to sleep with, in what position and how often (Bell 1999). They even suggested how one should wash one's hands and clean one's teeth on rising. This was not just simply good hygiene; the medical guidance offered was grounded in the concept of the six non-naturals. Some guides even included specific advice on the four humours, advising readers that:

> . . .ther be 4 humeros in man and 2 of thaim be frendes and 2 enemyes his 2 frendes be blode [sanguine] and fleume [phlegmatic] his 2 enemies ben colre [choleric] and malencolie and for they be enemies kinde hath prisoned thaim wher colre in galle [gall bladder] and malencolie in the milte [liver]. And if any of thaim breke prisone. . . they engendereth deadly sekenesse (C.U.L., MS Ii.6.17, ff. 5–6).

This very basic guide to humoral theory was often followed by an account of the seasons of the year, and their particular characteristics. Spring, a sanguine time, was when the sun melted the ice, trees smelt sweetly, birds grew new plumage and the sun 'enforce[d] them to synge'. (Aristotle 1528). Summer, on the other hand, was compared to a young man, who was hot and choleric of humour:

> Ffyr, Colour, Estas/and Juventus [young] Age,
> To-gidre Accorde / in heete and drynesse
> And Coleryk men/Citryn of visage (Steele 1894).

Indeed, authors described the four seasons in a specific order—spring, summer, autumn, winter—as they directly corresponded with the four humours and ages of man (Ibid.).

Humoral advice was necessary not only for eating and drinking (a phlegmatic man, for instance, was warned against the dangers of cold and wet foods such as lettuce), but also for all the non-naturals. Baths, for example, were recommended in some texts because they proved to be effective in purging the body by opening up the pores. In one prose version of the *Secreta*, known as *The Governance of Lordschipes* (c. 1400), the author begins by stating that 'bathes er on of Þe merueylles of Þys werld', as they could follow the four seasons: cold in winter, lukewarm in spring, hot in summer, and dry in harvest (Steele 1898). The author also suggests that bath houses should be built on elevated sites exposed to the wind, and have a furnace with hot flames and hot water. After relaxing in a bath, the reader was then advised to spend the rest of the day in joy and rest, as it 'is mykyl bettyr if a man haue disposicion ioy, gladnes . . . hope [and] triste [and] to laugh with ffrendys' (Ibid.). One way to achieve this was by using herbal and scented baths, as smell was a powerful therapeutic which, when inhaled, could raise the spirits and induce a general state of well-being. The bath house at Ely infirmary, for instance, not only boasted a piped water supply by 1288, but also was situated close to a fresh supply of herbs and flowers (Holton-Krayenbuhl 1997).

5.5 Spiritual Advice

The emphasis on moderation, balance and healthy living in *all* aspects of daily life, brings us to the final part of this chapter, which focuses on the Church's understanding of health and disease. The Church, for instance, had its own rules for human conduct, which focused on the avoidance of the Seven Deadly Sins and obedience to the Ten Commandments. The *Fasciculus morum*, a preacher's handbook composed by a Franciscan friar in England in the fourteenth century, for instance, is divided into seven parts, each analysing in depth one of the Deadly Sins. In Part VI, that which deals with Gluttony, may be found the following metaphor, replete with constant warnings of the dangers of worshiping the god of the belly and of gluttony, the kitchen:

> The bells ringing in it [i.e. the kitchen] are the kitchen boys who call out what roast or cooked dish is served. The altar is the dining table. Their chalice with it vials is the bowl with cups and tankards. The priests are their boon companions; their sacrifices, the slaughtered beasts and their roasted and boiled flesh; their incense, the smell and savor [sic] of the food. And they have two prayers: one before they are full, which goes, 'Oh, if only I had two stomachs!' the other when their belly nearly bursts: 'Ah, belly, have mercy; belly, mercy!' (Wenzel 1989).

The Church naturally tried, with varying degrees of success, to ensure that each and every individual body and soul remained free from the stains of sin after he or she had been baptised. But this was a seemingly impossible task, especially as the world was full of temptation and excess. As one medieval sermon put it:

> A man synneÞ in glotenye in dyvers miners, but Þe moste common maner is whan Þat a man takeÞ to meche mete or drynke, and specially when at a man falleÞ in dronkenship: for Þan he vanteÞ all is wittes and haÞ will and luste to do almaner synne, and namely lecherie (Ross 1940).

This advice evidently struck a chord, as other sermons also castigated the sinful man who lacked a balanced spiritual diet: he would live in such a constant state of metaphysical as well as real drunkenness that he would 'not see at Crist dwelleÞ in hem' (Ibid.). One fourteenth-century preacher compared each deadly sin to a state of intoxication; another wrote that too much drink 'blurs the senses, confuses the mind, stirs up lust, ties the tongue, poisons the blood, weakens all the limbs, and destroys one's health altogether' (Wenzel 1989). Certainly, too much wine had a desperate effect on the humoral balance, causing the body to overheat and the soul to be corrupted. Through constant abuse of the non-naturals, each of the deadly sins carried a humoral penalty.

Furthermore, as the author of *Fasciculus morum* noted:

> ... after diagnosing the sickness he [i.e. *Christus medicus*] gives the sick person a diet as he requires and prescribes what he should eat and what he should avoid .... Christ further heals us in many additional ways as if from physical illness: first, through the sweat of contrition, which one gets from hard exercise... Second, through the bloodletting of confession ... Third, through the diet of fasting and penance... Fourth, through the plaster or ointment of devout prayer. Fifth, through draining excessive bodily fluids... Six, through the surgical

> removal of evil companionship and the occasion of sin . . . And seventh, through the cautery of charity (Ibid.).

That Christ offered repentant souls a remedy for their own personal transgressions as well as the collective burden of Original Sin was a view shared by medieval theologians, who maintained that the holy medicine derived from Christ's flesh and blood had therapeutic effects upon both souls and bodies consumed with sin (Yoshikawa 2009). The mere sight of the eucharist was compared 'to a powerful electric current coursing through the body' (Rawcliffe 2008; Bynum 1987). Indeed, the doctrine of transubstantiation, which was formalised during the time of Grosseteste in twelfth century and imposed upon the laity in the thirteenth, held that during Mass Christ actually fed the spiritually sick with His own body (Rubin 1992). This was a particularly 'good medecyn to Þi soule', noted a thirteenth-century sermon, as 'Þe same body Þat died on Þe Crosse . . . is Þe same bodie on Þe Sacrament on Þe awtur in forme of brede' Moreover, just as 'Þe bodie is fed . . . with bodily brede', so 'Þe soule . . . lyeÞ with goostely foode' (Ross 1940).

The priest administered the sacrament during Mass. It seems that, unlike earthly food, there could be no danger of overindulgence when it came to consuming, or at least gazing upon, the body of Christ. As a ruling of the Fourth Lateran Council in 1215 made clear:

> Among other things that pertain to the salvation of the Christian people, the food . . . of God is above all necessary, because as the body is nourished by material food, so is the soul nourished by spiritual food (García and García 1981).

It was also during confession, which 'all the faithful' were expected to make at least once every year to their local parish priest, that sins were absolved and health restored (Ibid.). Sins were both the cause and symptom of disease, and confession achieved reconciliation with God, the Church, and the wider community (McNeill 1932). It is perhaps small wonder that Canon 22 of the Fourth Lateran Council is couched in medical terms, describing the parish priest himself as a physician for the soul who aids the 'recovery of bodily health' by hearing confessing and assisting the process of salvation. Sin manifested itself as disease in human beings, and the priest, described by the Council as a 'skilled doctor', was expected to hear confession and to give the sinner ('sick person') a suitable remedy so that he or she could recover (García and García 1981).

The literature of penitentials, or general priests' manuals, is replete with medical metaphors. These guides provided parish priests with the information they were expected to impart during confession (Hughes 1991). When he heard confession, the priest had to help the penitent to realise the cause of his or her actions, and as disease was thought to be the cause of sin, this meant evaluating the lifestyle of the individual concerned. What is more, as each of the deadly sins carried a humoral penalty, priests were expected to assess the humoral balance of their congregation. Grosseteste, for instance, himself advised in his *Templum Dei* (*The Temple of God*), which survives in over 90 manuscripts from the thirteenth to the fifteenth centuries, that clerics, before imposing penance, should give consideration to the individual's condition, gender, social status, age, and, not least, complexion. He argued that complexion had a bearing on the ability to withstand sin, and so he recommended that a priest should attempt an appropriate diagnosis (Grosseteste 1984).

Grosseteste himself certainly used medical language to full effect when driving his spiritual messages home. Elsewhere in the *Templum Dei*, for instance, he encouraged the reader to consider 'God as the physician, the sinner as infirm and wounded, the seven petitions [in the *Pater Noster*] as lamentations of the infirm to whom the physician gives preparations, medicine, and, after health and confirmation of health, joy to himself and others' (English trans. Loewen 2013). Further, in a letter probably written at the papal curia in Lyons in 1245, in which he sets out his position on the visitation of the Dean and Chapter of Lincoln Cathedral, he notes the role of physicians of the body and soul, who treat both the sick and healthy, and describes the healing powers of spiritual medicine:

> Now, a wise physician visits not only the sick but also his healthy patients, administering medicine that will ward off illness and preserve good health ... So, too, the prelate, the physician of souls, visits not only those who are spiritually sick, but those whose spiritual health is thriving, so that he may administer spiritual medicine as a protection from future illness and strengthen those he finds in good health (Grosseteste 2010).

For Grosseteste, good health, once discovered, should be revealed as it could itself 'serve as a medicine both to heal illness and to preserve good health in others' (Ibid.).

5.6 Conclusion

Clearly, medical ideas, and what might be termed the 'scientific and professional' tradition, shaped medieval understandings of health and healing; and in turn, so too did Christian ideas influence both the definition and understanding of health at every level of medieval society. Indeed, what should we now make of Fulbert of Chartres protestations that human medicine, represented by the famous ancient authority Galen, was less effective than divine healing, represented by Christ? As Iona McCleery has convincingly argued, medical historians, in particular, are now beginning to interpret these types of complaints through a lens of a more symbiotic relationship between medicine and religion (2014). In other words, although there were two types of medicine in the past, the dividing line between them was at times blurred and not as clear cut as some contemporaries might have liked to make out. Furthermore, running alongside this 'blurring' ran the development of a personal sense of responsibility for physical and spiritual health, as evident in the dissemination, translation and adaptations of medical advice manuals such as the *Regimen*. In short, the theory and practice of healing adopted in age of Bishop Grosseteste was complex and demonstrated a sophisticated level of medical, religious and scientific understanding, if indeed these three terms can be separated in this essentially anachronistic way.

References

Adamson, M. W. (1995). *Medieval dietetics: Food and drink in Regimen Sanitatis literature from 800 to 1400*. New York: Peter Lang.

Aquinas, T. (1982). *Theological texts* (T. Gilby, Trans.). Durham, NC: Labyrinth Press.

Arbesmann, R. (1954). The concept of 'Christus medicus'. in St. Augustine. *Traditio, 10*, 1–28.

Aristotle (Pseudo). (1528, 1970 ed.). *Secreta Secretorum* (R. Copland, Trans.). Amsterdam: De Capo Press.

Augustine of Hippo. (1972). *Concerning the City of God against the Pagans* (H. Bettenson, Trans.). Harmondsworth: Penguin.

Bell, R. (1999). *How to do it: Guides to good living for Renaissance Italians*. Chicago, IL: University of Chicago Press.

Bonfield, C. (2006). The *Regimen sanitatis* and its dissemination in England, c. 1348–1550. Unpublished PhD Thesis, University of East Anglia.

Bonfield, C. (2013). Therapeutic regimes for bodily health in medieval english hospitals. In L. Abreu & S. Sheard (Eds.), *Hospital life: Theory and practice from the medieval to the modern* (pp. 21–48). Oxford: Peter Lang.

Bowers, R. (1975). Choral Institutions within the English Church: Their constitution and development, 1340–1500. Unpublished PhD Thesis, UEA.

Brown, P. (1988). *The body and society: Men, women and sexual renunciation in early Christianity*. New York: Columbia University Press.

Carlin, M. (1989). Medieval english hospitals. In L. Granshaw & R. Porter (Eds.), *The hospital in history* (pp. 21–39). London: Routledge.

Crombie, A. (1971). *Robert Grosseteste and the origins of experimental science, 1100–1700*. Oxford: Clarendon.

Dugdale, W. (1817–1830). *Monasticon anglicanum* (J. Calley, H. Ellis, & B. Bandiel, Eds., 6 vols). London: Longman, Hurst, Rees, Orme & Brown & Alexander H. Thompson.

Ell, S. R. (1981). The two medicines. *Janus, 68*, 15–23.

Fulbert of Chatres. (1865). *Hymnus seu prosa de sancto Pantaleon (Hymn to St Panaleon)*. In *Patrologiae latinae cursus completes* (vol. CXII, col. 341). (J. P. Migne, Ed.). Paris: Garnier Frates.

García y García, A. (Ed.). (1981). *Constitutiones Concilii quarti Lateranensis una cum Commentariis glossatorum*. Vatican City: Biblioteca Apostolica Vaticana.

García-Ballester, L. (2002). *Galen and Galenism: Theory and medical practice from antiquity to the European renaissance* (J. Arrizabalaga, C. L. Cifuentes, F. Salmon, Eds.). Aldershot: Ashgate.

Getz, F. (1998). *Medicine in the english Middle ages*. Princeton, NJ: Princeton University Press.

Gil-Sotres, P. (1998). The regimens of health. In M. D. Grmek, B. Fantini, & A. Shugaar (Eds.), *Western medical thought from antiquity to the Middle ages* (pp. 291–318). London: Harvard University Press.

Grignaschi, M. (1982). Remarques sur la formation et l'interprétation du *Sirr Al-Asrar*. In W. F. Ryan & C. B. Schmitt (Eds.), *Pseudo-Aristotle, The Secret of Secrets: Sources and influences*. London: Warburg Institute Studies.

Grosseteste, R. (1861, reprint 2012). *Roberti Grosseteste Episcopi Quondam Lincolniensis Epistolae* (H. R. Luard, Ed.). Cambridge: Cambridge University Press.

Grosseteste, R. (1984). *Templum Dei* (J. Goering, Mantello, & A. C. Frank, Ed.). Toronto: Pontifical Institute of Medieval Studies.

Grosseteste, R. (2010). *The letters of Robert Grosseteste, Bishop of Lincoln* (F. A. C. Mantello & J. Goering, Trans.). Toronto: University of Toronto Press.

Gruner, O. C. (1930). *A treatise on the Canon of medicine of Avicenna incorporating a translation of the first book*. London: Luzar & Co.

Hardingham, G. (1985). The *Regimen Sanitatis* in late medieval England. Unpublished DPhil. Thesis: University of London.

Harrison, P. (2007). *The fall of man and the foundations of science*. Cambridge: Cambridge University Press.

Hippocrates. (1931). *Nature of man. Regimen of health. Humours. Aphorisms. Regimen 1–3. Dreams* (vol. IV, W. H. S. Jones, Trans.). Cambridge, MA: Harvard University Press.

Holton-Krayenbuhl, A. (1997). The infirmary complex at Ely. *Archaeology Journal, 154*, 118–172.

Horden, P. (2007). A non-natural environment: Medicine without doctors and the medieval European hospital. In B. S. Bowers (Ed.), *The medieval hospital and medical practice* (pp. 133–146). Aldershot: Ashgate.

Hughes, J. (1991). The administration of confession in the diocese of York in the fourteenth century. In D. M. Smith (Ed.), *Studies in clergy and ministry in medieval England* (pp. 87–163). York: Borthwick Publications.

Jamison, C. (1952). *The history of the Royal Hospital of St. Katherine by the Tower of London*. Oxford: Oxford University Press.

Jouanna, J. (2001). *Hippocrates* (M. B. Devoise, Trans.). Baltimore: John Hopkins University Press.

Jouanna, J. (2012). *Greek Medicine from Hippocrates to Galen: Selected Papers* (N. Allies, Trans.). Leiden: Brill.

Lindberg, D. C. (1992). *The beginnings of Western science: The European scientific tradition in philosophical, religious and institutional contexts, 600 BC–AD 1450*. Chicago, IL: University of Chicago Press.

Littré, É. (Ed.). (1839–1861). *Œuvres complètes d'Hippocrate*, (10 vols, vol. VI). Paris: Ballière (1962) (Reprint. Amsterdam: Adolf M. Hakkert).

Loewen, P. (2013). *Music in early Franciscan thought*. Leiden: Brill.

Manzalaoui, M. (1954). *The Secreta Secretorum in English thought and literature from the fourteenth century to the seventeenth century with a preliminary survey of the origins of the secreta*. Unpublished DPhil, University of Oxford.

Manzalaoui, M. (1977). *Secretum secretorum: Nine english versions*. Oxford: Oxford University Press.

McCleery, I. (2014). 'Christ more powerful that Galen?' The relationship between medicine and miracles. In M. Mesley & L. E. Wilson (Eds.), *Contextualizing miracles in the Christian West, 1100–1500: New historical approaches* (pp. 127–154). Oxford: Medium Ævum.

McNeill, J. T. (1932). Medicine for Sin as prescribed in the Penitentials. *Church History, 1*, 14–26.

Murray, J. (1991). Sexuality and spirituality: The intersection of medieval theology and medicine. *Fides et Historia, 23*, 20–36.

Nicoud, M. (2007). *Les régimes de santé au Moyen Âge: naissance et diffusion d'une écriture mé dicale (13e-15e siècle)* (Vol. 2). Rome: Ecole Française de Rome.

Nutton, V. (1995). Medicine in the Greek World, 800–50 BC. In L. I. Conrad et al. (Eds.), *The Western medical tradition* (pp. 19–22). Cambridge: Cambridge University Press.

Packard, F. R. (1922). History of the school of Salernum. In J. Packard, F. R. Garrison, & F. H. Harington (Eds.), *The school of Salernum: Regimen sanitatis Salernitanum, the English version by Sir John Harington*. London: Oxford University Press.

Paravicini-Bagliani, A. (2000). *The Pope's body* (D. S. Peterson, trans.). Chicago: University of Chicago Press.

Park, K. (1992). Medicine and society in medieval Europe, 500–1500. In A. Wear (Ed.), *Medicine in society: Historical essays* (pp. 59–90). Cambridge: Cambridge University Press.

Park, K., & Henderson, J. (1991). 'The first hospital among Christians': The Ospedale di Santa Maria Nuova in Early Sixteenth-Century Florence. *Medical History, 35*, 164–188.

Paynell, T. (Trans.). (1528). *Regimen sanitatis Salerni*, Sig. Aiij^{r}. London.

Rawcliffe, C. (1995). *Medicine and society in later medieval Europe*. Stroud: Alan Sutton Pub.

Rawcliffe, C. (1999). Medicine for the soul: The medieval english hospital and the quest for spiritual health. In J. R. Hinnells & R. Porter (Eds.), *Religion, health and suffering* (pp. 316–338). Abingdon: Routledge.

Rawcliffe, C. (2002). Curing bodies and healing souls: Pilgrimage and the sick in medieval East Anglia. In C. Morris & P. Roberts (Eds.), *Pilgrimage: The english experience from Becket to Bunyan* (pp. 108–140). Cambridge: Cambridge University Press.

Rawcliffe, C. (2008). Christ the physician walks the wards: Celestial therapeutics in the medieval hospital. In M. Davies & A. Prescott (Eds.), *London and the Kingdom: Essays in honour of Caroline Baron* (pp. 78–97). Donington: Paul Watkins Publishing.

Risse, G. B. (1999). *Mending bodies, saving souls: A history of hospitals*. Oxford: Oxford University Press.

Robinson, M. (2000). Johannes Hispalensis and the manuscript tradition: The history surrounding the time of John of Seville and the spread of his Work. Unpublished PhD, University of Exeter.

Ross, W. O. (Ed.). (1940). *Middle English Sermons* (no. 209, p. 9, 56). London: Early English Text Society.

Rubin, M. (1992). What did the Eucharist mean to thirteenth-century villagers? In P. R. Coss & S. D. Lloyd (Eds.), *Thirteenth century England IV* (pp. 47–56). Woodbridge: Boydell and Brewer.

Slack, P. (1979). Mirrors of health and treasures of poor men: The use of vernacular medical literature in Tudor England. In C. Webster (Ed.), *Health, medicine and mortality in the 16th century* (pp. 237–274). Cambridge: Cambridge University Press.

Southern, R. (1986). *Robert Grosseteste: The growth of an english mind in medieval Europe*. Oxford: Clarendon.

Steele, R. (Ed.). (1894). *Lydgate & Burgh's Secrees of old Philisoffres, A Version of the 'Secreta Secretorum' edited from Sloane* (no. 66, MS. 2464, stanza 194, ll. 1352–54, 43). Oxford: Early English Text Society (reprint 1973).

Steele, R. (Ed.). (1898). *Three prose versions of the Secreta Secretorum* (Vol. 74, p. 82). Oxford: Early English Text Society.

Thompson, A. H. (1937). *The history of the hospital and the new college of the annunciation of St Mary in the Newarke, Leicester*. Leicester: E. Backus.

Touwaide, A. (2014). Galen. In T. F. Glick, S. Livesey, & F. Wallis (Eds.), *Medieval science, technology and medicine: An Encyclopaedia* (pp. 179–182). New York: Routledge.

Walker Bynum, C. (1987). *Holy feast and holy fast: The religious significance of food to medieval women*. Berkeley: University of California Press.

Wenzel, S. (Ed. & Trans.). (1989). *Fasiculus Morum: A fourteenth-century Preacher's handbook*. Pennsylvania: Penn State University Press.

Yoshikawa, N. (2009). Holy medicine and diseases of the soul: Henry of Lancaster and *Le Livre de Seyntz Medicines*. *Medical History, 53*, 397–414.

Ziegler, J. (2001). Medicine and immortality in terrestrial paradise. In P. Biller & J. Ziegle (Eds.), *Religion and medicine and the Middle ages* (pp. 201–242). York: York Medieval Press.

Manuscripts

British Library, MS Cotton Cleopatra C V, ff. 25^{r} -28^{v}.

Cambridge University Library, MS Ii. 6.17, ff. 5–6.

Chapter 6
The Corruption of the Elements: The Science of Ritual Impurity in the Early Thirteenth Century

Sean Murphy

6.1 The Science of Ritual Impurity

William of Auvergne (d. 1249), writing in *De Legibus* (between 1220 and 1240?)—a long treatise on the non-moral laws of ancient Judaism and related matters—identifies six kinds of leprosy according to appearance, six kinds according to the site of infection, and six kinds according to color.[1] This is in addition to his observations on the three kinds of corruption caused by leprosy, as well as his distinction between two basic kinds: 'balding' leprosy and 'gnawing' leprosy. William's concern with leprosy is very much a concern with matter—the bodies, the clothing, the buildings that contract leprosy. He theorizes about the causes of leprosy, about its diagnosis, about its treatment, extending his analysis even to the proper definition of the term. And his concern with leprosy is also very much a concern with the spirit. He suspects, for example, ancient inducements to idolatry in the treatment of leprous clothes and houses, and finds a divinely mandated need for ritual correction to such idolatry. William assumes the moral neutrality of bodily leprosy in most cases, with due acknowledgement of the possibility that bodily leprosy occurs as a 'disease inflicted for sin by the wrath of God.' Above all else, he asserts the principle that the impurity of 'spiritual leprosy' is far greater than that of 'bodily leprosy.' We could perhaps call leprosy of the body a matter of natural

[1] Precise references to all matters mentioned in this introduction are provided in the notes that follow. The only biography of William of Auvergne is Valois (1880). A recent biographical sketch can be found in Murphy (2012). On William and his works, also see Morenzoni and Tilliette (2005). The *De legibus* is the second and much longer section of a single treatise, *De fide et legibus*. 1674. The only printed edition is in Guilielmus Alvernus (Guilielmus 1963). (The treatise has never been translated.)

S. Murphy (✉)
Department of Liberal Studies, Western Washington University, Bellingham, WA, USA
e-mail: Sean.murphy@wwu.edu

J.P. Cunningham, M. Hocknull (eds.), *Robert Grosseteste and the pursuit of Religious and Scientific Learning in the Middle Ages*, Studies in the History of Philosophy of Mind 18, DOI 10.1007/978-3-319-33468-4_6

science, leprosy of the spirit a matter of theology, but William would not. That distinction, reasonable as it seems, fails to capture William's point of view, according to which the material and spiritual are composite elements in a single conceptualization. Any sustained consideration by a medieval author of ritual purity laws in Leviticus (or related texts) invites a combined analysis of the material and the spiritual, the temporal and the eternal—in short, the 'scientific' and the 'theological.' For William, this is the case whether he is discussing leprosy or menstruation or other putative sources of impurity. We have long recognized that medieval commentaries on certain topics—the days of creation, for example—are a potential locus of combined scientific and theological analysis; something comparable can be seen perhaps in medieval treatments of ancient concerns with ritual purity, impurity, and purification.

6.2 Leprosy and Leviticus in *De Legibus*

William's treatment of leprosy across two chapters (here, following the chapter divisions in the 1674 Paris edition) of his *De legibus* is directly inspired by the legal mandates governing leprosy in Leviticus, Chaps. 13 and 14, where the biblical author identifies the signs of 'leprosy' in human bodies, clothing, and houses, and specifies the rituals required for purification.[2] This is evident especially in William's attention to the levitical rituals for purification of leprosy in the human body: First, the seven-fold sprinkling of the leper with water mixed with the blood of a sacrificed sparrow (Lev. 14:1–7). Then, on the eighth day after the sprinkling, the twofold anointing of the right ear, the right thumb, and the right big toe with the blood and oil of sacrifice (Lev. 14:10–18). William does not, however, provide a line-by-line commentary on Leviticus 13–14—nothing of the sort—nor, apparently, does he feel bound to address, by any other means, all the particulars of 'leprosy' described in Leviticus. He is selective in his topics and elaborates them, when so inclined, freely beyond the text of Leviticus.

On the various kinds of leprosy, for example, William's distinction between 'balding' and 'gnawing' leprosy in the 'hairs of the head and other [bodily] hair' is his own (*De legibus*, 42.2).[3] Certainly, Leviticus includes loss or change in hair among the indicators of a leprous condition, but the biblical text lacks the analytic

[2] That 'leprosy' is an inadequate and potentially misleading translation of the Hebrew term, 'tsar'at' has been widely noted. (William himself makes the point, as discussed below, that the term 'leprosy' is not used properly across Lev. 13–14.) For one concise, recent statement of the problem, see Coogan (2001) where the translators recommend 'surface affliction' as a better English translation of 'tsar'at.' For a more developed, though essentially compatible, discussion, see Milgrom (1991); here, the term 'scale disease' is preferred. The history of the translation of the Hebrew term into Greek, from Greek into Latin, and from Latin into English is well summarized in Rawcliffe (2006).

[3] All translations of *De legibus* are my own.

precision of William's distinction. And where Leviticus proceeds, without explicit logical or material linkage, from bodily leprosy to leprosy of clothing and, eventually, houses, William, in compressed sequence, links leprosy of hair to skin to clothing to stones and walls, all as similar manifestations of 'gnawing' disease or affliction (Ibid.). Such linkage amounts to a scientific theory, albeit in rudimentary form, of leprous affliction across a variety of materials. William's further divisions among kinds of leprosy grouped according to appearance, according to the site of infection, and according to color derive rather more directly from a close reading of Leviticus 13–14. This is made clear in William's prelude to the groupings: 'The kinds of leprosy are also distinguished according to the definition of the Law in this way' (*De legibus* 43.1). Even in this case, however, William brings an analytical and categorical precision to bear on his source material, reducing the welter of leprous signs in these two chapters of Leviticus to three crisp lists of essential points of comparison in the diagnosis of leprous conditions.

The literal value of Leviticus as a guide to leprosy is consistently affirmed by William: 'It ought in no way to be doubted that the signs that the Law establishes are *true signs* of the afflictions that the Law judged to be unclean and leprous' (*De legibus* 42.2). At the same time, however, William's study of leprosy in Leviticus is informed throughout by his interest in contemporary scientific-medical theory, as well as evidence derived from his own experience. His wide-ranging scientific interests lead William sometimes to elaborations, sometimes to corrections of the received material in Leviticus 13–14.

His elaborations include the enumeration of the 'three corruptions of leprosy'—rot, stench, and discoloration. And, shortly thereafter, the hypothesis that Leviticus commands the application of the blood and oil of sacrifice specifically to the right ear, right thumb, and right big toe, 'because leprosy is first diagnosed and felt there, that is, in those digits—and this because of their sensitivity, as is said...so it was commanded to be done such that there was the same order of purification as there was of infection' (*De legibus* 42.1). To which he adds this consideration: 'And perhaps in the extremity of the right ear some sign of leprosy customarily appeared, because that part is very much less sensitive' (*De legibus* 42.1). When introducing his distinction between 'balding' and 'gnawing' leprosy, moreover, William notes that the former 'is called a true leprosy among the medical people,' before elaborating on the latter. William's knowledge of medical theory, however, does not imply acknowledgement of its worth: 'Concerning discolorations, however, and patchiness in the skin, whether these are signs of true leprosy ought to be determined more by experience than by the art of medicine,' adding that, 'those who have lived with lepers a long time are accustomed to judge them' (*De legibus* 42.2).[4] The value William places on experience (of what we might call empirical

[4] On this last point, William is likely referring to the non-leprous members of leprosaria, that is, of European institutions for lepers (religious houses, hospitals, etc.) established in significant numbers from the early twelfth century onward. For recent research on medieval leprosaria, as well as the diagnosis of leprosy, both medical and para-medical (see Touati 1998; Rawcliffe 2006; Demaitre 2007).

observation) is also evident in his personal witness to leprosy 'in stones and walls,' an account which quickly unites a single observation with a degree of scientific theorizing. 'And we were asked about this many years ago, about a certain stone, which was devoured by itself in this way in a certain abbey, and they called it cancerous. It is not surprising, however, if this should be in a single stone out of an excess of dryness, because this happens in a mass of stones, that is, in the wall, as well as in the mortar from an excess of the same dryness. We see this same thing happen in wood, and this affliction can occur naturally in an entire wall, just as in an entire stone, because, having been produced by the same cause, it happens equally in the entire wall just as in one of its stones' (*De legibus* 43.1).

William's principal correction to the treatment of leprosy in Leviticus—a correction, I suppose, based on both his theoretical and empirical interests—comes right at the start of his only dedicated examination of the topic.[5] 'After this we speak about leprosy, where, first, you should know that the term 'leprosy' is not used here [i.e., in Leviticus] according to its proper meaning. For leprosy is properly a deadly itch, like a widespread cancer, creeping along with most secret pincers and spreading corruption through the whole body. That the Law, however, uses the term 'leprosy' loosely for many things is apparent from the plain text itself [...]' (*De legibus* 42.2). What then follows is a review of some of the multiple conditions described as 'leprosy' in Leviticus, at the end of which William concludes: 'Therefore, the term and definition of 'leprosy' is to be applied to nothing else here than what made a person loathsome (*deformem*) and, by the leprosy, marked out as unacceptable for cohabitation' (*De legibus* 42.2). According to the science of leprosy, then, Leviticus uses the term 'leprosy' in imprecise and, strictly speaking, improper ways. William considers this point worth emphasizing before all else. At the same time, William also wants to make clear that such imprecision is rationally explicable given the implied objectives (as William sees them) of purity laws governing 'leprosy.' Among other reasons William suggests for the wide and loose application of 'leprosy' to a variety of non-leprous conditions is this: 'God wanted the camp of his people not only to be clean but to appear clean, for the sake of his honor and glory, and so that his people would be pre-eminent in this respect among the other peoples. He also wanted his people to be on guard against not only the contagion of this disease, but even the suspicion of contagion' (*De legibus* 42.2).

So far, I have indicated ways in which William elaborates and corrects the levitical treatment of leprosy on the basis of scientific theory and his own experience. Interventions of this sort are occasional not systematic. This suits his own general approach to Leviticus 13–14, which, as I have said, avoids anything like a systematic commentary on chapter and verse. This is not to say, however, that

[5] William first mentions leprosy—this in Chap. 10—in relation to the sacrifice of the red calf, where he compares the parallel elements in these two separate rites of purification. At the start of the next chapter, however, he returns to leprosy and introduces the topic as if speaking about it for the first time.

William has no particular emphasis in his discussion of leprosy. He certainly does. He focuses, above all, on the levitical rituals of purification for leprosy, because they confirm, in his view, a scientific account of the *causes* of leprosy. And it is especially with respect to a theory of material causation that William's scientific interest takes him beyond the confines of Leviticus. His account of leprous causation is concise, emphatic: '[E]very single kind of leprosy is contracted radically and basically from one of the four elements through the humor corresponding to that element' (*De legibus* 43.1). Elsewhere, William observes that 'leprosy is mainly from a corruption of the blood' (*De legibus* 42.1). There is nothing, of course, about elements or humors in the text of Leviticus, not even an echo of those scientific theories of physical and physiological fundamentals. Nor could there be. These are entirely William's elaborations—modest but incisive—on the authoritative source for leprosy laws. Nevertheless, William sees such theories confirmed by the ritual of purification for leprosy mandated in that body of laws. The ritual, as he sees it, implies the underlying material phenomena. '[F]our materials are also used in the cleansing of a leper and a leprous house, namely cedar wood, hyssop, twice-dyed scarlet, and a living sparrow, from which the aspergillum is made in such a way that the cedar wood is like its handle, and the scarlet is the thread and binding material by which the branch of hyssop and the single living sparrow are bound at the same time, and these two, namely the branch of hyssop and the sparrow, are dipped at the same time, just like an aspergillum, in living water and the blood of the other sacrificed sparrow, so that whoever is to be cleaned is sprinkled with it' (*De legibus* 43.1). In his analysis of the sacrifice of the red calf (an analysis that immediately precedes and partly overlaps with his discussion of leprosy), William had already noted that 'the four elements were, in a way, signified by the materials burned' with the calf (*De legibus* 41.2). 'By the cedar, earth, because it comes forth from the earth and is of the substance and solidity of earth. By hyssop and water, because hyssop seems to be begotten and nourished by the rain alone, on account of which it grows among stones and in stony places. By the odor of both, air, especially according to those who believe that odor is nothing but vapor, a vapor, I say, released from the body emitting the odor, which is why, Isidore says, aroma is said like 'airoma.' And by the scarlet thread, fire, as was said' (*De legibus* 41.2). The sparrow, added to the ritual for leprosy, provides, as William sees it, a further, express signification of air (*De legibus* 43.1). And, so, the ritual purification for leprosy, he concludes, matches element to element: 'For this reason, this rite was provided by God, the creator of the elements, so that an uncleanness contracted from the elements, would be cleansed in this rite by the same' (*De legibus* 43.1).[6]

For William, then, every kind of leprosy derives from a corruption of the elements, and the materials used in the levitical ritual of purification for leprosy

[6] This repeats the point already made by William at 42.1. 'Therefore, it was suggested, according to the letter, by these four things that, from whatever of the four elements leprosy had been contracted, the very author of the elements, through a sprinkling of this kind and those other things which were added to it, cleansed the leprosy [...].'

signify a cleansing and restoration of the elements by an incorrupt form of the elements. This is science at the level of physics. Corrupt elements, as observers, including William, well know, corrupt the very bodies of which they are constituent. This, in the case of the human body, includes the three corruptions listed by William: rot, stench, and discoloration. At this level too, that is, the physiological, the purification ritual signifies the remedy for corruption: 'Another [purification is] from the cedar, which always repels rot from itself. [Another] from the hyssop, because it purges rotten humors and cleans away stains from the face, it spreads a sweet odor. Rightly are they used against the rot of leprosy. The incorruptible cedar and the purifying hyssop are fittingly used against the rotten humor and also against the stench and stains of that affliction. The juice of the hyssop, moreover, is believed to wash away stains of the face' (*De legibus* 43.1-2).

6.3 Science and the Literal Sense

William's evident interest in analyzing levitical leprosy from the perspective of scientific theory and experience arises from his thoroughgoing commitment, in *De legibus*, to the primacy of the literal sense in the interpretation of the Law of Moses. In fact, William's two chapters on leprosy in Leviticus are but one, relatively small part of a far grander exploration and defense of the rationality, goodness, and justice of the non-moral commands of the Law interpreted according to the letter. This feature of *De legibus*, signaled as early as 1974 by Beryl Smalley, is one which I have discussed at length in two recent articles—one on sex-related impurities in *De legibus*, a second on William's condemnation of contemporary 'pagan' idolatries (Smalley 1974; Murphy 2013, 2014). I will not rehearse all the relevant evidence here; a single, representative passage from the first chapter of *De legibus* conveys his position perfectly well: 'It is evident, therefore, from all these considerations that the Law of Moses was promulgated with God as its author and founder. That is why there is nothing in it that is useless, nothing pointless, nothing absurd. Therefore, there is nothing in it, whether command or prohibition, whether law or story, that does not have a rational explanation and sufficient reason, whether secret or manifest' (*De legibus* 25.1). I have also demonstrated elsewhere that William's defense of the literal value of even non-moral elements of the Law is founded on his dominant concern with contemporary idolatry, including sorcery and pagan practices, a concern that justifies, in William's mind, the continued consideration and sometime admissibility of an indeterminate but seemingly large number of non-moral commands.[7] This is made clear in, among other places, a key passage that concludes the section of *De legibus* on specific commands of the Law. '[I]t should be clear to you, and rightly so, that those things which seem to be absurd in

[7] On William's concern with idolatry in *De legibus* (see Murphy 2013, 2014). The same is discussed in (Smalley 1974), and briefly mentioned in (Smith 2005).

the law are to be understood according to the letter, and that there are just and rational reasons for its precepts and its prohibitions, and that in past times the simplicity and ignorance of the people, and the proximity and nearness of idolatries required precepts and prohibitions of this sort. We also make clear to you, that the greater part of those exist even among us, as, for example, commands against the observance of superstitions, which we enumerated just above' (*De legibus* 46.1-2).

All of this means that William has no truck with the existing tradition among Christian scholars of non-literal commentary on Leviticus. Consider the *Glossa Ordinaria* (c. 1115), for example, in which 'leprosy' is interpreted primarily as 'sin' or 'iniquity' or 'transgression' against the Law or Gospel, sometimes as 'heresy,' and in which literal readings of Leviticus 13–14 have no play at all.[8] Or the Leviticus commentary of Ralph of Flaix (fl. mid-twelfth century), which Beryl Smalley described as 'the standard commentary on Leviticus up to the mid-thirteenth century at least' (Smalley 1968), where leprosy symbolizes sin in all its varieties, nothing more, nothing less. Here, one sees the hegemony of the allegorical and the moral over the literal in a work explicitly motivated by Ralph's concern about the impact of 'Jewish' interpretation on his fellow Christians (Ralph of Flaix 1677).[9] And Ralph's influence extends from Peter the Chanter to Hugh of St. Cher to William of Middleton.[10] William of Auvergne, swimming against just such currents, devoted two full chapters of *De legibus* to a critical reflection on the practice of biblical interpretation, including an extended condemnation of excess, abuse, and basic incompetence in the practice of non-literal interpretation (*De legibus*, Chaps. 16 and 17).[11]

---

[8] *Bibliorum Sacrorum Glossa Ordinaria*, Venice (1603). Some representative examples of the commentary on Leviticus 13–14: the 'wound of leprosy' is 'transgression of the Law,' which makes the 'leper' guilty of violating all of the Law if he violates it in one respect (13:2); 'leprosy is false teaching, lepers heretics' (Isidore on 13:2); the 'whitened hairs' of leprosy are the 'open thought of sin' or the 'conscience of the heretics' (13:3); 'leprosy' is 'sin' and 'iniquity' (13:5–6); the cleansing of the leper is 'baptism or penance' (13:6); 'an inveterate leprosy' is said of one who has 'grown old in sin' (13:11); 'leprosy' is 'a violation of the law' (13:12); the 'white or red color in the bald head or bald forehead' is what is done 'against the rule of the Law or against the Gospel' (13:42); 'all the time that he is a leper and unclean' is when the leper is 'not yet perfected by penitence' (13:46); the 'two living sparrows' are an 'incorrupt mind and a firm faith' (14:4); the 'cedar wood, scarlet, and hyssop' are the 'prayers of the saints' (14:4).

[9] For Ralph's concern about 'Jewish' tendencies in Christian interpretation of Leviticus, see p. 48. His commentary on Lev. 13–14 is at pp. 130–150 in the 1677 edition. Ralph's introduction to Lev. 13, at p. 130, amply captures the aim of the twenty pages that follow: 'Leprosos itaque lex nominat, non eos qui peccaverunt, sed qui peccati sui poenitudinem nullam gerunt.'

[10] There is some orientation to commentaries on Leviticus to c. 1250, including a very brief mention of leprosy in Smalley (1974).

[11] The condemnation begins at 48.2, where William treats his 'fourth kind of signification,' that is, of a likeness between two or more things, none of which was intended to signify the other(s). A sample from 48.2: 'If the divine expositors and doctors spoke in this way about their allegorical and tropological interpretations, as well as their anagogical interpretations, and the Scriptures were satisfactory [in this respect], they would not have offended the understanding of their listeners or readers. But because it is said that such a thing signifies such a thing, and is a figure or prophecy or

William's considered literalism in his treatment of so much of Leviticus, including the chapters on leprosy under discussion here, implies, I think, a degree of realism with respect to ritual impurity. That is, Leviticus 13–14 encapsulates legitimate concerns with the physical reality of leprous conditions (and their impact on a community sanctified to God). I have suggested that literalism (or a high estimation of and attention to literal sense) helps to explain William's turn to contemporary scientific theory in his discussion of Leviticus 13–14. I think it equally plausible that William's invocation of scientific theory in this context indicates an at least limited realism with respect to the impurity associated with leprous conditions. Such realism is at least suggested by William's use of contemporary scientific theory. When William invokes contemporary theories about leprous conditions, he furthers the notion that leprous conditions are, in themselves, worthy of attention in a context of moral and theological analysis, that there is something genuinely wrong with such conditions, that is, with the physical and medical states that such conditions describe, and not simply with the moral conditions that the physical conditions are elsewhere taken to symbolize. This is consistent with William's inclusion elsewhere in *De legibus* of a scientific rationale, among other considerations—the threat of idolatry, the use of menstrual blood for sorcery, pollution of the tabernacle—for the reality of menstrual impurity.[12]

So what exactly, one might ask, was William reading of the medical-scientific literature on leprosy available before the mid-thirteenth century? That question is difficult, perhaps impossible, to answer. William, because of his overriding concern (as mentioned above) with idolatry and sorcery, mentions several books known to him that included sections on magical practices, astronomy, elemental transformation, and the mixing of kinds. These books no doubt included some measure of what we would call natural science, but William does not refer to them as works of natural philosophy or science, but instead condemns them as the 'books of the sorcerers.'[13] These include the *Neumich*, for example, and at least six others.[14] But these are clearly not the books William has in mind, when he writes about leprosy. And there is no mention, in any context, of works on medical science or, more broadly, on the elements considered in their own right or as they relate to the bodily humors.

That having been said, I think it instructive, at least for understanding William's perspective in *De legibus*, to consider influential examples of medical-scientific

parable of such a thing, when it seems that the one is not done or said in order to signify the other, they seriously offend their listeners.' On William's theory of biblical interpretation (see Smalley 1974; Smith 2005); on theory and practice (see Dahan 2005).

[12] In William's numbered list of six reasons for the commands about menstrual impurity in Leviticus, the second is that sexual intercourse during menstruation causes the corruption of the offspring conceived during the act; the sixth is the generally corrupting effects of menstrual blood. See *De legibus* 36.2. This is discussed in Murphy (2013).

[13] See, for example, *De legibus* 31.1 and 35.1.

[14] *Neumich* is mentioned at *De legibus* 43.2. For a summary list of the 'books of the sorcerers' mentioned by William and the contexts in which they occur (see Murphy 2014).

works on leprosy written near William's time. For present purposes, I consider three points of comparison: Gilbert the Englishman (c. 1180–c. 1250), an exact contemporary of William, whose *Compendium Medicinae* includes a lengthy consideration of leprosy; Bartholomew the Englishman (d. 1250), also an exact contemporary, whose *De propretatibus rerum*, an encyclopedic work in nineteen books, includes a single chapter on leprosy in Book 7, on diseases; and Theodoric of Cervia (1205–1298), whose widely copied *Surgery* likewise includes a long chapter on leprosy (Gilbertus Anglicus 1510; Bartholomaeus Anglicus 1975–1988; Surgery of Theodoric 1960). In all three cases, there are some specific parallels with William's discussion in *De legibus*. Gilbert and Bartholomew, for example, assert, like William, that leprosy is caused by a corruption of the humors; Theodoric likewise (Gilbertus 1510; Bartholomaeus 1975; Theodoric 1960). And where William asserts that leprosy is first felt and diagnosed in the digits, specifically the right thumb and right big toe, Gilbert, too, describes the loss of sensation in the digits of both hands and feet as symptomatic, though he describes it manifesting especially in the smallest and next-to-smallest digits (Gilbertus 1510). Generally, however, William has much less to say about the science of leprosy than any of these authors. This should not be surprising. For William, as I have said, the literal sense is paramount in the interpretation of Leviticus, but, even when he limits himself to the literal sense, the literal sense in no way limits him to strictly *material* considerations, to considerations of physics and physiology. For William, an interpretation of the Law according to the letter inevitably includes a consideration of moral and theological implications of the literal. The point here is not simply that valid non-literal interpretation (whether allegorical or tropological or anagogical or otherwise categorized) must rest soundly on a legitimate literal interpretation, though William clearly does think that is the case (see *De legibus*, Chaps. 16–17). Instead, the moral and theological are aspects of the literal itself—they are, in fact, the most important aspects for William. And so, when it comes to William's discussion of the impurity of leprosy in Leviticus 13–14, his interest in the literal sense, in the reality of ritual impurity, explains his appeal to medical-scientific theory, and also explains why there is so much less about both in his discussion of leprosy than in contemporary scientific literature on the same. Moral and theological considerations simply demand more attention than scientific ones.

At the level of physics, for example, he sees in leprosy the results of a corruption of the elements—this is a non-moral observation. But he also sees the possibility that the elements themselves, considered more generally, can be corrupted by sin. This point is made with respect to the sacrifice of the red cow, where the material elements ritually burned with the cow are supposed to recall the four elements and the necessity of redeeming the elements themselves. 'Moreover, these four are burned in commemoration of the final (*novissimae*) purification of the lower world, which is to happen through fire in dread of sins, and for the sake of a certain purification [of the elements] themselves, because, by their wrong (*abusione*), they [i.e. sins] were contracted by the very elements. [...] For this reason, therefore, this sacrifice was offered to the most high God from the four elements, just as to the author of the elements, and because [it is] to the author and for this purpose, whatever uncleanness was contracted by any of the four elements was understood to be expiated through this sacrifice' (*De legibus* 41.2). The observation about the sacrifice of the red cow is, in the very context, immediately applied to leprosy. In fact, the very reason William introduces leprosy at that point is because the materials used for ritual purification of leprosy are almost identical to those used in the sacrifice of the red cow (Ibid.). William takes this ethical-physical comparison between the ritual sacrifice of the red cow and the ritual purification of leprosy a step further, when he observes that the materials used for the former must be burned, while identical materials used for the latter are not. He reads this difference in treatment of the elements as a demonstration of the greater power of corruption sin has over the elements. (Sin, mainly of idolatry, being the reason for the sacrifice of the red cow.) This prompts him to a further distinction between the far greater 'impurity of spiritual leprosy' (idolatry) and the lesser 'impurity of bodily leprosy' (*De legibus* 43.1-2).[15]

But William also notes—based assumedly on the biblical stories, if nothing else, of Miriam (Num. 12:10–16), Gehazi (2 Kings 5:20–27), and Uzziah (2 Chron. 26:16–23)—that leprosy itself, as a disease in a human body, is sometimes a punishment from God for sin (*De legibus* 42.1).[16] This, too, though a moral observation about the divinely visited consequences of human sin, remains part of the literal analysis of levitical leprosy.

And, then, of principal importance for William's overall argument in *De legibus*: certain elements in the ritual purification of leprosy are themselves included by the Lawgiver as an antidote to idolatry. About the application of the term 'leprosy' and the necessary ritual purification to 'a scar from the burning of the flesh or skin,' for example, William suggests that 'he did this, because he abhorred the scar of a burn more than any other kind of wound, because of the idolatrous insult by which fire was worshipped. For fire worshippers, in the act of worship, made burns in their

[15] 'However, in this case, they are used whole and uninjured, whereas in the other case, that is, in the above named sacrifice, they are burned up: either for this reason—so that the uncleanness of sins, which is spiritual, is shown to be greater than the uncleanness of leprosy, which is corporeal; or for this reason—so that the three elements polluted through the sins of human beings are shown, in the end, to be purified with fire, and, for that reason, the twice-dyed scarlet, which in every case is the color of fire, is added, so that it suggests that the final purification must be done through fire. [. . ..] Furthermore, these four materials are used so that the leper realizes that the uncleanness of his leprosy is less corrupting to the three elements, [and] will truly realize how much less important is the uncleanness of bodily leprosy than the uncleanness of spiritual leprosy, and learn how much easier its purification is—while this one is seen to be purified by the sprinkling of the priest, that one is purified only by a powerful fire.'

[16] 'Consider, also, that three liquids were used in this rite: obviously, the living water, that is, bubbling and flowing water, which is of greater purity than standing or dead water—this therefore was used for the washing because leprosy is unclean; oil for cleansing because leprosy is dead; and blood for atonement, *because sometimes the disease of leprosy was inflicted for sin by the wrath of God,* or because leprosy is mainly from a corruption of the blood, such that by the blood of clean animals the corruption and infection of blood was cleansed.' (Emphasis added.)

flesh and skin. Just as other idolaters made cuts or engravings' (*De legibus* 42.2). For the application of 'leprosy' to clothing and houses, William imagines the following rationale: 'Therefore the Law dreaded these afflictions in clothes and homes, as much because of the health of its inhabitants, as because of idolatry; for gods of household and place [*penates et lares*] were worshipped because of these afflictions, either so that they removed them or lest they inflicted them. God, therefore, wanted, so that there would be recourse to his priests for these afflictions, that they be flooded with scrapings and burnings and washings and expiations; indeed, he wanted thereby to bring about through his priests some kind of expiation, so that demons of this sort were exterminated; and he wanted these cleansings to be done by a new rite, so that they were distanced from the rites of superstition.' (*De legibus* 43.1).

6.4 The Rationality of Ritual Law

De legibus demonstrates, in many places, William's profound affinity for Moses Maimonides' (1135–1204) rationalizing interpretation of non-moral commands of the Law. In writing, elsewhere, about menstrual impurity in *De legibus*, I have noted the general parallels between Maimonides' and William's thought on ritual law, as well as specific ones on menstrual impurity.[17] Nevertheless, I ought to caution that these admittedly striking parallels include significant divergences in detail—differences great enough to indicate, at least, William's independent-minded use of Maimonides, or perhaps, to put it more strongly, his intent to improve and sometimes correct positions advocated by Maimonides.[18] In the case of their respective treatment of the impurity and purification of leprosy in Leviticus 13–14, William and Maimonides are worlds apart. William's employment, in general, of medical-scientific theory in his examination of the literal sense, including moral and theological aspects, of Leviticus 13–14 is nowhere evident in Maimonides' *Guide of the Perplexed*. Where William emphasizes the physical, not moral causes of leprosy, Maimonides specifies with absolute confidence that 'it is a punishment for slander' (Maimonides 1963). As for the materials used in the rite of purification for leprosy, materials for which we have seen William detail a direct correspondence to the elements, Maimonides expresses total ignorance and uncertainty: 'The reason why purification from it was effected by means of cedar

[17] William's relationship to Maimonides is briefly investigated in Smalley (1974). There is a fuller treatment of the same in Guttmann (1889). Further details of William's debt to Maimonides, in general, and with respect to menstrual impurity are noted in Murphy (2013). For a partial account of the conduit of text and translation by which Maimonides, in Latin, shaped William's own approach to the Law of Moses, see Freudenthal (1988).

[18] On differences between Maimonides and William, see Murphy (2013). Further differences are noted in Guttmann (1889). For a brief comment on William's uses of Maimonides in the context of William's own relationship to Christian exegetical theory, see Dahan (2005).

wood, hyssop, scarlet thread, and two birds, is given in the *Midrashim*; but it does not fit with our purpose, and up to now I do not know the reason for any of these things; nor why cedar wood, hyssop, and scarlet thread were used in the ceremony of the red heifer [. . .]'(Ibid).[19]

However much William diverges from Maimonides with respect to leprosy, he is certainly closer to Maimonides, if only because of his interest in the topic, than to contemporary university-trained theologians—in England or France, among the mendicants or the seculars. I have yet to find, in the works of Alexander Nequam (1157–1217), Stephen Langton (c. 1150–1228), Thomas of Chobham (d. after 1233), or Alexander of Hales (d. 1245)—to name just a selection of theologians of William's generation—anything like William's interest in leprosy or, more broadly, ritual impurity, or, more broadly still, the non-moral commands of the Law.[20] John of la Rochelle (d. 1245), whose *Tractatus de praeceptis et legibus*, is included in Part III of Alexander of Hales' *Summa*, wrote at some length about the 'ceremonial precepts' of the Law, often quoting William of Auvergne's *De legibus*, while fundamentally opposing William's own commitment to a literal reading of the non-moral commands.[21] But even John of La Rochelle has nothing to say about leprosy. A generation beyond William, Albert the Great (d. 1280) devoted the second tractate of the second book in his *De causis proprietatum elementorum* to the 'corruption of the elements,' but Albert showed no interest there in Leviticus, or leprosy or, broadly, the corruption of the elements by moral causes or the ritual impurity occasioned by elemental corruption (Albert the Great 2010). I have no investment in the claim that William was unique among his contemporaries in this regard, but I have to admit that my search for some kind of parallel is so far without result.

And, finally, a word about Grosseteste. I think it safe to say that readers who appreciate Robert Grosseteste on his merits are likely to appreciate William of Auvergne, who, as theologian and bishop, also cut his own path. Were these men acquainted? To some extent, yes, as is made certain by Grosseteste's single surviving letter to William (Mantello and Goering 2010).[22] Their personal relationship, whatever it was, hardly matters to my point, because, with respect to levitical

[19] The *Mishneh Torah*, Maimonides' comprehensive summary of the Law, both written and oral, in fourteen books, provides, in Book 9, a detailed description of the purification of the leper, with consequences for a variety of mistakes possible in that ritual, but no rationale for the purification or the materials involved (see Maimonides 1950, 1954). Treatise 3 reviews Leviticus 13–14 in far greater detail than William does, but only with the practical purpose of explaining the correct implementation of the rules, not, again, with any apparent interest in exploring the rationale for leprosy-related impurity and purification.

[20] The works I have considered so far include: Neckam (1863); Stephen Langton, *Questiones* based on the contents summarized in Powicke (1928), Chobham (1968), and Alexander of Hales (1924–1948).

[21] On John of La Rochelle's authorship of this section of the *Summa* of Alexander of Hales, and on John's use of and critical response to William, see Smalley (1974).

[22] James McEvoy (2000), describes William as 'Grosseteste's friend,' but without citing this letter or any other evidence.

leprosy, these two had nothing in common. Grosseteste, as far as I know, never gave the subject attention, and there is no surprise in that, given his fundamental attitude towards ritual laws. It is only among authors like William, someone for whom the ritual commands of the Law have intrinsic rationality and morality, that such an analysis would be undertaken. But Robert Grosseteste, in his fascinating *De cessatione legalium*, takes a different view of the Law.[23] 'And because it is already clear that the fullest and most humble obedience consists in observing the law of deeds, or the positive law, it is manifestly appropriate that the positive law be added to the natural law both before the written Law and in the written Law. On account of this is resolved the charge of those who disparage the Mosaic Law as lacking a rationale for many of its commands. For, they say, there is no reason why someone ought not plow with an ox and an ass or wear clothes woven of wool and linen, and it did not suit, they say, the supreme wisdom of God to give the sort of commands that have no rationale. They do not understand, however that the rationality of testing and achieving perfect obedience consists in the observance of indifferent mandates that of themselves lack rationality' (Grosseteste 1986, 2012). For an author like William's contemporary fellow bishop, Robert Grosseteste, then, someone for whom such commands have, at best, only extrinsic rationality and morality, the science and theology of ritual impurity and purity is a largely closed book.

References

Albert the Great. (2010). *On the causes of the properties of the elements* (I. M. Resnick, Trans.). Milwaukee: Marquette University Press.

Alexander of Hales. (1924–1948). *Summa theologica* (Vol. 4). Quaracchi: Typographia Collegii S. Bonaventurae.

Bartholomaeus Anglicus. (1975–1988). *On the properties of things: John Trevisa's translation of Bartholomaeus Anglicus' De Proprietatibus Rerum* (Vol. 3, M. C. Seymour, Ed.). Oxford: Clarendon Press.

Coogan, M. D. (Ed.). (2001). *The New Oxford annotated Bible (NRSV)* (3rd ed.). Oxford: Oxford University Press.

Dahan, G. (2005). L'exégèse de la Bible chez Guillaume d'Auvergne. In F. Morenzoni & J.-Y. Tilliette (Eds.), *Autour de Guillaume d'Auvergne* (pp. 237–270). Turnhout: Brepols.

Demaitre, L. (2007). *Leprosy in premodern medicine: A Malady of the whole body*. Baltimore: Johns Hopkins University Press.

Freudenthal, G. (1988). Pour le dossier de la traduction latine médiévale du *Guide des égarés*. *Revue des Études juives, 147*, 167–172.

Gilbertus Anglicus. (1510). *Compendium Medicinae*, Book 7. Lyon.

Grosseteste, R. (1986). *De Cessatione Legalium* (R. C. Dales & E. B. King, Eds.). Oxford: Clarendon Press.

Grosseteste, R. (2012). *On the Cessation of the Laws* (S. M. Hildebrand, Trans.). Washington, DC: Catholic University of America Press.

Guilielmus Alvernus. (1963). *Opera Omnia* (Vols. 2). Paris; reprinted Frankfurt am Main.

[23] There is a brief consideration of *De cessatione legalium* in Murphy (2007).

Guttmann, J. (1889). Guillaume d'Auvergne et la littérature juive. *Revue des Études juives, 18*, 243–255.
Maimonides, M. (1950). *The code of Maimonides, book nine: The book of offerings* (H. Danby, Trans.). New Haven: Yale University Press.
Maimonides, M. (1954). *The code of Maimonides, book ten: The book of cleanness* (H. Danby, Trans.). New Haven: Yale University Press.
Maimonides, M. (1963). *The guide of the perplexed* (S. Pines, Trans.). Chicago: University of Chicago Press.
Mantello, F. A. C., & Goering, J. (2010). *The Letters of Robert Grosseteste, Bishop of Lincoln*. Toronto: Pontifical Institute of Medieval Studies.
McEvoy, J. (2000). *Robert Grosseteste*. Oxford: Oxford University Press.
Milgrom, J. (1991). *The Anchor Bible: Leviticus 1–16: A new translation with introduction and commentary*. New York: Doubleday.
Morenzoni, F., & Tilliette, J.-Y. (Eds.). (2005). *Autour de Guillaume d'Auvergne (d.1249)*. Turnhout: Brepols.
Murphy, S. (2007). Concern about Judaizing in Academic Treatises on the Law, c. 1130-c. 1230. *Speculum, 82*, 560–594.
Murphy, S. E. (2012). William of Auvergne. In D. Thomas & A. Mallett (Eds.), *Christian-Muslim relations: A bibliographical history* (pp. 288–294). Leiden: Brill.
Murphy, S. E. (2013). On the enduring impurity of menstrual blood and semen: Leviticus 15 in William of Auvergne's *De Legibus*. In J. Martínez-Gázquez & J. Tolan (Eds.), *Ritus Infidelium: Miradas interconfesionales sobre las prácticas religiosas en la Edad Media* (pp. 191–208). Madrid: Casa de Velazquez.
Murphy, S. E. (2014). Pagans past and present: Righteousness and idolatry in academic discussions of ancient religion, c. 1130-c. 1230. In S. Knaeble & S. Wagner (Eds.), *Gott und die Heiden: Mittelalterliche Funktionen und Semantiken der Heiden* (pp. 147–167). Berlin: LIT Verlag.
Neckam, A. (1863). De naturis rerum. In T. Wright (Ed.), *The Chronicles and Memorials of Great Britain and Ireland during the Middle Ages* (pp. 1–357). London: Longman, Roberts & Green.
Powicke, F. M. (1928). *Stephen Langton*. Oxford: Oxford University Press.
Ralph of Flaix. (1677). *Commentatiorum in Leviticum Libri XX*. In M. de la Bigne (Ed.), *Maxima Bibliotheca Veterum Patrum* (Vol. 17). Lyon.
Rawcliffe, C. (2006). *Leprosy in Medieval England*. Woodbridge: Boydell Press.
Smalley, B. (1968). Ralph of Flaix on Leviticus. *Récherches de théologie ancienne et médiévale, 35*, 35–82.
Smalley, B. (1974). William of Auvergne, John of La Rochelle, and St. Thomas Aquinas on the Old Law. In É. Gilson (Ed.), *St. Thomas Aquinas, 1274–1974: Commemorative Studies* (Vol. 2, pp. 11–71). Toronto: Pontifical Institute of Medieval Studies.
Smith, L. (2005). William of Auvergne and the Law of the Jews and the Muslims. In T. J. Heffernan & T. E. Burman (Eds.), *Scripture and Pluralism: Reading the Bible in the Religiously Plural Worlds of the Middle Ages and Renaissance* (pp. 123–142). Leiden: Brill.
Theodoric of Chobham. (1960). *The Surgery of Theodoric* (Vols. 2, E. Campbell & J. Colton, Eds.). New York: Appleton-Century Crofts.
Thomas of Chobham. (1968). *Summa Confessorum* (*Analecta Mediaevalia Namurcensia* 25) (F. Broomfield, Ed.). Louvain & Paris: Éditions Nauwelaerts.
Touati, F.-O. (1998). *Maladie et société au Moyen Âge: La lèpre, les lépreux et les léproseries dans la province ecclésiastique de Sens jusqu'au milieu du XIVe siècle*. Paris: L'Harmattan.
Valois, N. (1880). *Guillaume D'Auvergne, Évêque de Paris (1228–1249): Sa vie et ses ouvrages*. Paris.

Part III
Robert Grosseteste and Roger Bacon

Chapter 7
From *Sapientes antiqui* at Lincoln to the New *Sapientes moderni* at Paris c. 1260–1280: Roger Bacon's Two Circles of Scholars

Jeremiah Hackett

7.1 Roger Bacon's Two Circles of Scholars

A modern study, by Brian Clegg, *The First Scientist: A Life of Roger Bacon* (London 2003), illustrates the need to re-write the record on the life and works of Roger Bacon. Thankfully, the exaggerations of this book have been clinically exposed by the late John North.[1] With the publication of her recent book, Amanda Power has done a great service to Roger Bacon studies (Power 2013). Here, and in her paper 'A Mirror for Every Age: The Reputation of Roger Bacon,' Power has placed Roger Bacon back in the context of the thirteenth century where one is dealing with the medieval Roger Bacon and not with the new seventeenth- or nineteenth-century images (Power 2006).

In this section, I will investigate what I call 'Roger Bacon's two Circles of Scholars.' First I will speak about Bacon's early pre-1260 Circle of Scholars (*Sapientes antiquae*), then I will speak about Roger Bacon's Circle of Scholars at Paris c. 1260–1292, the (*Sapientes moderni*) among whom we must locate a *Sapientissimus*, possibly Gerard of Huy O. F. M., who is a very great Biblical Scholar. These two circles are quite distinct, both historically and geographically.

I will not expressly address the issues involved in his early Parisian Commentaries, other than comment on the recent hypothesis of Sylvia Donati about the possible non-authenticity of three of the eight Aristotelean Commentaries from

[1] TLS Review, March 28, 2003, 30: 'Every hero becomes a bore at last,' said Emerson. Roger Bacon could never have been called a bore, but the question is whether he qualifies as a hero. To our Victorian fore-bearers, this eclectic thirteenth-century Franciscan Friar was the true founder of experimental science, and Brian Clegg is determined to put him back on that pedestal, with a suitably updated label, 'the West's first true scientist.'

J. Hackett (✉)
University of South Carolina, Columbia, SC, USA
e-mail: HACKETTJ@mailbox.sc.edu

J.P. Cunningham, M. Hocknull (eds.), *Robert Grosseteste and the pursuit of Religious and Scientific Learning in the Middle Ages*, Studies in the History of Philosophy of Mind 18, DOI 10.1007/978-3-319-33468-4_7

tha1240s (Donati 2013). I believe that her proposal to exclude the *Quaestiones in Physicam I–IV, Questiones in Metaphysicam I–IV and Quaestiones in Metaphysicam IX* from the canon of Bacon's Aristotelian Commentaries from the 1240's appears to have a solid basis in manuscript description, style and doctrine. That still leaves five important volumes. In this section, I will focus entirely on Roger Bacon's works for Pope Clement IV, who occupied the *apostolica sedis* from 1265 to 1268.

There are many problematic issues in regard to our knowledge of Bacon's life, works, and teaching for the period 1260–1292, and I will deal here with just some of these issues. I will begin with the one item that has caused many problems for interpreters, namely, the problem of Chronology in regard to the life and works. I will then proceed to identify Bacon's concern with the *Sapientes antiqui* whom he claims to have seen and to have visited. [Text 2, par. 2] Finally, I will identify the members of Roger Bacon's Circle of Scholars at Paris in the Period 1260–1280. Included here will be a brief discussion of recent scholarship on Roger Bacon as a Franciscan Friar.[2] After 1280 he is back at the Franciscan *Studium* in Oxford.

7.2 Bacon's Life: Conflicting Chronologies and Texts

The evidence for my chronology can be found in the appendix to this section. The chronology established by Theodore Crowley, advocated by David C. Lindberg, and followed more recently by Yael Kedar is based one text alone. I believe that this chronology needs some adjustment. The one crucial foundational text for this chronology beginning with the words *Multum laboravi* has been excerpted from its context in *Opus tertium*, Chap. XX. This text is usually read in isolation and out of the context of the subject matter, that is, the education of Bacon's own young student, the *Iuvenis Iohannis*, who is about 20/21 years old in 1267–1268, and who has been a student with Bacon for about seven years. [TEXT 1, TEXT 2] It is clear from the evidence from the *Opus tertium* and related works that by 1267–1268, Roger Bacon was already an old person (a *senex*). In addressing the Pope in 1268, he speaks of 'us old men' (*nos senes*).

Roger Bacon could have been born as early as 1210. This date was argued for in the nineteenth century by Charles Jourdain (1874); he could have been born c. 1214 as has been argued by the renowned Franciscan Scholar, Arthur George Little (1914); this latter position is accepted by Thomas Maloney (1988), George Molland (2004) and Jeremiah Hackett (2013). However, the standard modern view is that of Theodore Crowley (1950) and David C. Lindberg (1996). They argue on the basis of just one text from the *Opus tertium* [Text 1] that Bacon was born forty-seven years before the writing of the *Opus tertium* in 1267. They calculate this on the basis of his statement that for the past forty years after he first learned the *Alphabetum*, he

[2] See the recent and forthcoming work (Johnson 2012). See also below nn. 59, 60. Bacon.

has been working on languages and sciences *in studio*, and that he had expenses as others commonly have. The problem then centers on which *alphabetum* is the object of Bacon's concern in Chap. XX, and what is the scope of *in studio* in the context of his remarks on the education of his own young student, the young John.

Theodore Crowley has also argued that Bacon could have been at Oxford in one of the Grammar schools directed by the Chancellor of the University, and Crowley may well be correct in this supposition (Ibid.). This would mean that Bacon was born c. 1220, educated at Oxford c. 1234–1241, and Professor at Paris c. 1242–1247/1248. It is sometimes held that c. 1248 he returned to Oxford, where he attended the lectures of Adam Marsh OFM. He would then have become a Franciscan Friar c. 1256/1257.

Here we run into a number of problems. First, Chap. XX of the *Opus tertium* is concerned with instruction in languages and mathematics, and the word *Alphabetum*, has been taken by Crowley and Lindberg to mean the first alphabet at the age of about 7, when Bacon was technically a *Puer*. A few paragraphs later in Opus Tertium (1859), Chap. XX the word *Alphabetum* is explicitly used to speak about the *Alphabetum philosophiae*, especially the basic knowledge of mathematics. Second, Bacon uses the standard terms such as *Adolescens, Iuvenis, Senex.*[3] And since he is talking about a youth who is learning mathematics and is talking about his own study of language and mathematics to indicate the he once was in the position of the *iuvenis Iohannes*, he is hardly speaking about a *Puer*. As a *iuvenis*, he would first have been taught by a grammar-master (*Grammaticus*) before proceeding onto the study of mathematics around the age of 11. Third, the dating of the birth at c. 1220 could not account for the following remark from the *Compendium studii theologiae*: 'Even the books of logic were not received and taught until late in the day. For Blessed Edmund, Archbishop of Canterbury was the first to teach the *Sophistical Elenchs* [of Aristotle] in my times (*temporibus meis*), and I both saw Master Hugh, who first taught the *Posterior Analytics* [of Aristotle], and I perused the words [in his book]' (Maloney 1988) [Text 3].

Sir Richard Southern holds that Blessed Edmund departed Oxford in 1222, and Bernard G. Dodd, the expert on logic, dates the teaching of Master Hugh to c. 1210 (Southern 1992; Dodd 1982).[4] Further, who is the great expert on natural philosophy and *perspectiva* mentioned in the *Tractatus de experiential in communi* who mentored Bacon *a iuventut*? The only known expert on natural philosophy and *perspectiva* during Bacon's early years at Oxford was Robert Grosseteste.

Now it is important to read Bacon's own words: sometimes he simply states that he has seen some of the ancients, when for example as we will see below he states many times *Nam vidimus* with regard to Robert Grosseteste and select scholars at Lincoln, and at Oxford. But of course, the important issue is the force of the term

[3] Speaking about the *Iuvenis Iohannes*, Bacon speaks of his in different contexts as *puer, iuvenis, asolescens*.

[4] See Dodd (1982) on the difficulties in establishing these dates for the teaching of Aristotle at Oxford.

Nam vidimus. There is only one text where he notes a particular conversation with another scholar, that is, with Adam Marsh OFM. Otherwise, he always uses the term *Nam vidimus*. The proper sense of the word perceive in Bacon is not the loose sense of a general glance at someone, but rather a direct encounter with an individual. He states that he witnessed some Franciscans questioning Master Adam Marsh OFM concerning the nature of the Agent Intellect. [TEXT 4] This, most likely was some time before 1257 and after Bacon ceased teaching in the Arts at Paris c. 1248. He tells us that he had seen Thomas of Wales, Bishop of St. David in Wales, but the latter died in 1255, and prior to that he had been a Bishop in Wales. And then his hero Robert Grosseteste passed away in 1253. Thus, he had to have seen these three scholars at some time before 1251, since he was in Paris at that time and seems to have been there until 1257 and later. Thus, we must look to the late 1220s to 1250 as the possible time for Bacon having seen the *Sapientes antiqui*.

Roger Bacon had a life-long concern with government and with the education of the Prince (Hackett 2006). Matthew Paris tells how Friar Robert Bacon of the Order of Preachers, in a speech before the King at Oxford, June 24, 1233, denounced the royal favorites, the Bishop of Winchester (Pierre des Roches) and Pierre de Rivleaux. A young *Clericus de curia regis*, one Roger Bacon made a witty remark about rocks. A. G. Little notes that while we do not have evidence that our Roger Bacon 'was ever a clerk of the royal court,' 'he had some knowledge of the inner workings of a chancery' (Little 1914). [Text 5] Still, we know that later in Paris c. 1265, Bacon moved in Ambassadorial Circles, and had intimate knowledge of the household of the brother of the King of France, Alphons of Poitiers. It would appear that after 1280 at Oxford, the edition of the *Secretum secretorum*, the very important *Mirror for Princes* was written for a Royal patron.[5] Further, we will note below his close connections with Master Raymond of Laon, an official of Cardinal Guy le Gros de Foulque, who in February 1265 was elected Pope Clement IV.

It was the firm conviction of the late James A. Weisheipl that Roger Bacon began his teaching in the Arts at Paris c. 1237, and in this I am inclined to agree. Certainly, I do not think he began to teach there much later than 1240. Now, it is standard lore that Bacon ceased to teach at Paris c. 1247–1248, and that he returned to Oxford from about 1248–1257. Sometime around 1256/1257 he joined the Franciscan Order. Bacon writes about the twenty years when I especially worked in the arts and sciences *neglecto sensu vulgi*. [Text 2] This phrase, however, is taken by most scholars to mean he had lectured on the texts of Aristotle *et sequaces eius vulgariter or per modum scholasticum* until 1247/1248. And then from 1248 to 1268 there is the new 'scientific' work *neglecto sensu vulgi*. But as we will see, this is an impossible hypothesis since Bacon tells us that from about 1256 to 1267 he did not do any professional academic work.

[5] The *Secretum secretorum* is a work that offers advice on statecraft and in the thirteenth century was thought to have been written by Aristotle for Alexander the Great, when in fact, it was a mid-twelfth century Latin translation of the tenth century Arabic work *Kitab sir al-asrar*.

But since Bacon distinguishes himself as one of the *Sapientes* in opposition to the leaders of the *vulgus philosophorum et theologorum* at Paris, and since in 1271 he talks of himself as having been a long experienced scholar of Aristotle, Avicenna and Averroes (*sicut ego expertus sum omnino*) [Text 6] I doubt he ever thought of himself as one of the *vulgus*. My argument, as I will indicate below, is that the term *neglecto sensu vulgi* has a specific determinate sense in Roger Bacon. It indicates work first in Languages, namely the grammar and logic of that language, and more importantly, it indicates a competence in the *Quadrivium*. This is precisely what Bacon states. For Bacon, *mathematics is the alphabetum philosophiae*. Note the words of the text: he had expenses, he organized scientific research, he organized schools. As we will see below, he certainly did not perform this task as a Franciscan Friar beginning c. 1257, nor did he do so as a *puer*. Writing in the *Opus tertium*, Bacon himself states that almost twenty years ago he was the *Magister Regens* at the inception of new masters in matters dealing with the *Quadrivium*, and none but he was fully competent in geometry. [Text 6a] This could have been as late as 1250–1252. That he was in Paris in 1251 is clear from his reference in *Opus maius IV* to have seen the leader of the Pastoreaux Rebels (Bacon 1964, *Opus maius*). Further, as Alain de Libera has argued, and as Thomas S. Maloney confirmed, the method and subject matter of the *Summulae dialectics* (*Summa Logica*) is more appropriate and typical of the late 1240s than the early 1240s (Maloney 2009).[6] Further, Bacon states that he heard Richard Rufus of Cornwall 'stultissimus' solemnly lecture at Paris after he had previously lectured at Oxford (1250–1253), that is, from 1253 to 1256. [Text 7] But of course by this stage, Bacon is thinking of become a Franciscan friar and by c. 1256–1257 he has become a Franciscan friar. The accumulated evidence here suggests that we must push the date of birth back before c. 1220. Further, I believe I have offered good reasons to question the common belief that Bacon was at Oxford from 1248 to 1257 where he attended the theology lectures of Master Adam Marsh O. F. M. It is likely that he did visit Oxford c. 1248 but he was back in Paris in 1251 and again 1253–1256, and remained there until possibly 1280. It is not impossible that his two years of rest from teaching took place at Oxford.

Now, as I will argue towards the conclusion of this chapter, I am convinced that Bacon was indeed a very sincere and committed Franciscan, but he was a Franciscan in a mold similar to but yet different from Bonaventure and his friend Richard Rufus of Cornwall. Still, he shared much in terms of theological method with his English colleague, John Pecham. The latter represents Bacon's philosophical and theological interests, with the possible exception of Bacon's deep commitment to the applications of *astrologia* to human affairs (Hackett 2003).

It is clear from the *Opus tertium* that apart from Richard Rufus of Cornwall (stand in for Bonaventure?) and Alexander of Hales, the main object of Bacon's

---

[6] Alain de Libera, however says that it is 'probable' that the work was given a final redaction in Oxford around 1250, and mention of a redaction implies an earlier and initial composition at Paris. De Libera sees it as an Oxford influenced work presented at Paris, between 1245 and 50. See xvii–xxii for a discussion of a possible argument by Bacon on this matter c. 1252.

criticism is the *vulgus philosophantium* at Paris, and the *Capita eorum*, namely, Albertus Coloniensis. The more I read these texts, the more I see that Roger Bacon is proposing a 'Research Program' to the Pope in Science, Philosophy and Theology that is directly defined over against the 'Research Program' of Albert of Cologne, and his followers at Paris, including the regular teachers of Philosophy such as Siger of Brabant and Boethius of Dacia, and Albert's student, Thomas D'Aquino. [Text 8] We must turn to Bacon's own training and to his debts to those he calls the ancient wise ones.

7.3 Bacon and the *Sapientes antiqui*

Now Roger Bacon often writes about the circle of scholars around Robert Grosseteste. In both the *Opus maius, and the Opus tertium*, he states: 'For we have seen some of the ancient wise persons who worked in languages such as the aforementioned Lord Robert, the translator and Bishop, and Thomas the Venerable Bishop of St. David, recently deceased, and Brother Adam Marsh and Master Hermann, the translator and certain other wise scholars.' [Text 9]

Again, in the *Compendium studii philosophiae* (1271), he states that the modern *Seculares* who c. 1267 teach theology have dismissed the old ways, and are drawn solely to honors and riches. 'And so they totally dismiss the ways of the ancient wise teachers some of whom we have seen in our own times, such as, Lord Robert, once Bishop of Lincoln, of holy memory, Lord Thomas, Bishop of St. David in Wales, Brother Adam Marsh and Master Robert Marsh, and Masters William Lupus [the treasurer of Lincoln Cathedral] and Master William of Sherwood.' [Text 9]

These men who flourished in the 1230s and by the 1240s, were a major part of Grosseteste's administrative staff as indicated by the following quote from Sir Richard Southern: 'In his general plan, the group of friars in his household were his missionaries, and his Archdeacons were his chief local agents comparable to Royal Sheriffs. To this office he appointed men who were his close collaborators in his learned enterprises and administration-such men as John of Basingstoke, Thomas Wallensis, William Lupus, William of Arundel, Richard of Gravesend and Robert Marsh. If he had his way, episcopal government would have been the strongest ruling force in England' (Southern 1992). Many of these men are the very ones that Bacon claims he has seen (*Nam vidimus*).

When we add to this the fact established by A. B. Emden that members of a Bacon family resided in the 1240s at a *domus scholarum*, a graduate residence at Oxford, and possessed a copy of Avicenna's *Healing*, and when one of them, Nicholas Bacon was appointed a diocesan official in 1244/1245 (Emden 1966)[7]

[7] The association of Nicholas and Peter Bacon with this small graduate household, and Nicholas's presumed ownership of a copy of the treatises of Avicenna and other Arab Philosophers invite speculation whether Nicholas and Peter may not have been related to the distinguished

by Robert Grosseteste, one must think seriously about Roger Bacon's connections with Oxford and Lincoln. As we will see below, writing to the Pope c. 1268, he speaks of not being able to contact his brother the scholar in England.

Bacon states explicitly to the Pope that when he was in the other form of life as a Master of Arts he had spent about 2000 Parisian librae on books, experiences, and on travel to visit the *Sapientes*. He tells us that he visited all the *Sapientes*. Since he praises Grosseteste, his household, and Master Adam Marsh as the greater clerics of the world and as the greater *Sapientes*, we must assume that he used this money to visit such persons. The other wise ones mentioned are Albert the Great (whom he could have seen at Paris c. 1245–1248), whom he would have met at Paris c. 1245–1248. And then there is the Biblical Scholar who is referred to as *Sapientissimus* who flourished in Paris in the late 1260s and 1270s.

Now, I do not doubt that Bacon had visited Oxford, and I believe he may even have visited Lincoln. But when he did so is still a mystery. Did it take place in the 1230s prior to his move to Paris to teach in the Arts? Could it have taken place after 1251? This latter hypothesis as we have just seen is impossible. Thus, there is reason to think that he met some of these scholars before 1248 or at possibly some of them between 1248 and 1251.

But how can we account for his knowledge of the works of Robert Grosseteste? Scholars hold that he must have learned about them in the 1250s when Grosseteste's Library was given to the Franciscan Studium at Oxford following Grosseteste's death in 1253. But this position which was proposed by the late James McEvoy is difficult to sustain. A period at Oxford or at home somewhere in England c. 1248–1251 would account for his contact with Master Adam Marsh and for his knowledge of Grosseteste's scientific works such as *De iride, De cometis, De lineis, etc*. One has to assume that Adam Marsh had access to these works. Such an encounter would account for Bacon's great knowledge of *Perspectiva*, especially the Optics of Ibn al-Haytham. Bacon notes that the subject was taught only at Oxford, that just on two separate occasions. And since this required knowledge of Ibn al-Haytham (al-Hacen's) *Optics*, it must have been in the late 1240s.[8] If he is correct about the claim that optics was taught only at Oxford prior to 1270, then, Bacon, the comprehensive master of Optics, must have learned his craft at Oxford. [Text 10]

---

contemporary bearing their surname, Fr. Roger Bacon, O. F. M., among whose many interests, the works of these philosophers were certainly one. Emden is of the view that Master Nicholas Bacon may have been the 'same man as Nicholas Bacon, clerk, who was instituted in 1244 or 1245 by Bishop Grosseteste as rector of the moiety of Stoke Rochford, Lincolnshire.' Also, A. G. Little had suggested that the Master Thomas Bacon who was suggested by Adam Marsh O. F. M. as a *socius* to Richard Rufus of Cornwall O. F. M. in 1252 may have been a brother of Roger Bacon.

[8] On the dating of *De aspectibus*, see Smith (2001). Commenting on the dispute about dating, Smith notes: 'The earliest incontestable evidence for its circulation is to be found in Bartholomeus Anglicus' *De proprietatibus rerum*, where *De aspectibus* is quoted several times. This work probably dates to the late 1240s.

We come now to the big unasked and unanswered Question: when and where did Bacon get such a good training in Mathematics, and above all his skilled training in grammar and logic? (Pinborg 1979) We have just seen that he already had training in mathematics and *perspectiva*. But where did he get his training in Greek?

It was the considered belief of S. A. Hirsch that Roger Bacon's command of Greek, that included knowledge of grammar, orthography, idiom, etymology was acquired in England from mature teachers of Greek such as Nicholaus Graecus (1914).[9] There is not time here to develop this topic, but the arguments of S. A. Hirsch seem to be very strong. This thesis is borne out by the great interest in Greek and Hebrew exhibited by Bacon c. 1272 in the *Compendium studii philosophiae* and in other related texts of that time. [Text 11]. This raises a question about Bacon's concerns: was he working with the named *Sapientissimus* on the Biblical Text at the Franciscan House of Studies in the late 1260s? At any rate, his interest in Greek at this time is professional and serious. And so, there is reason to think that he may have acquired this knowledge from scholars associated with the circle of scholars influenced by Grosseteste.

7.4 Bacon and His Patron, Cardinal Guy le Gros de Foulque (Guido Fucoldi)

We come now to the high point of Bacon's life, his encounter with the man who would become Pope Clement IV. Before looking at the chronology, allow me once again to draw on the wisdom of the late John North. In his important essay *Roger Bacon and the Saracens*, he paints a picture of the concerns of Roger Bacon that serves as an antidote to the over the top speculations of nineteenth-century historians and philosophers (North 1999). He places Roger Bacon in the context of the worlds of Islam and the Latin West in the mid-thirteenth century. While at times he seems more concerned with warfare in England, Italy and France, the world of Islam especially is close to Bacon's concerns. And it is in this context that the figure of Pope Clement IV takes on much significance. Bacon thought that the latter would be the good Pope who would lead the charge to prevent the expansion of Islam, especially after the destruction of Bagdad by the Mongols in 1258, and the loss of Damascus to the Mamlucks in 1260, where 'the old axis of Saladin had been re-instated.' Now, Bacon is truly interested in geo-politics and warfare and this interest is closely tied into his great interest in the *Secretum secretorum* and his moral philosophy.

1260 is a very important year for Roger Bacon. First, the war between the King and the Barons begins and lasts until 1264, and the man who would be Pope in 1265, Cardinal Guy le Gros de Foulque, was ambassador of the Papal Court to

[9] Hirsch points to the close connection between Bacon's Greek Grammar and the treatment of etymologies in the 1271 *Compendium studii philosophiae*.

England, but had been detained in France and was not allowed to enter England. In this war, Bacon's family suffered the costs of ransom, and he lost contact with his brother the scholar. Where did this scholar brother reside? Was it at Oxford? Further, any chance that Roger Bacon might continue his scholarly pursuits was finished by the statutes of the Council of Narbonne under the presidency of Bonaventure. It should be noted here that Cardinal le Gros de Foulque was Archbishop of Narbonne in 1260. No writing especially *De antichristo* could be published without the permission of the superiors, especially Bonaventure. Caught in this bind, what does Bacon do? He does an end-run around his superiors. That is never good policy. Given the statutes of Narbonne, it had practical consequences, namely, a time of isolation on bread and water. To return to the matter of Islam, North remarks:

> The Saracens were for Bacon 'a sect in one of the principal nations,' a sect bound by the law of Mohammed. How then could he reconcile himself to paying homage to the Saracens, when their religious views were in direct conflict with those of his own church? (Ibid.).

We come now to the re-birth of Roger Bacon as an active scholar c. 1266. We have the *Mandatum* that Pope Clement IV sent to Bacon in June 1266. Scholars are in agreement that in July 1266, Bacon received a directive or *Mandatum* from Pope Clement IV to write a work on philosophy and on other matters:

> To our dear son, Brother Roger, called Bacon, of the Order of Friars Minor. We have received your devoted letters gladly. And indeed we have attended carefully to the explanation of them which our beloved son, Sir William, called Bonecour, related orally to us, as faithfully as possible. So that we can obtain a clearer idea of what you intend, we command you by apostolic letters notwithstanding [non obstante] the contrary instruction of any prelate, to send to as soon as you can a fair copy of that work, which, when we were in a lesser office [Cardinal-Legate], we asked you to communicate to our beloved son Raymond of Laon, and explain in your explicit writings to us the remedies that you think we should adopt to address those issues that you have described on the occasion of such great danger, and do this quickly and as secretly as possible. [Text 12]

This then is the Papal Mandate issued in June, 1266. Yet, as is clear, Bacon had earlier contact with Cardinal Guy le Gros de Foulque (Guido Fucoldi), sometime prior to his becoming Pope in 1265. What did the Cardinal think of Bacon's ideas? How important was his encouragement as a motive for Bacon to begin writing? In the Gasquet Fragment, which is an introduction to the *Opus* maius, Bacon states explicitly that he had received a prior, first Mandatum from Pope Clement IV when the latter was in 'a lesser office, namely that of Cardinal.' This fact has been downplayed in the scholarship. Bacon had received an explicit *Mandatum* to write from Cardinal Le Gros de Folque. In this introduction to the *Opus maius*, Bacon writes: 'Certainly, your Magnificence was aware, since both Mandates asserted it, that I was under obligation by the strictest precept that I not communicate any writing which I made in this state of life [as a Franciscan Friar], just as all our congregation is known thus to be firmly obliged, and so I utterly shrank from writing.' [Text 13] This is a very important text, in that he provides us with a true picture of Bacon's actual absence from active work in the arts and sciences. By 1267, he had been an exile for about ten years. Further, he had written nothing.

Again, he speaks about the many impediments placed on him by his Franciscan superiors. In the *Opus tertium*, he compares his plight with that of the great Cicero: 'First, therefore in the Second Work, after the manner of the Letter of Cicero when he was called back after exile, and humbled himself and congratulated the Roman Senate, considering myself now for ten years an exile with respect to my fame for study'(*Opus tertium* 7). [Text 14]

When did the Cardinal send Bacon the first Mandatum? I believe it was in the period c. 1261–1262. Writing in the *Opus tertium* Bacon states that he began his instruction of his student Johannes, which has now lasted for six or seven years, since he first received a mandate from Cardinal Le Gros de Foulque. He also states that he began the composition of his central work, the *De multiplicatio specierum* when he first received the first Mandatum from the Cardinal. Further, he states that it is ten years since he received that Mandatum (*Opus tertium* 38).[10] Bacon began the educational preparation of his messenger, the *Iuvenis Iohannes*, on receipt of this first Mandatum from Cardinal Guy le Gros de Folque. Thus, sometime in the very early years of the 1260s, probably around 1260–1261, Bacon began his writing projects.

The big question therefore arises: When did Bacon write the *Opus tertium*? It is important to get a verifiable result on this since much of the chronology depends on this fact. After all as we saw above, scholars have tended to give 1267 as the time for the writing of the *Opus tertium*. This simply does not make any sense: Bacon tells us in the *Opus tertium* that he did not begin to finally write the *Opus maius* until after Epiphany, 1267, and if one takes into account that he has to draft the *De multiplicatio specierum*, the *Perspectiva*, *Communia naturalium* and the *Moralis philosophia*, we must push the writing of the *Opus tertiium* up to the years 1268–1269. Moreover, recent Bacon scholarship on 'Roger Bacon's *Communia Naturalium*: A thirteenth-century Philosopher's Workshop,' has come to a conclusion that this latter work was composed sometime between 1269 and 1270 (not later than 1271), and as I prove there, Bacon uses material from the *Opus tertium* in Part IV of Book two of the *Communia naturalium* (Bernardini and Rodolfi 2014; Hackett 2014). Hence, it is likely that this material was written sometime between 1268, when Pope Clement IV was still alive, and 1269.

He completed his edition of the *Secretum secretorum* at the Franciscan Studium in Oxford. By 1292, he had completed his *Compendium studii theologiae*, and he probably died at Oxford sometime after 1292.

[10] *Sed laboravi per annos decem [on the De multiplicatione specierum], quantumcunque potui vacare, et discussi Omnia ut potui, redigens in scriptum a tempore mandate vestre.* [This raises another issue. Elsewhere, Bacon states that he began teaching the Young John some six or seven years ago, after he received the first Mandatum from Cardinal Le Gros de Foulque. It would follow from his statement about the ongoing work on species that either the mandate was given in 1258 or he continued working on the text for ten years after 1261. At any rate, it clearly shows that he did these works for the Cardinal, later Pope Clement IV, in his spare time from his normal duties as a Franciscan Friar.

7.5 Bacon in 1266: An Exile with Respect to His Previous High Reputation as a Master in Arts

In view of the fact that Bacon was an old man without institutional support, how did he achieve so much writing? How did he do it? Did he have help? I am certainly convinced that he did have helpers. But who were they? Did they include the young theologian, moralist and astronomer Peter of Limoges? Yet, he is not mentioned among the great mathematicians praised by Bacon. What do we know about those in the Franciscan house of study and the University of Paris who might have shared Bacon's concerns? But first, we must ask: What kind of work are the *Opera* for Pope Clement IV? Second, why did he present the Pope with scientific treatises? *Cui bono*? Was Bacon writing for himself alone or was he a representative of a group of theologians, philosophers and scientists at Paris who had an agenda that differed considerably from the normal scholastic method of the regular teaching based on the Sentence-Commentaries? I believe that we must take Bacon at his own word when he states that he had nothing to present to the Pope but one or two chapters from different sciences. He was now for about ten years an emeritus Professor who had other duties as a Franciscan Friar.

7.6 The Uses of Mathematics in Opus maius, Parts IV–VII: Science Interpreted Moraliter.[11]

Parts IV–VI, and Sect. 4 of Part VII of the *Opus maius* deal explicitly with the applications of Mathematics to the natural, human and divine worlds. If one abstracts for a moment from Bacon's important concerns with sacred languages and with speculative grammar, one must notice that the applications of mathematics furnishes the main theme of Bacon's later writings. Bacon's opposition to the works of the young boys of both Orders, Dominican and Franciscan, has to do with the ignorance of the applications of mathematics even in theology. 'Of these sciences the gate and key is mathematics, which the Saints discovered at the beginning of the world, as I shall show, and which has always been used by all the saints and sages more than all the other sciences. Neglect of this branch now for thirty or forty years has destroyed the whole system of studies of the Latins. Since he who is ignorant of this cannot know the other sciences, they do not perceive their own ignorance, and therefore do not seek a remedy. And on the contrary the knowledge of this science prepares the mind and elevates it to certain knowledge of all things...' (*Opus maius* Part IV, 97). Again, he states, there is the example of Pythagoras, Ptolemy and Boethius. Further, 'For since there are three essential parts of philosophy, as

[11] I am, for reasons of space, omitting an account of Bacon's knowledge of Logic and Signs, the English provenance of which is acknowledged by, Alain de Libera, Thomas Maloney, Irène Rosier-Catach, Jan Pinborg.

Aristotle says in the sixth book of the *Metaphysics*, mathematics, natural and divine, the mathematical is of no small importance in grasping the knowledge of the other two parts, as Ptolemy states in the first chapter of the *Almagest*' (*Opus maius*, Part IV, 98–99).˙ Bacon covers a wide area in his application of mathematics, physics, astrology, theology, church history, geography.

For the purposes of this chapter, however, I will focus on the section in *Opus maius IV* dealing with Astrology/Scientia Experimentalis: the reason for this is to indicate that Bacon may well be the one condemned in regard to the charge of 'Astral Determinism' in the Parisian Condemnations of 1270, 1277. While Bacon is not named explicitly, it is clear that some of the propositions cover his doctrine of universal univocal causation based on the effects of the heavenly bodies on human bodies and on temperament. And while Bacon does attempt to save freedom of the Will, it is also clear that he does believe in an astral determination of individual personality and in the astral determinism of sects and religions. Or better, he clearly attempts to save Astronomy-Astrology for Statecraft and for the Church by means of argument and authority: the matter is so important to him that he treats of it in many places, most notably in *Opus maius IV*, *Opus maiux VI, part three, the Introduction and Notes to the Secretum secretorum*, *De secretis operibus naturae et de nullitate magiae*.

Writing on Astrology in the Middle Ages and Renaissance, the renowned authority on the subject, Graziella Federici Vescovini states:

> Roger Bacon was one of the staunchest defenders of the theory of the birth of religions brought about by the great conjunctions. He Bacon agrees with the astrological theory of religions and while he strongly advocates Freedom of the Will, he does so at the cost of opposing the latter to a more strictly deterministic world based on the influence of the stars. Proposition 68, 70, 76, 94 deal with this issue of Fatalism based on 'Astral Determinism' (Vescovini 2011, 2014).

I have argued that Proposition 101–107 beginning with the claim that no agency is open to alternatives; indeed, all agency is determined to one outcome clearly find warrant in Roger Bacon's works (Hackett 2000). The diversity of heavenly positions determines one's fate. Further, not only is one's physical complexion is determined but so also one's spiritual well-being is determined. Health, infirmity, life and death are determined by the stars such that the heavenly bodies influence human destiny.

Bacon *scientia experimentalis,* Part III*, and his Moralis philosophia,* Part IV, which Bacon calls the most important part of his Moral Philosophy, are a major attempt to save Mathematics (Astronomy-Astrology) in regard to contingent changes on earth and in regard to the human organism. It is the basis for his theory of prediction of the future and for his astro-sociology of world-religions. A related approach to Bacon can be found in the works of his younger contemporary, and I believe his helper, Master Peter of Limoges. This is clear from the sermon *De antichristo* (Bériou 1986).[12] Both John North and Graziella Federici Vescovini

[12] See Bériou (1986). This work reference to a *De antichristo* by Pierre de Limoges, who was a Master in Arts at Paris, and then, a theologian, in the 1260s indicates that he had with Roger Bacon a common concern with astronomy/astrology.

have documented Bacon's use and dependence of Abu'Mashr's *Liber conjunctionum*, and I have argued for his additional uses of Abu's Mashr's *Introduction to Astronomy*, Pseudo-Ptolemy, *Centiloquium*, Pseudo-Ptolemy, *De disposition sphaerae*.

To sum up: Bacon defends a radical opposition between inviolate Freedom of the Will, and physical laws of nature that are determinate and necessary. Further, all influence of *species in medio* are physical (Hackett 2011). There are no spiritual species in nature. In this, Bacon believes the *vulgus in philosophy and theology* is just confused. Bacon believes in Astral Determinism both in respect of physical and psychological reality. This has been shown in regard to the doctrine of species by Yael Kedar in her fine Ph.D. work at Haifa University (Kedar 2009). The importance of all of this for the education of the prince as seen in the Pseudo-Aristotle, *Secretum secretorum*, Part VII of the *Opus maius* is connected with the defense of astrology.

To appreciate how toxic Bacon's ideas may have appeared to some contemporaries one need only turn to the *Errores philosophorum* of Aegidius Romanum (Giles of Rome, OESA 1944). Written c. 1268–1271, this work is a head on attack on Greco-Arabian necessetarianism. I have already noted the same reaction to Bacon's concerns by the Master-General, Bonaventure.[13]

Giles of Rome's work was written between 1269 and 1272 and is exactly contemporary with Roger Bacon's writings for Pope Clement IV. Condemned here are the natural philosophy and metaphysics of Aristotle, Avicenna, Averroes, al-Ghazali, al-Kindi and Maimonides. With the exception possibly of Maimonides, all of these thinkers are defended by Roger Bacon. When one turns to al-Kindi and his theory of universal determinism, radiation and magic, one notices the extent of Roger Bacon's dependence on al-Kindi. Aegidius claims (1) 'al-Kindi erred in asserting that the future depends simply and without qualification upon the state of the super-celestial bodies. Hence in this same book in the chapter on the rays of the stars, he states that "one who knew fully the state of the heavenly bodies would have complete knowledge of both past and future."' (2) 'Again, he erred in believing that the effects of all causes in the world extended to every individual.' (3) Ulterius, erravit credens omnia de necessitate contingent.

These errors of the philosopher al-Kindi could be directed entirely against Roger Bacon, and in the context of 1270 they were probably directed against Bacon and his circle, especially since this circle took the work of mathematics so seriously and because by means of the *scientia experimentalis*, the older Bacon and the younger Magisters Peter of Maricourt (Picardus) and perhaps Magister Petrus de Limoges, attempted to defend this new form of scientific study in the context of the more verbal and dialectical education in the arts. The latter had interests similar to Bacon in regard to the importance of astronomy/astrology, and to their uses in Theology.

[13] See note 34 above.

7.7 The Applications of Mathematics: Perspectiva

Richard Newhauser has demonstrated the significant use by Master Peter of Limoges of Roger Bacon's moral uses of the study of *Perspectiva*. He shows the extent to which this Master drew heavily on the third part of Bacon's *Opus maius*, Part V, the *Perspectiva* (Newhauser 2012).[14] In *Roma magistra mundi*, the essays in honor of Fr. Leonard E. Boyle, I provided reasons for thinking that Paris MS BN Lat. 7434 contains a version of the *Perspectiva* that pre-dates the text of the *Perspectiva* in the *Opus maius*. Further, I believe that it contains the oldest text of the *Perspectiva*.[15]

Just recently, A. Mark Smith has shed new light on the nature of Bacon's *Perspectiva*. Smith comments as follows:

> Although Grosseteste's effort to submit the physics of light to geometrical analysis was only partly successful, he inspired Roger Bacon to bring that effort to fruition. It is perhaps no exaggeration, in fact, to say that in sharing the same Augustinian theological leanings, the same desire for broad learning, and the same enthusiasm for applying mathematics to the analysis of natural philosophy, Bacon was, in a sense, Grosseteste's alter ego. But Bacon had one clear advantage over Grosseteste: he could draw on a much wider array of sources in carrying out his program (Smith 2015).

As Bacon notes in the *Opus tertium*, there was no knowledge of *Perspectiva* at Paris in the 1260s; the subject had been taught but twice, and that was at the University of Oxford. Smith does a great job in situating the three parts of the *Perspectiva*. He devotes more care to part one on the relation of optics and the psychological process of perception than Lindberg did. Of course, this was the big bone of contention between both scholars: Was Bacon primarily interested in providing a geometrical theory of vision after the manner of Ibn al-Haytham or was the concern with optics subordinated to providing for a more comprehensive and better theory of perception and knowledge. In regard to Roger Bacon, I believe that Smith has now shown that the latter position is the correct one in the case of Bacon. He shows that this is not so in the case of Witelo. Here, much more than in Bacon and in Pecham, one is dealing primarily with a detailed mathematical treatment of optics. Indeed, Smith is quite correct in noticing that with respect to the mathematics, Bacon does not make advances in the arguments. He simply falls back on Euclid in his attempt to solve the problem of the angle of incidence. But of course, Bacon was writing an introductory *persuasio* to encourage the serious study of optics in the part of the curriculum of the medieval university, called the *Quadrivium*.[16]

---

[14] See Newhauser (2012) Introduction, xi–xxiii for Peter's reliance on both Roger Bacon and John Pecham, but especially, Peter's use of part three of Bacon's *Perspectiva*. Quite significant here is the fact that Peter nowhere mentions Bacon by name.

[15] See Hackett (1998). I have prepared an edition of this text and am working on a translation of this text.

[16] Might it not have been the case that it was Bacon's intention by means of his works for Pope Clement IV to influence the progress of science at the Papal *Studium* in Viterbo. As we will see below, the influence of his works there in the mid-1270s has been acknowledged by David C. Lindberg.

Indeed, Bacon's advocacy for optics was very successful. In the next decade, Bacon's confrere, John Pecham wrote the textbook for the university teaching of optics, the *Perspectiva communis*. It is a masterpiece of educational pedagogy. Thus, Bacon's plea for the improved study of science in the medieval university was not a failure; it was a massive success.

I will argue that Bacon's intense theological concerns condition the manner in which he sets up the parameters of his scientific account of optics in Part I and Part II of the text.

7.8 The Third Part of the *Perspectiva*

One issue has bothered students of Bacon's optics. Why does Bacon make such a spirited defense of a combination of an intromission and extramission theory of vision. Now, of course, one could argue that he is combining elements of Neoplatonic Aristotelianism with the straight geometrical optics of Ibn al-Haytham. There is some truth in that view. However, a review of the Third part of the *Perspectiva*, that is the part which deals with the spiritual uses of optics, the part indeed which so much appealed to the great Master of Preaching, the theologian at the Sorbonne, Peter of Limoges as can be seen from his *Tractatus de oculo morali* indicates another reason for the combining of intromissions and extramission. Here one can find Bacon's rationale for his integration of both an intromission theory of vision and an extramission theory of vision.

Following his remarks on the need for optics in order to know the natures of things, Bacon writes about the preservation of the spiritual pupil of the eye, that is, the soul.

> In a comment on this, Bacon lists the seven spiritual gifts, the seven virtues, three theological and four cardinal virtues, seven gifts of the Holy Spirit and seven petitions of the Our Father, the eight beatitudes (here, the eyelids provide a convenient eight member to the seven parts of the eye, 'so that to the eight spiritual guardians there correspond the same number of corporeal ones' (Lindberg 1996).

David C. Lindberg saw Bacon's combining of the intromission/extramission theories of vision as a sign of confusion in optical theory. I can see why he would think so, but I am not so sure that that alone is the issue. Could it not be the case that Bacon's theological motivation, namely, the need to find a model for the relation of grace and free will requires him to have a perceptual theory that unites intromission and extramission to match what is required in theological doctrine. Is it not the case that Bacon finds a convenient symbiosis between natural science and the requirements of theology? Indeed, it is the case. That Bacon is motivated by theological concern is clear from what follows in his text:

> It has been said that not only is intromission (of species) required for vision, but also the extramission and cooperation of its own for vision, but also the extramission and

cooperation of its own power and species. Similarly, spiritual vision requires not only that the soul should be the recipient from without of divine Grace and Powers, but also that it should cooperate by its own power. For consent and the exercise of free will are required, along with the grace of God, if we are to see and gain the state of salvation (Ibid.) [Text 15].

In other words, Bacon's fusion of extramission with intromission theory of vision is motivated by the need to find a natural analogue in vision and perception for the relation between Grace and human freedom of the will.

Further, citing the relation between the need for proper distance for vision and spiritual distance, Bacon draws a very tight parallel between perspectival perception and spiritual perception, in this case drawing closely from Ibn al-Haytham.

Since corporeal vision is of three kinds—namely, sense alone [direct impact of the species of light and color], knowledge of the universal [vague individual] and knowledge by syllogism—it is likewise necessary for mankind to have a threefold [spiritual] vision. For by sense alone we gain an insufficient grasp of a few things, such as light and color; and this cognition is weak, revealing whether things exist and what they are. But by knowledge we grasp what kind they are and what qualities they possess; whether the light of the sun or the moon, whether white or black. By syllogism we grasp everything associated with light and color according to all twenty common sensibles. Therefore, the first cognition is weak, the second is more perfect, and the third is most perfect. So it is that in spiritual vision; for what a man knows by his own sense alone is very modest, since he lacks the other two kinds of cognition, [the first of which is] through teachers, from youth to old age, for we can always learn them from those who are wiser than ourselves. And [if cognition is by sense alone] we are also without the third kind of cognition, which occurs through divine illumination (Ibid.).

But what is also significant about *cognitio per syllogismum* is that it has nothing whatsoever to do with the common Aristotelian use of the term Syllogism as in formal argument. Bacon is referring to the use of the term in Ibn al-Haytham's optics where it refers to direct intuitive perception of singulars. This is the kind of confirmatory use of intuitive experience that in the *Opus maius*, Part VI, Bacon holds is needed to confirm the rational teachings of books and teachers. It has much significance for epistemology in that it starts a tradition that in Duns Scotus and William of Ockham and their contemporaries will become central, the intuitive cognition of singulars (South 2002). This is the source of the doctrine of experience and intuitive cognition in William of Ockham which has been clearly outlined by Peter King (2003).[17]

The remaining two sections deal with the subdivision of vision into direct, reflected and refracted vision and to its uses in moral persuasion. It also deals with the application of mirrors to the technology of war.

[17] For the beginning of this account of experience in Roger Bacon, see Hackett (2008–2009).

7.9 The Utility of Mathematics: Opus maius VI: De scientia Experimentali

In this section, I will deal with only one aspect of Bacon's concerns in his *Opus maius,* Part VI. The matter of his relation to the Magister Petrus de Maricourt, Picardus, has been carefully examined recently by Sylvia Nagel (2012). I will examine Bacon's stated concern in the aforementioned text about the *nova translatio* of Aristotle's *Meteorologica.*

Bacon is quite clear: he does not have time to complete a formal treatise on the rainbow, halo and related matters. But why in the first place is Bacon so exercised about the rainbow and halo? What is the big deal? Here we must return to Robert Grosseteste, and the matter of the translations of Aristotle, especially the translation of the *Book of the Meteorologies* of Aristotle.

7.10 Roger Bacon's Criticism of the Translators, Especially the Translation of William of Moerbecke

Throughout the *Opus maius* and elsewhere, Bacon talks about the importance of Book Three of Aristotle's *Metheorologica* on the rainbow and halo. Further, the account of the rainbow and halo provides Bacon with the *exemplum* for the first prerogative of his *Scientia experimentalis*, the confirmation by adequate experiences of the rational claims of the other parts of Natural Philosophy and Optics. But why pick Book three of Aristotle's *Metheorologica*? Surely, there are lots of other examples from Ibn al-Haythan, Ptolemy and others. What is going on here? Does it perhaps have to do with difficulties in the translation of Aristotle from Greek into Latin? (*Perspectiva* 324–325).

The late Joseph Brams once observed that Bacon had spoken strongly against *Guillelmus Flemengus*, translator, and that his remarks are compromised by 'le ton ironique et le gout de l'exaggération.' Yet, he saw Bacon's remarks as an issue of some embarrassment for modern scholars. The newly published critical edition and study of William of Moerbecke's *nova translation of the Meteorologica* of Aristotle in the *Aristoteles Latinus* series by Gudrun Vuillemin-Diem makes it possible for the first time to provide a trustworthy evaluation of Bacon's remarks in *Opus maius VI* on the rainbow (Aristoteles Latinus 2008, X, 2.1/2.2, *Meteorologica*). Further, her study of Moerbecke's translation methods, especially in his translation(s) of the *Meteorologica* allow us to prove that Bacon's strong remarks in 1271–1272 about Moerbecke as a translator were no exaggeration. They expressed a serious concern with the first two versions of the translation, which did, indeed, have serious difficulties.

What are the dates of the three successive translations and which scholars at Paris first used them? G1a the uncorrected version was done in 1260. It is likely that Bacon knew this version c. 1266–1267, though he may also have known version

G1b, the second version. This latter version was known to Thomas Aquinas c. 1269–1271, and GT, the third version, was known at the Papal Court sometime after 1270. It is the view of Gudrun Vuillemin-Diem that Thomas Aquinas is the first authentic user of the second version of the translation.[18] Or to be more precise, he may have used G1b (1267), and he certainly knew the third version form 1270. Unfortunately, we lack the commentary of Thomas Aquinas on Book Three of the *Meterorologica*. However, if Thomas Aquinas is the first authentic use of the second version, G1b, what can we say about Roger Bacon's use of either the first or second version? We can state that Roger Bacon knew the *translatio nova* already c. 1266–1267, and may have known the second version G1b, thereby preceding Thomas Aquinas as the *first to comment* on the new translation of the *Meteorologica*. We can also say quite categorically that Roger Bacon, drawing on his knowledge of Aristotelian commentators from Grosseteste, Adam of Buekfield and others, wrote on the first prerogative on experimental science as a correction of Moerbecke's account of the rainbow: this is what he means by his appeal to the practice of Robert Grosseteste and the difficulties of the perverse translations of Aristotle. Like Grosseteste, he intends to use his own experience and other authors, most notably Seneca and writers on optics to correct and advance the study of the rainbow. Bacon tries to correct Moerbecke's Aristotle by a new treatment of the figure of the rainbow and he shows how an adequate use of instruments such as the astrolabe one can give precise measurements of the possible altitude of the rainbow (42°). He also studies the question of the objectivity of rainbows and other secondary stars, the role of vision and the nature of the colors. The question naturally arises concerning Bacon's move to write on philosophical topics c. 1260. Did the new translation of Aristotle move him to return to Natural Philosophy and Logic in order to offer a broad criticism of the new readings of the text of Aristotle?

But the question remains; could Bacon, the writer of a Greek Grammar and the commentator in *Compendium studii philosophiae* of the meaning of Greek words, have been wrong and just simply prejudiced in a negative manner alone about Moerbecke; was it a fit of pique that he did not get to translate the *Meteorologica*? Was it just a case of corporate professional rivalry? At any rate, according to the editor of the latter text in the *Aristoteles Latinus*, it is clear even in version two (1267), that Moerbecke had trouble with the geometrical figure of the rainbow. A careful review of all three versions by the editor proves that there are major problems in Moerbecke's first two versions: numerous omissions of words, especially syntactical words, mechanical translation from the Greek with assimilation mistakes, additions by Moerbecke himself, difficulties with unusual (that is, scientific) Greek words, confusion of single words such as those for cloud and fog, significant failure in reading the text, numerous textual conjectures. Above all, there was a significant problem with the diagram of the rainbow.

[18] Ibid., Aristoteles Latinus, X, *Meteorologica Translatio Guillelmi De Morbeka*, 2.1: 349–350.

We can conclude then that Roger Bacon, onetime Master in Arts, writer of a Greek Grammar, knew a weak text when he saw one. Later in the mid-1277s, Bacon's text from the *Opus maius* on the rainbow and halo c. 1267, was in use at the Papal Centre of Studies: As David C. Lindberg states: Witelo's theory of the rainbow parallels Bacon's on a number of points, some of which are original with Bacon (1971). The result is that we must eschew the suspicion of Bacon that has been common among some Robert Grosseteste scholars, and we must take Roger Bacon at his word especially when there is corroborating evidence for doing so. Bacon raised a very important and serious epistemological issue when he held that even when the experimental universal is established, there remains an issue of certifying the results of mere arguments. Thus, experience was not only an important source of knowledge; it was also the key to the verification of rational claims.

Finally, where does Bacon fit into the program of Bonaventure's *De reductio atrium ad theologiam*? Recently, Timothy Johnson has provided a convincing answer to this matter. First, he critiqued Dieter Haartrup's attempt to make Bacon the author of this early Bonaventure programmatic work (Johnson 2009). Second, he proves that the *Opera* for Pope Clement IV are typical Franciscan Wisdom Scriptures, and shows that Bacon's more this-worldly' spirituality, discovering God in the midst of everyday life including scientific discovery, differs from the program of Bonaventure and Pecham.[19] They both warned against the '*hospites scientiae*.' They worried that Bacon's concerns would lead the Friars to a more worldly concern. And yet, Timothy Johnson proves in a forthcoming paper that Bacon's masterpiece account of the mystical meaning of devotion was encouraged by *Il Poverello*. And so, after all the seventeenth to twentieth-century imagery of Bacon as the first scientist, we have come full circle to the real Medieval Franciscan theologian who has a deep interest in languages and mathematical science after the manner of his heroes, Robert Grosseteste and Adam Marsh. Here, we must take seriously Bacon's reference to the *Sapientissimus* who was a master of the Biblical Text. As we saw above, Bacon was also concerned with the status of the text of Aristotle. But Bacon's primary concerns after 1257 are with Franciscan Wisdom, and especially with the status of the Biblical Text and with issues in Theology.

7.11 Conclusion

I set out to determine the manner in which Roger Bacon identifies his two circles of scholars. There is the circle of the ancient wise ones such as Robert Grosseteste, Adam Marsh and scholars at Oxford and Lincoln. After 1260 Bacon moves in the context of the Franciscan house of Studies at Paris. Bacon was both a product of

[19] Idem, *Wisdom has built her house:* she has set up her seven pillars: Roger Bacon, Franciscan Wisdom, and the Conversion to the Sciences, (forthcoming) (see also Johnson 2014).

his own ancient teachers at Oxford and a committed Franciscan at the stadium in Paris.

Appendix

Conflicting Chronologies:

[Text 1] Opus Tertium, ed. Brewer, pp. 65–67: Multum laboravi in scientiis et linguis, et posui jam quadraginta annos postquam didici primo alphabetum, et fui semper studiosus; et praeter duos annos de istis quadraginta fui semper in studio; et habui expensas multas, sicut alii communiter.... Similiter de figuris et numeris in geometria et arithmetica, sine quibus nihil sciri potest de potestate philosophiae, ut opera quae scripsi probant...Nam hoc est alphabetum philosophiae; ut nunquam possit homo aliquid dignum scire, postquam harum scientiarum ignorant potestatem. Et hoc factum est contra dies Antichristi, ut tollatur tota sapientia philosophiae, et per consequens theologiae quantum est in expositione Scripturae. Nam textus ipse, et expositiones sanctorum sunt plenae numeris, et figuris, et caeteris mathematicis consequentibus ad haec, ut ego probo in Majori Opere, comparando mathematicam ad theologiam ...Et scitis figuris et numeris possumus omnia scire de facili; quia tota sapientia exit ab eis sicut a radicibus, et per haec declaratur, sicut patet ex iis quae mitto.

[Text 2] Ibid. 58–59: Quarta ratio est propter meipsum, quia jam a iuventute laboravi in scientiis, et linguis, et omnibus praedictis multipliciter; et collegi multa utilia, et ordinavi de personis. Nam **quaesivi amicitiam omnium sapientum inter Latinos, et feci juvenes instrui in linguis, et figuris, et numeris, et tabulis, et instrumentis, et in multis necessariis. Et examinavi omnia quae hic necessaria sunt, et scio qualiter procedendum est, et quibus auxiliis, et quae sunt impedimenta; sed non possum procedere propter defectum expensarum praedictarum**.

Nam per viginti annos quibus specialiter laboravi in studio sapientiae, neglecto sensu vulgi, plus quam duo millia librarum ego posui in his, propter libros secretos, et experientias varias, et linguas, et instrumenta, et tabulas, et alia, tum ad quaerendum amicitias sapientum, tum propter instruendos adjutores in linguis, in figuris, in numeris, et tabulis, et instrumentis, et multis aliis.

[Texts 3] *Compendium studii theologiae*, ed. Maloney, 46: Nam Beatus Edmundus, Cantuariae Archiepiscopus, primus legit Oxoniae librum *Elenchorum* temporibus meis: et vidi magistrum Hugonem, quo primo legit librum *Posteriorum*, et librum eius conspexi."

Tractatus de experiential in communi, ed. Hackett, p. 293: "Nam in translatione libri "Meteorologicorum" pervulgata apud Latinos usque nunc, dicitur quod a radiis lunae non sit iris nisi bis in quinquaginta annis **et maxime naturalis et perspectivus quem vidi voluit et hoc verum salvare et causam eius reddere dum eius auditor a iuventute fueram constitutus**.

[Text 4] Ibid., Unde quando per tentationem et derisionem aliqui Minores praesumtuose quaesiverunt a fratre Adam, "Quid est intellectus agens?" Respondit, "Corvus Eliae"; volens per hoc dicere quod fuit Deus vel Angelus. Sed noluit exprimere, quia tentando et non propter sapientiam quaesiverunt."

[Text 5] A. G. Little, Roger Bacon Life and Works (1914), 2–3.

[Text 6] *Compendium studii philosophiae*, ed. Brewer, 468: Sic translatae sunt et scientiae communes, ut logica, naturalis philosophia, mathematica, ut nullus mortalis possit aliquid dignum de eis intelligere veraciter, **sicut ego expertus sum omnino.**

[Text 6a] *Opus tertium*, ed. Brewer, 139: Si igitur dignetur vestra gloria considerare quae nunc scribo, et in Primo Opere, poteritis conferre cum omni geometro et naturali, et neminem inveniretis qui vobis resistet. Adolescens quidem vobis in his omnibus poterit respondere, quia docui eum omnia, quae sunt de istis figurationibus corporum.

Sed fere viginti anni sunt quod egi intra principia multa magistrorum novorum de hac materia; sed nullus unquam inventus est in tota universitate qui terminos ipsos intelligeret; et ideo pluries feci lectionem magistri novi de veritate quod Aristoteles, et Averroes narrant, cum expositione vocabulorum, et tamen nullus potuit disputationi respondere.

[Text 7] *Compendium studii theologiae*, ed. Maloney, 87: Et optime novi pessimum et stultissimum istorum errorum <auctorem>, qui vocatus est Richardus Cornubiensis, famosissimus apud stultam multitudinem etc.

[Text 8] *Opus tertium*, ed. Brewer, 30–31: "Quod philosophia jam data sit Latinis, et completa, et composita in lingua Latina, **et est facta in tempore meo** et vulgata Parisius, et **pro auctore allegatur compositor ejus. Nam sicut Aristoteles, Avicenna, et Averroes allegantur in scholis sic et ipse: et adhuc vivit et habuit in vita sua auctoritatem, quod nunquam homo habuit in doctrina. …et de errore vulgi decepti per eum…Sed iste per modum authenticum scripsit libros suos, et ideo totum vulgus insanum allegat eum Parisius sicut Aristotelem, aut Avicennam, aut Averroem, et alios auctores**.

[Text 9] *Opus maius III*, ed. Bridges 88–89: "Nam vidimus aliquos de antiquis qui laboraverunt in linguis sicut fuit dominus Robertus praefatus translator et episcopus, et Thomas venerabilis ansistes Sancti David nuper defunctus, et frater Adam de Marisco et Magister Hermannus translator, et quidem alii sapientes.

Compendiuim studii philosophiae, ed. Brewer, 428: "Ita quod totaliter dimiserunt vias antiquorum sapientum, **quorum aliquos vidimus nostri temporibus; scilicet, dominum Robertum, quondam episcopum Lincolniensem, sanctae memoriae, et dominum Thomam, episcopum Sancti David in Wallia, et fratrem Adam de Marisco, et Magister Robertum de Marisco, et Magistros Willelmum Lupum, et Willielmum de Schyrewode, et aliquos alios eis similes, quorum vestigia moderni saeculares omnino dimiserunt**.

[Text 10] *Opus tertium*, ed. Brewer, 37: Haec autem scientia non est adhuc lecta Parisius, nec apud Latinos, nisi bis Oxoniae in Anglia…

[Text 11] *Compendium studii philosophiae*,, ed. Brewer, 430–465, and elsewhere. See S. A. Hirsch, ed. Bacon's Greek and Hebrew Grammar (OHI, Steele).

[Text 12] *Fr. Rogeri Bacon Opera*, ed. Brewer, p. 1. [Text 13] Card. Gasquet Fragment, 500: "Unde Raymundus de Lauduno qui vestre clementine locutus est de scripturis meis meum propositum nullatenus intellexit. **Magnificentie quidem vestre innotuit ut utramque mandatum pretendit quod precepto fui obligatus artissimo ne scritum in hoc statu a me factum communicarem, sicut et nostra tota congegatio firmiter noscitur obligari, et ideo componere penitus aborrebam.** Nam componi nihil potuit nisi scriptoribus traderetur, qui vellem nollem transcriberent pro ipsis vel amicis, et sic communicarent omnibus ut pluries vidi scripta secretissima per fraudem divulgari scriptorium, et inciderem in conscientiam de transgressione precept.

Praeterea cum non potui communicare amicis meis carissimis et coadiutoribus necessariis since quibus nichil possum, neglexi compositioni insistere scripturarum.

[Text 14] Anthony à Wood, *Historia et Antiquitates Universitatis Oxoniensis*, 138: "Prelati enim et fratres me jejuniis macerantes tuto custodiebant, nec aliquem ad me venire voluerunt, veriti ne scripta mea aliis quam summum pontifici et sibi ipsis pervenirent. This is corroborated by his remarks in *Opus tertium*, ed. Brewer, 15: "Et primum impedimentum fuit per eos, qui mihi praefuerunt, quibus cum nihil scripsistis in excusationem meam, et eis non potui revelare vestrum secretum…,

[Text 15] *Perspectiva*, ed. Lindberg, 324–325: Et dictum est quod ad visionem exigitur non solum ut fiat intus suscipiendo, sed extramittendo et cooperando per virtutem et speciem propriam. Similiter et visio spiritualis non solum requirit ut anima recipiat ab extra, scilicet a Deo gratias et virtutes, sed cooperetur per virtutem propriam. Nam motus liberi arbitrii et consensus requiruntur cum gratia Dei ad hoc ut videamus et consequamur statum salutis. See R. Newhauser, "Inter scientiam et populum," 702 for Peter of Limoges's uses of this text from Bacon.

[Text 16] *Compendiium studii philosophiae*, ed. Brewer, "Et sic de aliis. **Maxime iste Willelmus Flemingus qui nunc floret. Cum tamen notum est omnibus Parisius literatis, quod nullam novit scientiam in lingua Graeca, de qua presumit. Et ideo omnia transfert falsa et corrumpit sapientiam Latinorum. Solus enim Boethiius scivit de omnibus interpretationibus linguas sufficienter. Solus dominus Robertus, propter longitudinem vitae et vias mirabiles quibus usus est, prae aliis hominibus scivit scientias; quia Graecum et Hebraeum non scivit sufficienter ut per se transferret, sed habuit multos adjutores. Omnes autem alii ignoraverunt linguas et scientias et maxime hic Willelmus Flemingus, qui nihil novit dignum neque in scientiis neque in linguis; tamen omnes translationes factas promisit immutare et novas cudere varias. Sed eas VIDIMUS et SCIMUS esse omnino erroneas et vitandas.**

References

Aristoteles Latinus. (2008). *Meteorologica: Translatio Guillelmi Morbeka, edditit.* In G. - Vuillemin-Diem (Ed.). Brussels: Brepols.

Bacon, R. (1964). *Opus maius* (R. Bridges, Ed.). Oxford: Minerva Reprint.

Bériou, N. (1986). Pierre de Limoges et la fin du temps. *Mélanges de l'École française de Rome, 98*, 65–107.

Bernardini, P., & Rodolfi, A. (Eds.). (2014). *Roger Bacon's Communia Naturalium: A 13th century philosopher's workshop*. Florence: Edizioni Del Galuzzo.

Crowley, T. (1950). *Roger Bacon: The problem of the soul in his philosophical commentaries.* Louvain: Duffy & Co.

Dodd, B. G. (1982). Aristoteles Latinus. In N. Kretzmann, A. Kenny, & J. Pinborg (Eds.), *The Cambridge history of later medieval philosophy* (pp. 45–79). Cambridge: Cambridge University Press.

Donati, S. (2013). Pseudoepigrapha in the Opera hactenus inedita Rogeri Baconi? The commentaries on the physics and metaphysics. In J. Verger & O. Weijers (Eds.), *Les débuts de l'enseignement universitaire à Paris (1200–1245 environs)* (pp. 153–203). Turnout: Brepols.

Emden, A. B. (1966). Accounts relating to an early Oxford houses of scholars. *Oxoniensia, 31*, 77–81.

Giles of Rome. (1944). *Errores Philosophorum: Critical text with notes and introduction* (J. Koch, Ed., & J. O. Riedl, Trans.). Milwaukee, WI: Marquette University Press.

Hackett, J. (1998). The hand of Roger Bacon, the Writing of the Perspectiva and MS; Paris BN Lat. 7434. In J. Hamesse (Ed.), *Roma, magistra mundi* (pp. 323–336). Louvain-La-Neuve: Brepols.

Hackett, J. (2000). Aristotle, Astrologia, and Controversy at the University of Paris (1266–74). In J. Van Engen (Ed.), *Learning institutionalized: Teaching in the Medieval University* (pp. 69–110). Notre Dame, IN: University of Notre Dame Press.

Hackett, J. (2003). Agent intellect and intelligible species in Roger Bacon and John Pecham. In G. Marchetti, O. Rignani, & V. Sorge (Eds.), *Ratio et Superstitio: Essays in Honor of Graziella Federici Vescovini* (pp. 117–136). Louvain-La-Neuve: Textes et Études De Moyen Âge.

Hackett, J. (2006). Mirrors of princes, errors of philosophers: Roger Bacon and Giles of Rome (Aegidius Romanus) on the Education of the Government (the Prince). In H. B. Clarke & J. R. S. Phillips (Eds.), *Ireland, England and the Continent in the Middle ages and beyond: Essays in memory of a Turbulent Friar, F.X.Martin, O.S.A.* (pp. 105–127). Dublin: University College of Dublin Press.

Hackett, J. (2008–2009). Roger Bacon's concept of experience: A new beginning in medieval philosophy? *The Modern Schoolman, 86*(1 2), 123 146.

Hackett, J. (2011). Ego Expertus Sum: Roger Bacon's science and the origins of empiricism. In T. Bétanouïl & I. Draelants (Eds.), *Expertus Sum: L'expérience par les sens dans la philosophie naturelle médiévale (Actes du colloque international de Point-à-Mousson 5–7 fé vrier 2009)* (pp. 145–173). Florence: Edizioni del Galuzzo.

Hackett, J. (2013). Roger Bacon. In E. N. Zatta (Ed.), *Stanford Encyclopedia of philosophy.* Stanford, CA: Stanford University.

Hackett, J. (2014). Motion, time and *Aevum*. In P. Bernardini & A. Rodlfi (Eds.), *Roger Bacon's Communia naturalium. A 13th century philosopher's workshop* (pp. 191–216). Florence: Rodolfi.

Hirsch, S. A. (1914). Roger Bacon and philosophy. In A. G. Little (Ed.), *Roger Bacon essays* (pp. 101–152). Oxford: Clarendon.

Johnson, T. J. (2009). Back to Bacon: Dieter Hattrup and Bonaventure's authorship of the *De Reductione. Franciscan Studies, 67*, 133–147.

Johnson, T. J. (2012). Franciscan bodies and souls: Bonaventure and Bacon on scripture, preaching, and the cura corporis/cura animae. In T. J. Johnson (Ed.), *Franciscans and preaching: Every miracle from the beginning of the world came through words* (pp. 73–89). Leiden: Brill Academic Publishers.

Johnson, T. J. (2014). Place, analogy, and transcendence. In G. Melville, B. Schneidmüller, & S. Weinfurter (Eds.), *Innovationen durch Deuten und Gestalten* (pp. 83–96). Regensburg: Verlag Schnell-Steiner.
Jourdain, C. (1874). *Discussions De quelques Points de la Biographie de Roger Bacon*. Paris: Comtes rendus des séances de l'Academiedes inscriptions et belles-lettres.
Kedar, Y. (2009). *Species* as signs: Roger Bacon (1220–1292) on *Perspectiva* and *Grammatica.* Ph.D., University of Haifa.
King, P. (2003). Two concepts of experience. *Medieval Philosophy and Theology, 11*, 203–226.
Lindberg, D. C. (1996). *Roger Bacon and the origins of perspectiva in the Middle ages*. Oxford: Clarendon.
Lindberg, D. C. (1971). Lines of influence in thirteenth-century optics: Bacon, Witelo, and Pecham. *Speculum, 46*, 66–83.
Little, A. G. (1914). Introduction: On Roger Bacon's life and works. In A. G. Little (Ed.), *Roger Bacon essays* (pp. 1–33). Oxford: Clarendon.
Maloney, T. (1988). *Roger Bacon: Compendium of the study of theology*. Leiden: Brill.
Maloney, T. (2009). *Roger Bacon: The art and science of logic*. Toronto: Pontifical Institute Medieval Studies.
Molland, G. (2004). *Bacon, [Bakun] Roger, c. 1214-1292? Philosopher and Franciscan Friar* (Oxford dictionary of national biography). Oxford: Oxford University Press.
Nagel, S. (2012). Pietro Peregrinus, il sapiens-simplex eccelente di Ruggero Bacone. In E. Menestò (Ed.), *I Francescani E Le Scienze (Atti del XXXIX Convegno internazionale Assisi, 6–8 ottobre, 2011)* (pp. 21–47). Spoleto: Centro Italiano di Studi Sull'alto Medioevo.
Newhauser, R. (2012). *Peter of Limoges: The moral treatise on the eye*. Toronto: Pontifical Institute of Medieval Studies.
North, J. (1999). Roger Bacon and the Saracens. In G. F. Vescovini (Ed.), *Filosofia e scienza classica, arabo-latina medieval e l'éta moderna: Ciclo di seminari internazionali (26–27 gennaio 1996)* (pp. 129–160). Louvain: Brepols.
Pinborg, J. (1979). The english contribution to logic before Ockham. *Synthèse, 40*, 19–42.
Power, A. (2006). A mirror for every age: The reputation of Roger Bacon. *English Historical Review, 121*(492), 657–692.
Power, A. (2013). *Roger Bacon and the defence of Christendom* (Cambridge studies in medieval life and thought, Vol. 4). Cambridge: Cambridge University Press.
Smith, A. M. (2001). *Alhacen's theory of visual perception* (Vol. 91). Philadelphia: Transactions of the American Philosophical Society.
Smith, A. M. (2015). *From sight to light: The passage from ancient to modern optics*. Chicago, IL: The University of Chicago Press.
South, J. (2002). Duns Scotus and the knowledge of the individual revisited. *History of Philosophy Quarterly, 19*(2), 125–147.
Southern, R. W. (1992). *Robert Grosseteste: The growth of an english mind in medieval Europe* (2nd ed.). Oxford: Clarendon.
Vescovini, G. F. (2011). *Le Moyen Âge Magique: La magie entre religion et science aux xiiie et xive siècle*. Paris: J. Vrin.
Vescovini, G. F. (2014). The theological debate. In B. Dooley (Ed.), *A companion to astrology in the renaissance* (pp. 99–140). Leiden: Brill.

Chapter 8
The Theological Use of Science in Robert Grosseteste and Adam Marsh According to Roger Bacon: The Case Study of the Rainbow

Cecilia Panti

8.1 The Importance of Science for Theology

In two of my recent articles, I have presented an inquiry into the legacy of the scientific thought by Robert Grosseteste among early Franciscan scholars. In the first of them (Panti 2012), I tried to show that references to light, colour and optical phenomena in theological works and sermons by Grosseteste related to his teaching at the Franciscan school of Oxford were mainly intended as symbolic exemplifications for illustrating theological topics, such as the dogma of trinity and the nature of virtues and free will. These examples display a sort of 'technique' that Grosseteste probably wanted to transmit to his pupils as a methodological tool for preaching the *sacra doctrina*, and it seems that his pupil and friend Friar Adam Marsh shared the same methodology, at least according to what a few indirect sources suggest. In the second paper (Panti 2016), I addressed my research to how Grosseteste's scientific ideas were applied in the theological writings of mid-thirteenth century English Franciscans. A comparison of three approaches to the exegesis of the same verse of *Eccelsiasticus* (*Sirach* 43: 4, *tripliciter sol exurit montes*), respectively by Grosseteste, Friar Roger Bacon and Friar Thomas Docking, has shown that the three scholars made use of optics and mathematics in explaining why the sun is a threefold cause of heat on the top of mountains. Although Docking quotes extensively from Grosseteste's works and Bacon, in turn, demonstrates knowledge of both Docking's and Grosseteste's exegesis, the three scholars travelled along independent paths.[1] My analysis, eventually, challenged

[1] Friar Thomas Docking was the seventh lector of the Oxford Minors in the early sixties. He included scientific discussions in his exegetical works by quoting long passages from the

C. Panti (✉)
University of Rome Tor Vergata, Rome, Italy
e-mail: panti@lettere.uniroma2.it

J.P. Cunningham, M. Hocknull (eds.), *Robert Grosseteste and the pursuit of Religious and Scientific Learning in the Middle Ages*, Studies in the History of Philosophy of Mind 18, DOI 10.1007/978-3-319-33468-4_8

both Docking's real commitment to Grosseteste's methodology and Bacon's empathy with Docking's use of science for theology.

In this third paper, I wish to deepen Bacon's claim on the importance of science for theology for verifying how much it adheres to Grosseteste's view on the same subject. It is known, in fact, that in the works written for Pope Clement IV, Bacon asserts that his tenet is exemplified in the writings of Grosseteste and his friend and pupil friar Adam Marsh. Bacon indeed, presents himself as a representative of their tradition of teaching, as if their influence on him had been decisive in turning him to the interests associated with both scholars. These interests include foreign and ancient languages, the importance of mathematics, optics and experimental methodologies, and a renewed critical study of the Holy Scriptures.[2] For Bacon these three contexts are tightly linked, since a correct literal exegesis necessarily requires knowledge of languages and sciences. The present paper cannot take into account languages, and is limited to demonstrating the connection between theology and science, together with Bacon's affiliation to Grosseteste and Marsh. It is important to remark that the claims by Bacon concerning the use of philosophy and science for theology have been analyzed in depth in past and recent studies (Hackett 2012; Power 2013). Here, they will be examined only in their application to a relevant case study, namely the nature of the rainbow. I will try to demonstrate that in spite of Bacon's explicit assertion that Grosseteste's *De iride* reveals that sciences are fundamental for theology, Bacon's distance from Grosseteste's view is definite and clear, not only with regard to the nature of the rainbow, but also its utility in the *scientia divina*.

Before turning to this, it is important to underline that Bacon associated Grosseteste and Marsh in his references to their interests, though no treatise by Marsh survives for attesting his alleged scientific concerns and their exegetical utility.[3] Hence, the only way for verifying Bacon's words, apart from testing their accordance with Grosseteste's claims, is to confirm the reliability of Bacon's direct acquaintance with both scholars. The first part of this paper deals with this question, while the second part examines Bacon's theory of the rainbow in its theological framework. In the third section, I will consider the differences between his and Grosseteste's thought on the nature of the rainbow.

Lincolniensis. Although Bacon might have known Docking when he was at Oxford in 1247–1250 (Docking was likely among the pupils of Adam Marsh), he surely came across his exegetical works on a later occasion, in the sixties. See (Little 1943; Catto 1968, 1984).

[2] In all passages concerning Grosseteste and Marsh, unless differently specified, the English translations of Latin texts are mine.

[3] His only writings known to us are his letters, which contain no reference to these subjects. See (Lawrence 2006 & 2010).

8.2 Roger Bacon and His Knowledge of Robert Grosseteste and Adam Marsh

Although it seems undisputable that Bacon met both Grosseteste and Marsh, no documentary evidence testifies his familiarity with them or his having been their pupil. However, Bacon's numerous references to both scholars and his open affiliation with their alleged method of teaching suggest that his contacts could not be limited to accidental knowledge.[4]

A first possibility for placing this meeting is the late twenties or early thirties, which implies that Bacon was born in about 1214 and was in Oxford as a student of Arts.[5] The piece of evidence associating him with the town is, however, only a record attesting that in 1233, when he would have been, more or less, 20 years old, he acted as a cleric of the king's court.[6] By that time, he would have already fulfilled his studies in Arts and perhaps a basic instruction in Law, which would justify his position at the court. In 1233, Roger was neither a Franciscan, nor might he have been connected with the friars' school, where Grosseteste was teaching theology. At approximately the same year, Adam Marsh was entering the order. Adam made his profession a few years later, at Worcester, where he probably served his novitiate (Lawrence 2006 & 2010). If Bacon first arrived at Oxford in 1233, he would have had only an indirect knowledge of Grosseteste and Marsh. However, the reputation of Grosseteste and his friendship with the Oxford Franciscans and specifically with Adam Marsh might have nourished his later confidence in associating himself with their theological and mathematical interests. Moreover, the example by Marsh, a master of Arts from a noble family close to the king, who resigned his status and possessions for love of Francis, may have instilled in Bacon

[4] The knowledge that Bacon had of Grosseteste (and Marsh) is still a matter of debate. Past scholars commonly held that Bacon was a pupil of Grosseteste, while in recent times this view has been revised see (Southern 1986). The chronology of both scholars is too obscure for solving the question. In what follows, I try only to figure out whether Bacon's direct knowledge of both scholars is plausible in order to test the reliability of his claims on their common scientific and theological interests.

[5] On Bacon's chronology and for further bibliography see (Hackett 1997a, b; Power 2013).

[6] Matthew Paris is referring to an episode that happened in June of 1233 (Paris 1876), when the English barons refused to present themselves to King Henry III, who summoned them to Oxford. Matthew asserts (pp. 244–5) that Roger, a cleric of the *curia*, played his wit with a fresh sense of humor, for alerting the king of the bad influence exerted on the barons by Bishop Peter of Winchester. Matthew also mentions (p. 244) that another *Bacun*, namely Robert, a Dominican friar, was there and spoke to the king. The editor supposes that the name *Rogerus* is a mistake for *Robertus*, but the different status and position, and consequently identity, of the two men is clearly stated: the one was a *clericus*, the other a friar, who preached *verbum dei* and delivered an 'open' speech (*libera voce*) against Bishop Peter. This source for Bacon's biography is commonly overlooked or neglected by modern scholars.

the seed of his later conversion. Bacon himself belonged to a wealthy and noble family close to the king, and was early in life to become master of Arts at Paris.[7]

More importantly, the *Opus tertium* (Bacon 1859a) also tells us that Grosseteste was the master of Adam (*Opus tertium* c. 50, 186–7; quoted below). This noteworthy remark may signify that Marsh was in Grosseteste's classroom either as a student of Arts, before 1226, or as a student of theology, during his novitiate in 1232/33–1235. This last possibility matches perfectly with Bacon's presence at Oxford in 1233, and reinforces his assertions of the common objectives of the two scholars.

Apart from the event of 1233, it is likely that Bacon had been at Oxford as a student of Arts since about 1227 (Power 2013). Documentary evidence is missing, with the exception of two autobiographical notes, which point directly to Bacon's early scientific interests and specifically to the discussion on the rainbow at Oxford. In the *Opus tertium*, Bacon states:

> This science [*namely perspective*], has not been taught up to now among Latins except twice at Oxford, in England, and there are no more than three men who know its value (*Opus Tertium* c. 11, 37).

In the *Opus maius* (Bacon 1900), he adds:

> In fact, in the translation of the books on *Meteorology* divulged among Latins up to now, it is stated that a rainbow cannot be made by moon rays but twice every fifty years; and the greatest natural philosopher and expert on perspective, whom I saw, wanted both to save this truth and to explain its cause while I was his pupil in my youth (*Opus maius*, vol. 2 pt. 6, 173, addendum).[8]

If read together, these statements imply that the young Roger was in the classroom of an Oxonian master of Arts, an expert in optics, who lectured on Aristotle's *Meteorologica*.[9] Bacon knew Grosseteste's *De iride*, written in late twenties, which makes use of the *Meteorologica*, though it does not refer to lunar rainbows. Is it possible that this master was Grosseteste? A positive answer would imply that Bacon attended Grosseteste's last lessons in the Arts, and this, in turn, requires that Grosseteste started teaching theology as late as 1229, when he was asked to teach the Franciscans.[10] This matches with Bacon's biography only if we

---

[7] On Bacon's vocation and its cultural and spiritual background see (Power 2013).

[8] This passage is an addition referring to *Opus maius*, vol. 2, pars 6 (*De scientia experimentalis, Tractatus de experientia in communi*), 173, after line 18. See *Opus maius*, vol. 3, 181 (notes and additions). The transcription by Bridges presents the misreading *fuerit* instead of *fuerim*. This addition is transmitted only in the miscellaneous MS Vat. lat. 4091, at fol. 57v (fol. 52v old foliation). The paper quire containing it transmits abstracts from the *Opus maius* written by a fifteenth-century hand. I checked the MS in situ.

[9] The words by Bacon concerning the Oxonian teaching of the *Meteorologica* (Aristotle 2000) are reliable also because the public reading of Aristotle's works was forbidden in Paris at that time. Bacon had likely been the first master to lecture on Aristotle at Paris since 1237–1240. As regards the rainbow as a topic that requires competence on perspective, see below, § 2.

[10] The chronological extension of Grosseteste's teaching *in sacra doctrina* is a matter of debate. According to a traditional view, defended by James McEvoy (1982), Grosseteste taught theology as early as 1214. Richard W. Southern (1986), proposed a much later start, 1225, while Joseph

assume that he was born no later than 1214, because students of Arts were teenagers and in the late twenties Bacon should have been in his teens.

As regards Adam Marsh, he incepted as master of Arts at Oxford in 1226 and resigned his post in 1232 or 1233, when he took the Franciscan habit (Lawrence 2006 & 2010). Thus, although it seems more likely that Bacon might have been in his rather than in Grosseteste's classroom, and notwithstanding Bacon's assertion that Marsh was also a great mathematician, it is hard to accept that he might have referred to Marsh as *maximus perspectivus*.[11] Besides, in recalling his attendance at Marsh's scholarly discussions (that will be mentioned below), Bacon states that Adam was a friar and this likely indicates that Bacon is referring to discussions that happened in biennium 1248–1250, when Adam was teaching theology.

Bacon's insistence on associating Grosseteste and Marsh based on their interests in mathematics and perspective seems to refer, finally, to his alleged acquaintance with them in the late twenties. However, a difficulty in confirming this hypothesis is that Bacon made rare use of Grosseteste's writings in the works related to his teaching years at Paris.[12] In other words, if he had been Grosseteste's pupil, it seems that the teachings of his master had hardly any immediate effect on him. Yet, after Bacon left his post as Master of Arts, it seems that he had a sort of afterthought. Perhaps, his subsequent return to England in 1247 was a sort of second chance for him in rediscovering his cultural roots after years of disappointment at Paris. However it may be, if these data are collected together, they tell us that Bacon's supposed first stay at Oxford coincided with crucial changes in Grosseteste's and Marsh's lives, determined by the establishment of a growing Franciscan community and the development of their school.

The second possible occasion for Bacon to meet Marsh and Grosseteste might have taken place in a completely different milieu. In 1244/45, Bacon was a Master of Arts at Paris, and might have met them there, on their way to the council of Lyon.

Goering (1995) and James R. Ginther (2004), hold that Grosseteste's first and only chair of theology was in the quinquennium (1229–1235) at the Oxford Franciscan school. This last hypothesis is better also for the chronology of Grosseteste's scientific works, which can be placed within the years 1210–1230. See (Panti 2013).

[11] The passage from the *Communia mathematica* (Bacon 1940) lists three great mathematicians (*Communia mathematica* pt. 2, dist. 3, c. 3, 117–8), Robert Grosseteste, Adam Marsh and John *Bandoun*, likely John 'of London', who in turn has been identified with John of Tynemouth. See (Knorr 1990). In the *Opus tertium*, 34–35, Bacon refers to four excellent mathematicians: 'One needs the best mathematicians (...). But there aren't any, save for two perfect (*perfecti*) ones, namely, master Jo(hn) (of) London and master Peter of Maharncuria of Picardy. There are two other good (*boni*) ones, namely, master Campanus of Novaria, and master Nicholaus, tutor of Lord Almaricus de Monte Forti'. The identification of Tynemouth with John of London has been later reconsidered by Wilbur R. Knorr (1996).

[12] It is important to underline that the *Liber de sensu et sensato* (Bacon 1937) attributed to Bacon and written during his years as master of Arts, makes use of Grosseteste's *De colore*. See (Tachau 2014). Also Bacon's *Computus*, likely an early writing, makes use of the homonymous work by Grosseteste and quotes him by name. This is the only reference in which Grosseteste is not mentioned in the past tense. See (Bacon 1926).

Grosseteste was now a bishop and Adam a friar.[13] Two autobiographical notes by Bacon, respectively from the *Opus maius* and the *Opus tertium*, seem to attest to this meeting, although their interpretation is controversial:

> For when the University of Paris was convoked, I twice saw and heard the venerable William, Lord Bishop of Paris of Blessed memory, in the presence of all, teach that the agent intellect cannot be a *pars animae*, and the Lord Robert Bishop of Lincoln and Adam Marsh and elders of this rank supported the same teaching (*Opus maius* vol. 3, pt. 2, c. 5, 47; trans: Hackett 1996).

> However, it is false that the agent intellect is a part of the soul. (...) And all the old wise men, including some who are still alive, said that it was God. Hence, I twice heard the Venerable Bishop of the Church of Paris, the Lord William of Auvergne, with the university congregated before him, reprove those < who said that it was a part of the soul > and dispute with them. And he proved through certain reasons, which I give, that all of them were in error. In fact, the Lord Robert, the Bishop of Lincoln, and Brother Adam Marsh, very great clerics of the world and perfect in divine and human wisdom, supported this same teaching. As a result, when certain impudent Franciscans on account of derision and temptation asked Brother Adam 'What is the agent intellect?' he replied: 'The raven of Elias', wishing through this to say that it was God or an angel. But he did not wish to explain because they asked not in order to gain wisdom, but in order to embarrass him (*Opus tertium* c. 23, 74–5. trans: partly modified- Hackett 1996).

Both passages are placed in the context of a discussion on the nature of the agent intellect and, immediately after, of angels. According to Bacon, all old wise men such as William of Auvergne, Grosseteste and Marsh rightly defended the opinion that the agent intellect is not a part of the soul, but God or an angel; he mentions a couple of university convocations held at Paris and lead by William on this subject. The jest by Marsh in the second of these passages refers to the Biblical event of Elias nourished by ravens, namely angels, sent him by God (1 Kings 17: 2–6). Now, if his remark is read together with the episode of the double university convocation, it implies that Grosseteste and Marsh were present at one of these events, on their way to Lyon.[14] If this were true, the subtle statement by Marsh would signify that he, a Friar Minor, could not openly manifest his thought, because it was contrary to the teaching of his Parisian confreres, such as John de la Rochelle, who was the regent master at that time. Adam was already an outstanding scholar, because, on his coming back from Lyon, he was requested for one of the two Franciscan chairs of theology, left vacant at Paris exactly in these days, after the death of both John de la Rochelle and Alexander of Hales.[15]

If, however, these passages are divided into two parts, they simply mean that Grosseteste and Marsh on one side and William of Auvergne on the other shared the same opinion as regards the agent intellect, at least according to Bacon. But this, in

[13] Grosseteste was at Lyon to defend his right to visitation against the canons of Lincoln Cathedral. See (Srawley 1955).

[14] This interpretation was firstly proposed by Hackett (1996). On the Parisian episode and for further bibliographical references see (Panti 2012).

[15] Grosseteste wrote to the provincial minister of England requesting that other brothers were appointed to that post. See (Panti 2012).

turn, would mean that Bacon, at some point in his life, and surely before 1259, the year of Marsh's death, attended a discussion between Adam and other Franciscans, likely a dispute held during Adam's teaching at the school of Oxford, in the biennium 1248–1250. Thus, if we exclude the idea that Adam's remark refers to the Parisian convocation, the third and last possibility of Bacon's meeting with Adam (though not necessarily with Grosseteste) took place at Oxford in 1248, when Bacon returned to England, after he gave up his post at Paris. As far as we know, he might have been among the students of the Franciscan school exactly at the time of Adam's regency, namely by the end of 1247 up to 1250. This would mean that Roger had been a novice since then, and that he remained in this status until about 1257, the year which is commonly posited as his entrance into the Order. This long lasting novitiate is not easy to justify, especially because in this decade Bacon was involved with his independent scientific research. However, it must be considered that many of the novices were *magistri* in search of a theological education, like Marsh himself had been. Besides, we know that Bacon was back in Paris in 1250/51 (*Opus maius*, vol. 1, 401), exactly when Marsh resigned his teaching. Bacon's alleged contact with Marsh and the Oxford Franciscan School, accordingly, lasted only for the two years of Adam's teaching. If this were true, it is possible to justify another remark of the *Opus tertium* concerning Adam: namely, how he tackled the problem of the movement of angels. In fact, subsequent to the second passage on the Parisian convocation quoted above, Bacon enters into details on the nature of angels and their being nowhere and immobile and, in developing his reasoning, he includes a query he asked Adam (*Et cum quaesivi*). Again, Bacon reports the direct answer by Marsh (*respondit quod*), a further remark pronounced by him (*dixit*) and, finally, expressly states that Marsh's teachings were in agreement with those of Grosseteste, his master:

> And when I asked a very wise man, namely Brother Adam Marsh, how it was possible that the soul of Blessed Ambrose attended the funeral of Saint Martin, he answered that the corporeal distance is nothing for the soul. (...) In fact, if spiritual beings have no relation with the divisions of the corporal distance, a demon burning in person in the Hell is not missing from any other place; why would he not operate those things that are permitted to him, such as inducing men to commit sins? Indeed, Brother Adam said: 'As two sentences are not physically distant according to their property, likewise two spiritual beings, such as a human soul here and an evil spirit in Hell'. (...) However, some < scholars > tickle themselves in these and other things, for they are induced to this not by the power of reason but by their imagination, enjoying falsities more than truths. Consequently, they vilify this claim, provided it is true. Indeed, it stood from the consent of old wise men, such as Brother Adam and his master Robert Grosseteste, and others (*Opus tertium* c. 50, 186–7).

Bacon is here referring to the story reported in *De miraculis Sancti Martini* (I, 5, 918C-919A) by (Gregory of Tours 1879) that Ambrose apparently fell asleep while he was celebrating Mass in Milan, but, indeed, was miraculously present at the simultaneous funeral of Martin at Tours. This long passage shows a sense of familiarity that Bacon had with Marsh at some point in his life, likely when the latter was a teacher. This closeness justifies also Bacon's knowledge of Marsh's involvement with the program of translations from the Greek by Bishop Grosseteste and the affinity of the theological methods of the two men.

The following scheme sets out the chronological details discussed above:

1227/28–1233: Bacon might have been at Oxford as student of Arts. He was in the classroom of the *maximus perspectivus*, who lectured on *Meteorologica* and on the nature of lunar rainbow.

1227–1230: last scientific writings by Grosseteste (*Computus correctorius, De iride*, *De lineis*, *De natura locorum*), all known to Bacon and quoted in his late works (after c. 1256).

1226: Adam Marsh incepts as Master of Arts at Oxford.

1229/30–1235: Grosseteste teaches theology at the Franciscan school of Oxford.

1233: Roger Bacon, *clericus de curia*, displays his wit before Henry III at Oxford.

1232/33: Adam Marsh enters the Franciscans; if he spent his novitiate at Oxford, he must have studied theology under Grosseteste (Bacon asserts that Adam was a pupil of Grosseteste).

1235: Grosseteste is elected bishop of Lincoln; Marsh makes his profession at Worcester.

1236/40–1247/48: Bacon is Master of Arts at Paris.

1235–1253: Grosseteste is Bishop of Lincoln, he introduces in England Greek books and speakers.

1244–46: Grosseteste and Marsh are at Lyon; Marsh is requested for a chair of theology at Paris, perhaps Bacon meets them at a university convocation at Paris.

1247–1250: Marsh is Master of Theology at the Franciscan school of Oxford.

1247/48: Bacon resigns his post at Paris and starts his involvement in experiments and knowledge of languages; he returns to England, perhaps attends Marsh's lectures at Oxford.

1250: Marsh stops teaching theology; Bacon is back in France.

1253: Grosseteste dies.

1257 c.: Bacon enters the Franciscan Order, perhaps at Paris.

1259: Marsh dies.

One particular statement by Bacon concerning Grosseteste and Marsh verifies two occasions of his closeness to them at Oxford. This passage is from the *Compendium studii philosophiae* (Bacon 1859c), written in 1271. Bacon mentions by name a few masters who taught at Oxford or were related to Grosseteste's circle there. His list puts them in an approximate chronological order that covers almost exactly the two periods of his supposed stays at Oxford: 1227/30–1233/38 and the biennium 1248–1250 (chronological details concerning those masters are in italics within square brackets):

> For forty years [*since 1230*] the secular clergy have neglected the study of theology and philosophy along the true paths of those studies (...) to such an extent that they have completely left the paths of the wise men, some of whom I have seen in my own time, namely the lord Robert, formerly bishop of Lincoln of holy memory, the Lord Thomas, Bishop of St. David in Wales [*third master at the Franciscan school at Oxford, from 1238*], Brother Adam Marsh, master Robert Marsh [*brother of Adam, incepted in theology in 1250 at the presence of Grosseteste*], masters William Lupus [*archdeacon at Lincoln and master in Law and Arts in the early fifties*] and William of Sherwood [*formerly Parisian

master of logic; since 1249 master at Oxford], and others like them (*Compendium studii philosophie* c. 5, 428).[16]

In this passage, Bacon witnesses the profound change in the methodology of university learning that had occurred gradually in the last decades, mainly at Paris, and had driven towards the refusal of the study of languages and experimental methods and the dismissal of the sacred texts, because of the introduction of the 'Sentence-Method' for the study of theology. According to him, Oxford preserved the seeds of the right model of learning as long as the interests of Grosseteste and Marsh remained alive. For Bacon, the new approach corresponds to the destruction of the Christian sense of wisdom and imposes the need for an overall reform of teaching and learning. This is actually the cultural context from which Bacon develops the urgency of re-uniting sciences to theology, the union of which constitutes a fundamental aspect of the proposal of reform that he illustrated to the pope in his later writings. It is in expressing this new demand that the references to Grosseteste and Marsh and their methodology are collocated. Hence, it is in this context that Bacon will introduce the case study of the rainbow as a bright exemplification of his and their claim of the importance of the sciences for theology.

8.3 Final Cause, Efficient Cause and the Bible: Roger Bacon and the Example of the Rainbow

As already mentioned, Bacon abandoned his university career in 1247 and, apart from the evidence of his return at Paris in 1250, the chronology of his later life remains tentative. As far as can be inferred from his writings, he might have spent time in both Oxford and Paris as an independent scholar, who invested money in books and assistants for his new enthusiastic engagement with sciences, experiments and languages (*Opus tertium*, c. 17, 59.). References to Grosseteste and Marsh begin in his later writings, written when he was a friar, after c. 1257. Since then, his social status had changed totally. He freely embraced poverty, stopped his independent engagement in secular studies and in 1264 even his family lost its rank and was brought to impoverishment and exile (Power 2013). Yet, his commitment to a reform of learning was gradually developing, and when, finally, he was in contact with Cardinal Gui Folques, the future Clement IV, the impulse that urged him to promote his novelties was conveyed in the writings that he addressed to the pope in 1268. In the *Opus maius* Bacon connected his reform to the eschatological context of a renewal of Christendom, while in the *Opus minus* and *tertium* he summarized the core issues of this enterprise.

For the present purpose, it is sufficient to note that in the *Opus maius*, after Bacon has evidenced (in the first part of the work) how to eradicate the main causes

[16] On Thomas of Wales see (Costambeys 2004), On Robert Marsh see (Lawrence 2006). On William of Sherwood see (Kretzmann 1966); see also (*Opus tertium*, c. 3, 14).

of errors that prevent scholars from following the right path of wisdom, he then turns to the connection between philosophy and theology, the highest science. Here, he remarks that all wisdom comes from God and has its root in God's Word, the Holy Scriptures, which explain the divine plan of creation and redemption (*Opus maius*, vol. 1, parts 1 and 2, 1–44). Human beings can grasp all wisdom contained in the Holy Scriptures with the help of philosophy and science, which are 'included' in the Bible as well, in so far as they offer literal explanations concerning natures and properties of all natural things created by God. Yet, after asserting that God 'posited all creatures' in the Bible and that philosophy is entirely within the Holy Scripture (*Opus maius*, vol. 1, pars 2, c. 8, 43), Bacon adds a very important claim, namely that the Bible openly illustrates only the final cause (*causa finalis*) of everything.

Scriptures tell us about every creature; nonetheless, they openly reveal only its final cause, namely what is its aim according to God's plan. The Bible, indeed, leaves the efficient cause, that is to say why something happens, to be dealt with by philosophy. Bacon chooses the rainbow as one of the infinite possible examples for illustrating his central assertion:

> Among infinite examples, it is sufficient here to mention the rainbow. Aristotle the philosopher impedes us from understanding anything worthy of knowledge because of his obscurities, and I do not wonder that Avicenna did not understand the nature of the rainbow. The reason for this is that the philosophers were ignorant of the final cause of the rainbow; and in their ignorance of the end, they are ignorant of those things that pertain to the end, because the end imposes a necessity upon those things, as Aristotle states in *Physics*, book 2. The final cause of the rainbow is the dispersal of watery humidity, as it is manifest in *Genesis*, so that every time a rainbow appears there is a dispersal of clouds in infinite drops (. . .). And this evaporation of water cannot be done by means of the rainbow, unless it happens through sunrays. In fact, infinite rays congregate by means of many reflections and refractions and their congregation is the cause of the dispersal and evaporation of waters, and consequently a rainbow is generated because of many reflections (. . .). Yet, the final cause of the rainbow is given in the book of *Genesis*, when it is said 'I will posit my bow on the clouds of the sky, so that there will be no more deluge on earth'. From this, it is possible to investigate the efficient cause and how a rainbow is generated, which are not appropriately known by the philosophers, at least according to what is manifested in their books. And the same < argument > runs as regards every other creature (*Opus maius*, vol. 1, pt. 2, dist. 1, c. 3, 108; Aristotle, *Physics* 2, 9, 200a7-10; Gen. 9: 12–15).

Bacon is explicit on the importance of the example of the rainbow: given that its *final* cause is the dispersal of humidity, as *Genesis* asserts, this explains why Aristotle and pagan philosophers could not properly demonstrate the *efficient* cause of the rainbow, namely because they did not know the final cause given in the Bible. In fact, while the final cause is openly given, the efficient cause, which depends on it, must be investigated by means of philosophy. This clarifies 'how' philosophy is within the Holy Scriptures and why philosophers investigate in vain without the knowledge of the final cause. In the case of the rainbow, the dispersal of humidity (final cause) is due to the modification of mist and clouds into an infinite amount of drops, caused in turn by the congregation of an infinite amount of solar rays multiplied through many reflections and refractions. For Bacon, the congregation of sunrays in the mist has the mechanical effect of forming drops, which are the sign that humidity is dissolving. This is the true and only efficient cause of the rainbow, which is an appearance of reflected rays on each drop under special conditions. Yet, this mechanism cannot be appropriately grasped unless the final cause itself is previously known.

Bacon underlines, in addition, that the explanation of the efficient cause needs in turn the help of mathematics and perspective. The efficient cause, in fact, is always due to a radial action, i.e. the action of species:

> Every efficient cause acts by its own power, which it produces in the adjacent matter, as the light of the sun produces its power in the air and this power is light [*lumen*] diffused through the whole world from the light [*lux*] of the sun. This power is called similitude, image, species, and many other names, and it is produced by substance as well as by accident, spiritual as well as corporeal. (…) This species produces every action in the world, for it acts on sense, intellect, and all the matter in the world for the generation of things (*Opus maius* vol. 1, pt. 4, dist. 2, c. 1, 111).[17]

The well-known Baconian theory of *species* or immaterial rays, derived in turn from Grosseteste's *De lineis* and Alkindi's *De radiis*, implies that every efficient cause acts by its own power. All species are invisible except sunrays, which emanate from the sun in straight lines by infinite auto-multiplication of themselves (*multiplicatio specierum*). Consequently, the efficient cause has always a mathematical explanation, given the geometrical behaviour of the multiplied rays and species.[18] However, it may happen that mathematics needs, in turn, the help of other sciences to suit the explanation of a given phenomenon. This is, again, the case of the rainbow: the *scientia experimentalis* is necessary for solving specific geometrical problems, which necessitate a special arrangement of observations for a plain justification of the phenomenon itself. The experimental science supplies mathematics with specific experiences and practical figurations. Bacon mentions in his *Opus minus* (Bacon 1859b) a master dealing with these experiments on rainbows:

> However, experimental science must be mastered appropriately, at least because that problem of the rainbow and coloured circles has a difficulty as regards geometry; namely, the basic principle < of the phenomenon > and what comes after those examples are plainly described by means of a huge wisdom. Surely, those things concerning the rainbow and the coloured circle < of the Moon > kept me busy for a month before I could understand them by means of < geometrical > figures and experiments (…). And I am sure that nobody among the Latins but one, who is the most sapient of them, could correctly understand this subject (…) by means of wonderful experiences, which must be done not only during the day, but also by night, as regards the Moon rainbow and the coloured circles around it (*Opus minus* 317).

In this remark, Bacon underlines that the problem of the rainbow cannot be solved by geometry alone, but needs the support of appropriate experiences.[19] Only one sapient man could fulfil these last requirements and, on this occasion, Bacon is not thinking of Grosseteste, who was a mathematician and *perspectivus*, but Peter of Maricourt (*Petrus Peregrinus*), the highest authority in experimentations (*dominus experimentorum*) and author of the *Epistula de magnete* (Nagel 2012).

---

[17] On Bacon's theory of species, its sources and application see (Lindberg 1983).

[18] See (Grosseteste 1912a). On the geometrical basis of Bacon's physics see (Panti 2014).

[19] On the *exemplum* of the rainbow as a case study for experimental science, see (Hackett 1996).

Here, therefore, he is not speaking about the teacher of *perspectiva*, whose lessons he attended in his youth, as also the *Opus tertium* clearly confirms:

> (…) only experience, and not demonstrations, certifies this < as regards the rainbow>. Therefore, I posit the root of these experiences, which nobody among Latins can understand but one, namely Peter < of Maricourt > (*Opus tertium* c. 11, 37).

It is evident that the example of the rainbow is crucial for Bacon, who posited its solution in a wide framework that includes the position of philosophy within the Holy Scriptures and its divisions into different disciplines, such as mathematics, optics and experimental science, within the new organization of wisdom and sciences (Hackett 2012).

In the *Opus maius*, the setting of this framework goes on with a long and detailed section on how philosophical wisdom has been inserted within the divine project of creation and redemption from the beginning of the world: it was entirely revealed to the Patriarchs and Fathers, and even pagans and ancient philosophers received traces of it. This is the subject of the remaining sections of part 2, while part 3 of the *Opus* deals with languages, which are the first step in the progression towards wisdom and the fundamental instrument for reading appropriately the Holy Scriptures (*Opus maius*, vol. 1, xcvii–ci). With part 4, Bacon turns to mathematics, the second step to theology and the foundation stone of all other sciences; here, the argument of the rainbow is considered once more, and in this occurrence, the problem under discussion will be, again, its efficient cause. It is useful to reflect on how Bacon introduces it in the course of part 4 of the *Opus maius*.

Species, rays and their geometrical behavior are the core argument of a long discussion, which is intermingled with numerous examples (given in part 4, distinction 4) concerning apparently inexplicable effects of natural events. Several of these exemplifications, for instance those on climates and heat, are based on other sources, also on Grosseteste's late scientific works (*De lineis*, *De natura locorum*), which Bacon quotes without mentioning by name their author. Yet, after these examples, Bacon turns (in part 4, distinction 5) to a new set of practical exemplifications on the utility of mathematics in theology. The point of view is now the direct observation of natural events and their being narrated in the Holy Scriptures. After a quick recall of how the Bible pre-contains philosophy, and consequently mathematics, Bacon repeats his claim that the ancient Fathers and wise men knew mathematics and sciences. Here, eventually, we find one of the statements on the *viri famosissimi*, Grosseteste and Marsh:

> There were very famous men, such as Robert Grosseteste Bishop of Lincoln and Friar Adam Marsh and many others, who knew how to unfold the causes of all things by the power of mathematics and to illustrate appropriately both human and divine wisdom. The proof of this is manifest in the writings of those men, such as *On impressions*, *On rainbows*, *On comets*, *On the generation of heat*, *On the investigation of places of the earth*, *On heavens* and others, which both philosophy and theology employ (*Opus maius* vol. 1, pt. 4, dist. 1, c. 3, 108).[20]

[20] On the attribution of these works see (Thomson 1940; Panti 2001).

This claim not only illustrates the importance of science for theology as a specific aspect of Grosseteste's and Marsh's legacy and as part of the cultural reform that Bacon is presenting, but also works as a sort of 'bibliographical footnote' to the sources employed in the preceding scientific exemplifications. This claim, frequently cited for attesting the deference that Bacon had towards Grosseteste and Marsh and their methodology, is therefore inserted within a long discussion explaining that, without a profound knowledge of mathematics, the theologian cannot grasp the literal sense of Scriptures. Specifically, the exegete ignorant of mathematics cannot understand the right efficient cause of a phenomenon described in the Bible, the final cause of which is, however, openly expressed to the reader.

Bacon deplores the fact that Grosseteste's and Marsh's attitude has no more followers in his time: this matches perfectly also with the passage from the *Compendium studii philosophiae* cited above (end of § 1; *Compendium studii philosophie* c. 5, 428), explaining how their method had been cultivated at Oxford and how he himself adhered to it. However, Bacon does not say *how* the two masters applied sciences to the *sacra doctrina* and, as far as his subsequent examples testify, it should be questioned whether Bacon's *own* employment of science was really in line with that of those wise men.

The *Opus tertium* will condense Bacon's long line of reasoning within a single passage, which, nonetheless, contains what the *Opus maius* had expanded on and articulated; starting with the example of the rainbow:

> Philosophy makes nothing but explain natures and properties of natural things, which are contained in the Holy Scriptures from the highness of the heavens down to their <lowest> boundaries, and of all artifacts and moral claims, as I declare there [in the *Opus maius*] by means of the example of the rainbow. Consequently, this is the proper way of knowing the Holy Scriptures, and the way of the blessed men and ancient sages, such as the Bishop of Lincoln and Brother Adam, and others; so that, in such a way, the complete wisdom of philosophy is known in the divine book (*Opus tertium* c. 24, 82).

Thus, turning again at the *Opus maius*, Bacon goes on with offering effective examples of how mathematics is useful for knowing the world of creation, and lists seven ways of illustrating how a correct literal exegesis of the Bible should be conducted. Then, he develops two exemplifications that make clear how the final cause of a phenomenon is given in the Bible and how the corresponding efficient cause of this phenomenon can be individuated. The first example concerns Genesis 9: 14–15 and deals with the rainbow, the second is the exegesis of Ecclesiasticus 43: 4 (*tripliciter sol exurit montes*), namely why the sun is a threefold cause of heat on mountains (Panti 2016). Here is Bacon's explanation as regards the rainbow:

> From what has been said, it follows that the divine bow is ordered against deluge and abundance of waters. Therefore, it is necessary that, as long as this bow appears in the sky, there is an effective dissipation of watery humidity; and this is true. In fact, clouds are abundantly dissipated and there is persistent rain (...). However, the dispersal of watery humidity cannot happen but because of something having the capacity of consuming, and in the generation of a rainbow we find nothing except sunrays and clouds. The congregation of clouds is the material cause; accordingly, the diffusion of rays is the efficient cause. Nonetheless, incident rays [*i.e. direct rays*] cannot perform great and amazing operations,

> because they do not intersect each other. On the contrary, a convergence of < their > force is required for this, namely for educing an effective operation. Yet, a convergence cannot happen except by reflection and refraction < of rays>. Consequently, it is necessary that the rainbow be generated by means of infinite reflections and refractions in infinite drops falling down continuously, so that the truth of both < its > colours and figure be revealed by means of those multiplications according to < geometric > figures, angles and lines, and not by means of the diversity of the material (*materiae*) of the cloud, as it is written in the text of the Latins and as everyone believes, as I will demonstrate with valid experimentations (*Opus maius* vol. 2, pt. 6, 213).

Convergence of rays is due to their reflection, namely the coming back of the ray falling upon a reflecting surface, and to their refraction, namely the bending of the ray entering within a transparent body such as water, mist and clouds. This convergence, therefore, occurs in connection with an enormous multiplication of rays and, consequently, provokes the consumption of humidity, which 'condenses' in a countless amount of drops. Geometry and perspective demonstrate how such a multiplication and convergence happen and work, according to 'lines, angles and figures' as Bacon states recalling Grosseteste's homonymous *De lineis angulis et figuris* (Grosseteste 1912a). In addition to geometry, however, appropriate experiences show how the required effect happens under specific climatic circumstances, geographic locations and sight conditions.

In conclusion, the example of the rainbow, as well as that of the heat on mountains, are proposed by Bacon in connection with his recalling Grosseteste's and Marsh's writings and teaching as illustrations of how philosophy and sciences explain the literal sense of the Holy Scriptures: specifically, they exemplify how to look for the efficient cause of natural phenomena, the final cause of which is openly expressed in the Bible. This, for Bacon, is the proper exegetical way suggested by wise men of past generations.

8.4 The Efficient Cause of the Rainbow: Grosseteste's *De iride* and Its Use in Bacon's *Opus maius*

The treatise *On the Rainbow* (*De iride*) is one of Grosseteste's scientific *opuscula* 'useful for philosophy and theology' listed by Bacon. The Franciscan friar demonstrates his direct and deep knowledge of this openly praised treatise, in which, for the first time in medieval science, it is held that the rainbow is due to the refraction of solar rays upon mist and a cloud. Grosseteste justifies his tenet by means of the science of perspective, the nature and scope of which are clarified at the beginning of the treatise (Grosseteste 1912b). For Grosseteste, experts in optics (*perspectivi*) follow a scientific method different from that of natural philosophers (*physici*). The latter examine how (*quia*) something happens through the connection of secondary causes and factual observations, while the former know why (*propter quid*) it happens by investigating the efficient cause, i.e. the immediate cause of that phenomenon. Grosseteste says that Aristotle looked for the *quia*, while he wants to investigate the *propter quid* of the rainbow and, in doing so, he must consider the emission of both visual and luminous rays, which is the object of the science of perspective.[21]

This science, Grosseteste continues, is divided into three parts, according to the behavior of the ray. Direct ray is the object of the science of vision (*de visu*), reflected ray of the science of mirrors (*de speculis*) while refracted ray of a science that 'has remained untouched and unknown among Latins until the present time'. For Grosseteste, however, it is precisely this last component of *perspectiva* that explains the efficient cause of the rainbow, and this is done by means of mathematics, specifically of geometry, which measures the angular bending of the radial line (*species visibilis*) in its crossing different diaphanous media (*De iride*, 72–3).

The search for the efficient cause drives Grosseteste to criticize the Aristotelian solution to the problem, namely that the rainbow is due to the reflection of light upon a cloud, and toward assuming that the phenomenon is due to light refraction.

> Now a rainbow cannot be produced by means of solar rays passing in a straight line from the sun and falling into the concavity of a cloud, for they would make a continuous illumination in the cloud not in the shape of a bow, but in the shape of the opening on the side towards the sun through which the rays would enter the concavity of the cloud. Nor can a rainbow be produced by the reflection of the rays of the sun from the convexity of mist descending from the cloud as from a convex mirror, in such a way that the concavity of the cloud may receive the reflected rays and thus a rainbow appears, because if that were so the shape of all rainbows would not be an arc (*De iride*, 75–6; trans: Lindberg 1966).

For Grosseteste, the rainbow happens because of a threefold refraction:

> I maintain that the outside of a cloud is convex and the inside concave (...) and since the mist descends from the concavity of a cloud, it must be pyramidally convex at the top, descending to the earth, and therefore more condensed near the earth than in the higher part. Therefore there are altogether four transparent media through which a solar ray penetrates: <first> pure air containing the cloud; second, the cloud itself; third, the higher and rarer mist coming from the cloud; and fourth, the lower and denser part of the same mist. Therefore, (...) solar rays must be refracted first at the interface between the air and the cloud and then at the interface between the cloud and the mist. By these refractions the rays converge in the density of the mist and, being refracted there again as from the vertex of a pyramid, spread out not into a round pyramid, but into a figure like the curved surface of a round pyramid expanded opposite the sun (*De iride*, 76, Lindberg 1966).

As we have seen, Bacon pays open tribute to Grosseteste's *De iride* for its capacity to explain the efficient cause through mathematics (*Opus maius* vol. 1, pt. 4, dist. 1, c. 3, 108: *per potestatem mathematice...causas omnium explicare*), exactly because mathematics explicates the behavior of species or rays, as examined above (§ 2). Moreover, in a passage from the *Compendium studii philosophiae* (*Compendium studii philosophie*, c. 8, 469), Bacon praises Grosseteste for his

[21] The edition of *De iride* by Baur (Grosseteste 1912b) needs, in this passage (p. 72), an important correction, namely *quia* instead of *quid*. The right reading is given by the majority of the mss, among which the oldest ones: Madrid, 3314, f. 90r; Vatican Libr., Barb. lat. 165, f. 403. The difference between a knowledge *quia* and *propter quid* is given by Grosseteste in his commentary on the *Posterior Analytics* (Grosseteste 1981), at p. 189. On this subject see (Rossi 1996).

capacity for grasping the cause of every phenomenon 'by means of independent experiences, alternative sources and special sciences (*per experientiam propriam, et auctores alios, et per alias scientias*) a hundred times better than what the inaccurate Latin translations of Aristotle can offer'. However, despite Bacon's praise of all these aspects that makes the *De iride* an example to follow, in the *Opus maius* he openly and resolutely criticized Grosseteste's explanation of the rainbow.

To my knowledge, no modern scholar has paid adequate attention to why Bacon rejects the reasons put forward by Grosseteste after having commended his method. Both Carl Boyer and David Lindberg skip the problem in their respective inquiries into Grosseteste's and Bacon's theory of the rainbow. The former dismisses Bacon by labeling his theory 'retrogressive' and asserting that Bacon's lengthy explanation in the *Opus maius* is surely not better than that of Grosseteste. The latter, on his own, wonders how Bacon could have been 'so stupid as to reject obvious progress', and tries to justify why the English friar 'rejected Grosseteste's forward-looking appeal to refraction' and returned to a reflection theory which was reminiscent of the Aristotelian theory (Boyer 1954; Lindberg 1966; Crombie 1953; Eastwood 1966; Hackett 1998).

In my opinion, the answer must be looked for in Bacon's claims on the difference between the final and the efficient causes of a phenomenon and on how they are included in the Bible, as examined in the paragraph above. Given that the Bible only expresses the final cause of every phenomenon, the efficient cause, given by mathematics and its applications to nature, must justify it. As we have seen, for Bacon *reflections and refractions* of rays provoke, through their convergence, the dispersal and consumption of humidity by creating innumerable drops. Consequently, the rainbow must be only an appearance, a sign that drops are everywhere. Contrary to this, Grosseteste's explanation insists on the fact that the rainbow is a real thing, caused by rays entering the cloud and the mist. This, accordingly, would imply the stable presence of humidity, and not its dispersal.

In the long discussion on the rainbow proposed by Bacon in the *Opus maius* part 6, on experimental science (*Opus maius*, vol. 2, pt. 6), several chapters are devoted to experiences proving that the sun, the observer and the rainbow are tightly connected as regards every aspect of this phenomenon. Bacon points out, as Avicenna had already evidenced, that each observer sees a different bow according to his terrestrial location and movement, as well as according to the altitude and movement of the sun. It is evident, therefore, that there are as many rainbows as observers. This ability to move with the observer, Bacon continues, is not characteristic of images formed by refraction. A further observation supports the same point: crystalline stones produce colors by refraction and these colours are similar to those of the rainbow, but, differently from it, are located in the same place for all observers. There are, in conclusion, as many rainbows as observers, and each bow moves in correspondence to the movement of the sun. For Bacon, therefore, the rainbow is only a visual appearance, completely different from the iris in crystals:

> If it is said that solar rays passing through a crystal produce real and fixed colours, which produce a species and have objective reality, we must reply that the phenomena are different. The observer alone produces the bow, nor is there anything present except

> reflection. In the case of the crystal, however, there is a natural cause, namely, the ray and the corrugated stone, which has great diversity of surface, so that a diversity of colours results according to the angle at which the light falls (*Opus maius*, vol. 2, pt. 6, c. 8, 191–192).

In the case of the rainbow, in fact, colours are due ‘to the humours and colours of the eye’, for these colours ‘exist only in appearance’ being an effect caused by the roundness of each drop, acting as a convex mirror (*Opus maius*, vol. 2, pars 6, c. 10, 193 and c. 12, 197).

For Bacon the rainbow is or, better, *must* be just an optical effect, which accompanies the dispersal of humidity (final cause) into drops, due, in turn, to the convergence of sunlight (efficient cause) infinitely multiplied through reflections and refractions upon the clouds. The rainbow, in his account, is produced when sunlight falls on small drops ‘infinite in number’ and from each of these raindrops reflection occurs as from a spherical mirror, and since they fall without interval they seem from a distance to be continuous. Therefore the image of the sun seems continuous and not multiplex according to the multitude of drops (Lindberg 1966). The resulting rainbow fails to look like the sun, because small spherical mirrors distort size and shape and produce the appearance of colours where no colours really exist. Proof that they are only an appearance is their variability in location for different observers. The rainbow is produced from a different set of drops for each observer: as an observer moves, different drops successively serve as reflectors, so that the reflecting surface moves together with the observer. Sunrays penetrate and converge everywhere in the entire mist, but only drops opposite the sun and having the required distance and orientation reflect sunlight back to the observer (Ibid.). Hence, Grosseteste’s solution, which considers the rainbow as a real thing caused by a triple refraction of solar rays entering a cloud and the moisture beneath it with a mechanism similar to refraction within the crystal, is totally wrong, as Bacon openly declares (the passages quoted verbatim from the *De iride* are in italics):

> In the same way those < scholars are wrong > in saying that the rainbow is caused by refraction, so that it has *a figure like the curved surface of a round pyramid expanded opposite the sun*, for the reason that, for them, *it assumes the shape of an arc. And since the vertex of the aforementioned figure*, as they assert, *is near the earth and it is expanded opposite the sun, half the figure or more must fall on the surface of the earth and the remaining half or less onto a cloud opposite the sun.* But this is rejected, because it is proved that the rainbow cannot generate by means of refraction (...). They say, in fact, that *solar rays must be refracted first at the interface between the air and the cloud and then at the interface between the cloud and the mist. By these refractions the rays converge in the density of the mist*, in fact the more dense part is lower, being heavier; *and, being refracted there again as from the vertex of a pyramid, spread out not into a round pyramid*, as they say, *but into a figure like the curved surface of a round pyramid expanded opposite the sun.* It is evident, therefore, that they obtain this figure because of three refractions. However, in sprays there cannot be three but only one < refraction>, nonetheless the same figure occurs, such as that in the sky; thus, it is not the cause of this refraction. Why, finally, will the rays not form a round pyramid, but a figure like *the curved surface of a round pyramid*? For the latter is not in accord with the law of refraction, since refraction must produce a regular round pyramid (*Opus maius* vol. 3, pt. 6, c. 11, 195) (*De iride*, 77).

If the rainbow were produced by refracted rays within a cloud, it would be ‘a thing fixed in one place in the cloud, which would vary neither with the motion nor with the number of observers', exactly as it happens in the crystal. And this, in turn, would mean that no dispersal of humidity is occurring.

As Lindberg noted, Bacon does not reject the possibility of refraction by a cloud, for he treats the cloud as a lens in explaining the halo of the Moon in the *Opus majus* (Lindberg 1966). Bacon assumes that rays from the sun are refracted by a spherically shaped mist and, later in the *Opus tertium*, he asserts that refraction occurs even in individual drops. Thus, Lindberg underlines, when it made sense to use refraction as an explanatory principle, Bacon not only did so willingly, but went beyond Grosseteste to consider refraction in individual drops (Ibid.).

Still according to Lindberg, 'it is one thing to show the absurdity of the refraction theory of the rainbow; it is quite another to establish the reflection theory by showing its greater ability to account for the important phenomena, and Bacon was far more successful at the former than at the latter' (Ibid.). Now, if even Bacon's solution was far from being plain and clear - and, in fact, Bacon needs both geometry and experimentations to figure out a quite satisfactory explanation - why was he so resolute in denying every aspect of Grosseteste's theory? Could not he try to use, at least in part, the hypothesis of the refraction, as Theodoric of Freiberg did a few decades later? In my opinion, the answer to these questions lies in what Bacon had previously asserted on the relationship between the final and the efficient causes. Since the final cause is indubitable because it is openly expressed in the Bible, as seen above, the efficient cause must justify it. Thus, if the efficient cause is the convergence of rays, causing consumption of humidity into drops, the rainbow cannot be another effect; it must be, indeed, a *sign*, an *appearance* of the effect. This appearance is due to reflection of rays under special conditions, while refraction is excluded for the formation of the bow, given that the bending of the rays requires the permanence of the cloud and the moisture, similarly to what happens in the iris within a crystal.

As a matter of fact, neither the *maximus perspectivus*, be he Grosseteste or not, nor the *sapientissimus experimentator* Peter of Maricourt can explain properly the phenomenon of the rainbow unless their competences are joint and unless they look previously to the Bible, in search of the final cause.

8.5 Conclusions

In looking at how Roger Bacon refers to Grosseteste and Marsh, it emerges that his remarks are detailed and elusive at the same time. They are detailed as far as Bacon demonstrates knowledge of Grosseteste's scientific works and philosophical claims by Adam, referred to through the reconstruction of his oral teachings. Yet these allusions are elusive as far as they intend to qualify both masters as leaders of a tradition of thought that is, conversely, Bacon's own. In this respect, both wise men are 'icons' for corroborating Bacon's claims as regards the use of philosophy and science in theology. As we have seen in the case of the rainbow, Bacon considers Grosseteste a valid methodological guide for the relevance that he attributed to the science of perspective in the solution of specific natural problems, such as that of the rainbow. For both scholars optics is pivotal in understanding the efficient cause of every phenomenon, always explained in terms of the geometry of radial forces. However, for Bacon this science is fundamental in a broader and deeper sense, namely because its correct use gives evidence also to the final cause of the observed phenomenon, the aim of which is openly given in the Bible. Thus, Bacon does not hesitate in criticizing Grosseteste's solution, as it happens in the case of the rainbow, if it is in contrast with this last requirement.

The *De iride* by Grosseteste had a considerable diffusion, and in one of the earliest manuscripts that transmit it, the Vatican, Barb. Lat 165 (late thirteenth century), it is immediately followed by an addendum that develops further Grosseteste's theory of threefold refraction of sunlight by means of 'experiments and reasons' (*experimento et ratione*). The addendum insists on the role of the raindrops and proposes experiments to figure out why the motion of the observer is accompanied by an identical motion of the rainbow, which is apparently inexplicable if Grosseteste's solution is accepted, as Bacon has evidenced. Both the parallel movement of rainbow and observer and the role of raindrops, as we have seen, are fundamental for Bacon's rejection of Grosseteste's idea. Besides, this manuscript transmits also the *De fluxu maris*, a treatise on tides formerly attributed to Grosseteste or even to Marsh, but now firmly ascribed to Adam of Exeter, a pupil of Grosseteste, dear friend of Marsh and he himself a renewed Master of Arts, who wrote it *in scolis*, before entering the Franciscan Order at Oxford in about 1230 (Panti 2013). This addendum, which Greti Dinkova-Bruun and I are presently studying and editing, and the context of its manuscript transmission reinforce Bacon's claims concerning the teaching of the *perspectiva* at Oxford. Hence, they attest that the problem of the rainbow as well as those of tides, climatic variations and other problems of natural philosophy largely discussed by Bacon in his later writings, were part of the teaching curriculum in Arts and were, in one way or another, connected to Grosseteste and his circle of pupils and friends that Bacon, in his youth, likely met. By the way, one can easily notice that in his criticism of Grosseteste's theory of rainbows, Bacon uses the plural ('those who say', etc., see *Opus maius* vol. 3, pt. 6, c. 11, 195). Is it possible that he knew the defense of the refraction theory presented in the addendum, and that this very text is a proof that Grosseteste's theory was discussed in the Oxford school at the presence of Bacon? These questions need a further investigation that a chronology, as seen above, may help to answer, if corroborated by a deeper study of the texts.

However, Bacon was conscious that the *De iride* by Grosseteste was exactly the kind of scientific work fundamental for conducting a reliable literal exegesis of the Bible according to the criteria that he has formulated, and, not surprisingly, he mentions this work among the other useful writings by Grosseteste. Both Grosseteste and Bacon considered theology to be the study of the Bible; yet, they differed in applying mathematics and natural philosophy to the divine science. For the former, the sciences and above all the study of light offered the possibility to develop images, examples and allegories useful in enlightening the most complex tenets of the Christian faith. For Bacon sciences have much more space within the theological discourse than for Grosseteste. Mathematics, experimental science and all other sciences are the intellectual instruments that allow human beings to verify the profound correspondence between the laws governing the natural world and the divine project of creation as revealed in the Scriptures.

References

Aristotle. (2000). *Meteorology in the Arabico-Latin tradition* (P. L. Schoonheim, Ed.). Leiden: Brill.

Bacon, R. (1900). *Opus maius* (J. H. Bridges, Ed.). London: Williams & Norgate.

Bacon, R. (1859a). *Opus tertium* (J. S. Brewer, Ed.). London: Longman.

Bacon, R. (1859b). *Opus minus* (J. S. Brewer, Ed.). London: Longman.

Bacon, R. (1859c). *Compendium studii philosophie* (J. S. Brewer, Ed.). London: Longman.

Bacon, R. (1926). *Computus* (R. Steele, Ed.). Oxford: Clarendon.

Bacon, R. (1940). *Communia Mathematica* (R. Steele, Ed.). Oxford: Clarendon Press.

Bacon, R. (Attributed). (1937). *Liber de sensu et sensato* (R. Steele, Ed.). Oxford: Clarendon.

Boyer, C. B. (1954). Robert Grosseteste on the rainbow. *Osiris, 11*, 247–258.

Catto, J. I. (1968). New light on Thomas Docking. *Medieval and Renaissance Studies, 6*, 135–149.

Catto, J. I. (1984). Theology and theologians. 1220-1320. In J. I. Catto & R. Evans (Eds.), *The history of the University of Oxford: The early Oxford Schools* (Vol. 1, pp. 471–518). Oxford: Oxford University Press.

Costambeys, M. (2004). Thomas Wallensis. In *Oxford dictionary of national biography*. Oxford: Oxford University Press.

Crombie, A. C. (1953). *Robert Grosseteste and the origins of experimental science 1100-1700*. Oxford: Clarendon Press.

Eastwood, B. S. (1966). Robert Grosseteste's theory of the rainbow. *Archives internationales d'histoire des sciences, 19*, 323–324.

Ginther, J. R. (2004). *Master of the sacred page: A study of the theology of Robert Grosseteste, c. 1229/30–1235*, Aldershot: Ashgate.

Goering, J. (1995). When and where did Grosseteste study theology? In J. McEvoy (Ed.), *Robert Grosseteste. New perspectives on his thought and scholarship* (pp. 96–130). Turnhout: Brepols.

Gregory of Tours (Gregorius Turonensis). (1879). *De miraculis Sancti Martini Episcopi libri quatuor*. Patrologia Latina 71. Paris: J. P. Migne.

Grosseteste, R. (1981). *Commentarius in Posteriorum Analyticorum libros* (P. B. Rossi, Ed.). Florence: Olschki.

Grosseteste, R. (1912a). De lineis angulis et figuris. In L. Baur (Ed.), *Die philosophischen Werke des Robert Grosseteste, Bischofs von Lincoln* (pp. 59–65). Münster im W: Aschendorff.

Grosseteste, R. (1912b). De iride. In L. Baur (Ed.), *Die philosophischen Werke des Robert Grosseteste, Bischofs von Lincoln* (pp. 72–78). Munster im W: Aschendorff.

Hackett, J. (1996). Scientia experimentalis. From Robert Grosseteste to Roger Bacon. In J. McEvoy (Ed.), *Robert Grosseteste. New perspectives on his thought and scholarship* (pp. 89–120). Turnhout: Brepols.

Hackett, J. (1997a). Roger Bacon on Scientia experimentalis. In J. Hackett (Ed.), *Roger Bacon and the sciences: Commemorative essays* (pp. 277–316). Leiden: Brill.

Hackett, J. (1997b). Roger Bacon: His life, career and works. In J. Hackett (Ed.), *Roger Bacon and the sciences* (pp. 9–23). Leiden: Brill.

Hackett, J. (1998). Experientia, Experimentum and perception of objects in space: Roger Bacon. In J. Aertsen & A. Speer (Eds.), *Raum und Raumworstellungen im Mittelalter* (pp. 101–120). Berlin: Walter De Gruyter.

Hackett, J. (2012). Roger Bacon and the moralization of the science: From Perspectiva through Scientia experimentalis to Moralis philosophia. In Enrico Menestò (Ed.), *I francescani e le scienze. Atti del XXXIX Convegno internazionale. Assisi, 6–8 ottobre 2011* (pp. 369–392). Spoleto: Fonazione CISAM.

Knorr, W. R. (1990). John of Tynemouth alias John of London: Emerging portrait of a singular medieval mathematician. *The British Journal for the History of Science, 23*, 293–330.

Knorr, W. R. (1996). «Falsigraphus vs. adversarius»: Robert Grosseteste, John of Tynemouth, and Geometry in 13th-Century Oxford. In M. Folkerts (Ed.), *Mathematische Probleme im*

Mittelalter. Der lateinische und arabische Sprachbereich (pp. 333–359). Wiesbaden: Harrassowitz.

Kretzmann, N. (1966). *William of Sherwood's introduction to logic*. Minneapolis, MN: University of Minnesota Press.

Lawrence, C. H. (2006 & 2010). *The letters of Adam Marsh* (2 vols). Oxford: Clarendon.

Lindberg, D. C. (1966). Roger Bacon's theory of the rainbow: Progress or regress? *Isis, 57*, 235–248.

Lindberg, D. C. (1983). *Roger Bacon's Philosophy of nature, De multiplicatione specierum, De speculis comburentibus, Critical Edition, English Translation, Introduction and notes*. Oxford: Clarendon Press.

Little, A. G. (1943). Thomas Docking. In A. G. Little (Ed.), *Franciscan papers, lists and documents* (pp. 98–121). Manchester: Manchester University Press.

McEvoy, J. (1982). *The philosophy of Robert Grosseteste*. Oxford: Clarendon Press.

Nagel, S. (2012). Pietro Peregrino, il sapiens-simplex eccellente di Ruggero Bacone. In *I francscani e le scienze* (pp. 21–47).

Panti, C. (2001). *Moti, virtù e motori celesti nella cosmologia di Roberto Grossatesta. Studio ed edizione dei trattati «De sphera», «De cometis», «De motu supercelestium»*. Florence: SISMEL-Edizioni del Galluzzo.

Panti, C. (2012). Scienza e teologia agli esordi della scuola dei Minori di Oxford: Roberto Grossatesta, Adamo Marsh e Adamo di Exeter. In E. Menestò (Ed.), *I Francescani e le scienze. Atti del XXXIX Convegno internazionale. Assisi, 6–8 ottobre 2011* (pp. 311–351).

Panti, C. (2013). Robert Grosseteste and Adam of Exeter's physics of light. Remarks on the transmission, authenticity and chronology of Grosseteste's Scientific Opuscula. In J. Flood, J. R. Ginther, & J. W. Goering (Eds.), *Robert Grosseteste and his intellectual milieu. New editions and studies* (pp. 165–190). Toronto, ON: PIMS.

Panti, C. (2014). Natural continuity and the mathematical proofs against indivisibilism in Roger Bacon's *De celestibus (*Communia Naturalium, II). In P. Bernardini & A. Rodolfi (Eds.), *Roger Bacon's* Communia Naturalium. *A 13th century philosopher's workshop* (pp. 159–190). Florence: SISMEL-Edizioni del Galluzzo.

Panti, C. (2016). The theological use of science at early Oxford Franciscan School: Thomas Docking, Roger Bacon and Robert Grosseteste's works. In M. Robson & P. Zutshi (Eds.), *St Francis and his followers: Studies in memory of Dr John Moorman*. London: Taylor and Francis. (In print).

Paris, M. (1876). *Chronica Majora*, vol. 3 (H. Richards Luard, Ed.). London: Longman

Power, A. (2013). *Roger Bacon and the defence of Christendom*. Cambridge: Cambridge University Press.

Rossi, P. (1996). Robert Grosseteste and the object of scientific knowledge. In J. McEvoy (Ed.), *Robert Grosseteste: New perspectives on his thought and scholarship* (pp. 53–75). Turnhout: Brepols.

Southern, R. W. (1986). *Robert Grosseteste: The growth of an English mind in medieval Europe*. Oxford: Oxford University Press. (rpt. 1992).

Srawley, J. H. (1955). Grosseteste's administration of the diocese of Lincoln. In D. A. Callus (Ed.), *Robert Grosseteste: Scholar and Bishop* (pp. 171–177). Oxford: Clarendon Press.

Tachau, K. H. (2014). Scholastics on colors in the mind, colors in the world. In *Forme e oggetti della conoscenza nel XIV secolo. Studi in ricordo di Maria Elena Reina* (pp. 337–370). Florence: SISMEL Edizioni del Galluzzo.

Thomson, S. H. (1940). *The writings of Robert Grosseteste, Bishop of Lincoln (1235–1253)*. Cambridge: Cambridge University Press.

Chapter 9
Laying the Foundation for the Nomological Image of Nature: From Corporeity in Robert Grosseteste to Species in Roger Bacon

Yael Kedar

9.1 Introduction

Since the seventeenth century, the expression 'laws of nature' had become an essential element in the conceptual vocabulary of modern science. Kepler's laws of planetary motion, Galileo's law of falling bodies, and Newton's gravitational, inverse-square law are considered hallmarks of the scientific revolution. Historians and philosophers of science regard the concept of law constitutive of the structure and premises of early modern science, yet the questions of its origin and development are still moot.

Without entering into the question of what exactly a law of nature is,[1] it seems a matter of course that any attempt to find and formulate such laws assumes that nature is governed by laws. This assumption constitutes, in Catherine Wilson's phrasing, a 'nomological image of nature' (Wilson 2008). In this chapter I examine the nomological image, as it was developed by two thinkers of the thirteenth century, namely, Robert Grosseteste and Roger Bacon.

The nomological image of nature assumes that the explanatory terms of natural phenomena are universal, necessary and impersonal, describing neither the actions of the individual nor those of specific actors, but rather the factors linking objects and processes into a whole. This image of nature expresses unity and order,

[1] The literature on the question of the definition, validity and characteristics of laws of nature is extensive, and goes back to the problem of induction raised by Hume. The prominent scholarly works in the contemporary debate include, among others, Armstrong (1983), Van Frassen (1989), Giere (1999), Ward (2002), Mumford (2004), Maudlin (2007), Lang (2009).

Y. Kedar (✉)
University of Haifa, Haifa, Israel

Tel-Hai College, Qiryat Shmona, Israel
e-mail: yraizman@gmail.com

J.P. Cunningham, M. Hocknull (eds.), *Robert Grosseteste and the pursuit of Religious and Scientific Learning in the Middle Ages*, Studies in the History of Philosophy of Mind 18, DOI 10.1007/978-3-319-33468-4_9

which—at least in principle—can be described and formulated quantitatively. I contrast this image with the Aristotelian scheme, whose basic units were essences and 'natures', final causes, and the interplay between potentiality and actuality. The use of such terms places individual objects and phenomena at the center of attention and does not allow a view of nature as a system.

Note that we are discussing here only explanations which are considered natural, that is, which do not involve divine intervention, animistic factors or intelligences as causes. Natural explanations assume nature's self–sufficiency in its regular operation. Within these constraints, I find the Aristotelian conception of natural action intrinsic and therefore particular. A natural agent in the Aristotelian conception acts according to natural tendencies and considered either in itself or as a part of a species or genus. It is its specific nature that determines its linkage with other agents.

The contrast I discern between the seventeenth century nomological conception of nature versus the Aristotelian metaphysics of natures and essences is challenged by Nancy Cartwright who argues that:

> Most modern accounts in the philosophy of science take it that the attempts of the scientific revolution to banish natures from science were successful. The idea of natures operating in things to determine their behaviors was replaced by the concept of a law of nature (Cartwright 1992).

Cartwright contends that this account is distorted. Aristotelian-style natures were not replaced by laws of nature, and are still central to the modern explanatory program. Laws of nature, she adds, are in fact about natures and what they produce. Indeed modern science replaced occult powers by powers that are visible, and gave up the search for essences, but to this day it is concerned with natures: 'Our most wide-raging scientific knowledge is not knowledge of laws but knowledge of the *natures* of things' (Cartwright 1992). Natures, in opposition to laws, Cartwright clarifies, 'tell us what *can* happen, not what *will* happen' (Cartwright 1999). In other words, they inform us with the potential behavior of an entity, while the actualization of that behavior beyond certain limiting conditions remains unknown. This is the source of Cartwright's skepticism regarding the concept of laws of nature as universal and necessary. She claims that our laws of nature apply to specific, well-defined circumstances, and are severely limited in their scope. Our scientific successes do not portray a unified world of universal order, she claims, but rather 'a dappled world of mottled objects' (Ibid.).

Cartwright's argument exemplifies my assumption well: there is in fact an opposition between metaphysics of natures and metaphysics of laws; one cannot hold to a metaphysics of natures and at the same time adhere to the universality and necessity of laws. Cartwright's thesis serves to pinpoint the difference between the two kinds of explanation: the one relies on potentials and possibilities; the other is binding and unconditional.

I do not take a stand in this paper regarding the correctness or otherwise of Cartwright's description of how modern science works. My aim here is to follow the formation and development of the nomological conception of nature and to uncover the assumptions that made it possible. I will not assess its validity. I therefore do not make a claim about how nature behaves, be it according to natures or according to laws. I do, however, aim to clarify what each position warrants. More precisely, I describe and analyze a moment in the history of science in which a view of a unified, ordered nature was forming and inquire into what was it that made such a view possible.

Several ideas have been proposed concerning the source of the concept of law of nature in the seventeenth century. Among these one finds Cartesianism, medieval voluntarism and medieval mathematics.[2] Those opting for the medieval mathematics stance place Robert Grosseteste and Roger Bacon at the center of their argument. Crombie assumes that it was Grossreteste who by his metaphysics of light inaugurated the project of mathematization of nature and the move away from explanation by form to explanation by law (Crombie 1959). Schramm contends that Bacon was the first to invoke the concept 'law of nature' as a comprehensive, unconditionally binding and constitutive ordering of nature (Schramm 1981). Ruby argues that Bacon's use of 'law' in his science of optics for regularities in nature was 'indistinguishable from ours' (Ruby 1986).

These suggestions receive substantial support from the relevant texts. I will present just a few examples. In *De statu causarum*, Grosseteste remarked:

> An efficient action is said to be efficient in two ways: in one way from this, that a patient is the effect and consequence of an action, as passion universally follows action. In another way it is said that some actions cause the passion not because that passion is according to the nature of the thing causing that action, but because it is required according to natural law or positive law that passion follows such action (*De statu causarum*).[3]

Grosseteste distinguished between action caused by the nature of a thing and action which is required by a law of nature. The difference between the two modes of causation requires further analysis; however, the distinction is clearly laid down. In the *Hexamaeron*, Grosseteste commented:

> ... the command of bringing forth produced the plants from the earth in the beginning, and it was a sort of law of nature [*et veluti lex quedam nature fuit*], and was fixed in the earth, giving it the capability of breeding and bearing fruit in the future (Grosseteste 1996).

In another place in the same text, he noted:

> Eating meat was not granted to nature in a state of health by the law of nature [*quod cranium esus non nature sane, lege nature*], but in virtue of weakness, as a medicinal remedy (Ibid).

To be sure, these references to laws in the *Hexaemeron* are not quantitative. The invocation of laws in this manner does not bind phenomena together or provide strict causal explanations. However, Grosseteste did set up several law-like

---

[2] The studies supporting the claim of a Cartesian origin include, among others, Zilsel (1942), Needham (1956) and Henry (2004). Studies in favour of medieval voluntarism include, among others, Foster (1934), Oakley (1961), Milton (1981) and Klaaren (1985).

[3] 'Actio vero dicitur efficiens dupliciter: uno modo ex hoc, quod passio est effectus illatioque actionis, sicut ad actionem universaliter sequitur passio. Alio modo dicitur actio efficiens passionem, non quia illa passio sit secundum naturam rei efficientis illius actionis, sed quia secundum legem naturalem vel positivam debitum est, ut talem actionem talis passio consequatur.'

quantitative rules, such as that natural force is most active when propagated in a straight line, and that the force is strong in inverse proportion to the length of the line (McEvoy 1983), or that every agent multiplies its power spherically. In *De iride* Grosseteste formulated a law of refraction, according to which a ray passing from one medium to another medium, which is denser, will be refracted at an angle equal to half the angle of incidence; and the law of reflection, stating that the incidence and reflected angles are equal (Grosseteste, ed. Baur 1912a, b). Even though Grosseteste did not use often the term *law* regarding these rules, the theorems appear to be law-like formulations.[4]

By comparison, the terminology of laws in Bacon is significantly richer. He was the one to attach the term *law* to the formulas made by Grosseteste. He wrote of 'laws of refraction,' *leges istarum fractionum* (Bacon, ed. Bridges 1964 4.4.2); 'law of incidence and reflection at oblique angles,' *lege incidentiae et reflexionis ad angulos obliquos* (Ibid.); 'law that governs passage from the subtler to the denser substance,' *legem incessus a subtiliori in densius* (Bacon, trans. Lindberg 1983 2.4); 'laws of multiplication [of species],' *legibus multiplicationum* (Bacon, trans. Lindberg 1996 1.6.2 and 1.8.2); 'laws of material forms,' *leges formarum materialium* (Ibid. 1. 6.3); 'laws of material and corporeal things,' *leges rerum materialium et corporalium* (Ibid. 1.6.4); 'common laws of nature' (Bacon 1964 4.2.3); 'universal law of nature,' *lege nature universalis* (Bacon 1983 1.6); and 'law of particular nature,' *lege nature particularis* (Ibid.).

What is common to these expressions is that they all refer to descriptions of natural phenomena, mainly but not exclusively to the propagation of rays of light and their encounter with different mediums. Bacon, however, expanded the application of these geometrical laws to everything material and corporeal—*leges rerum materialium et corporalium.* If Grosseteste's law of refraction is essentially a geometrical law within the boundaries of Greek tradition, Bacon's set of laws of corporeal forms tells a different story.

It has been argued that prior to the seventeenth century the expression 'laws of nature' had been in use merely as a loose metaphor or a restricted statement of principles in mathematics, never as an explanatory proposition in natural philosophy. Mathematics, the argument continues, was considered to be detached from physical causation; it therefore could not provide physical explanation in terms of causes which was the aim of natural philosophy (Henry 2004). It is my contention that Bacon's concept of laws of nature was not a loose metaphor, but had explanatory and causal force regarding various physical phenomena other than the radiation of light. His laws display a necessary, causal connection between geometrical properties and physical effects. The laws that Bacon formulated were descriptive of

[4] Eastwood (1967) analyzes this law and argues it was a completely original formulation by Grosseteste and that no such law is to be found in any earlier known treatise on optics or natural science (p. 406). He finds another law formulated in *De iride*, which he calls 'the law for location of an image in reflection'. However, Grosseteste could easily find this law in Euclides' *Catoprics* and al Kindi's *De aspectibus*.

the behavior of species, which he thought were corporeal and physical (for example, species transmit heat).

Bacon stressed the universal, uniform and compulsory aspects of these laws. For example, in *De multiplicatione specierum* he wrote that 'the death of individuals (with respect to the present life) is necessary according to the law of universal nature.' (Bacon 1983 1.6) In discussing the law of refraction, which describes the passage of species from a body of one density to another of a different density, he wrote, 'if it [the species] does not fall perpendicularly, then of necessity it alters its direct advance, and makes an angle on entering the second body.' (Bacon 1964 4.2.2). The law of reflection has predictable and necessary consequences: 'If, therefore, a concave spherical mirror be placed against the sun, an infinite number of rays will converge to one point by reflection. And therefore of necessity fire is ignited when a concave mirror is placed against the sun' (Ibid.). The production of fire by following the law of reflection demonstrates the causal role of a geometrical law in the realm of physics.

There are many other examples; the evidence is clear: in the writings of both Grosseteste and Bacon there is a discourse of laws of nature. The metaphysical assumptions of a universe run by impersonal, universally binding principles, concordant with this discourse are present as well, alongside a tentative quantitative formulation.

I accept the claim made by Crombie, Schramm and Ruby that there was, in the thirteenth century, a concept of a quantitative, universal, law of nature, with Grosseteste and Bacon as the protagonists.[5] I leave aside for the moment the question of origin and influence; that is, I do not argue that the concept of a lawful nature upheld by Grosseteste and Bacon anticipated or influenced the seventeenth-century concept. Rather, I concentrate upon their respective philosophies of nature and pose the following questions: what was it that enabled Grosseteste and subsequently Bacon to conceive of an ordered nature governed by laws? How did they arrive at this outlook of nature as a lawful and unified system?

9.2 Grosseteste's 'Corporeity'

Grosseteste was not the originator of the doctrine of the first form. Aristotle had been typically understood to propound a notion of first or prime matter which is common to the four elements and therefore common to all physical things.[6] Since the transmutation of the elements one into another must be a continuous process

[5] Eastwood (1967) argues that it was Grosseteste's law of refraction is only semi-quantitative, because 'it is based on qualitative principles and might better be called the qualitative law of refraction'.

[6] Aristotle discussed this concept on several occasions. For example, in *Metaphysics* VII, 3, 1029a, he stated that prime matter is not a particular thing, not of a definite quantity and does not fall under any category; in *Physics* I, 9, 192a, he declared that it is outside the sphere of being and becoming.

(given that *ex nihilo nihil fit*), it therefore requires an underlying and surviving principle which Aristotle called 'first matter' (Hyman 1965). The notion of corporeal form was introduced by Simplicius of Cilicia (c. 490–560), in his commentary on *Physica* 1.7, 230, 21–9 (Wolfson 1929). In Simplicius, corporeal form meant something like the capacity of matter to receive dimensions. According to Avicebron (c. 1021–1058), the founder of the hylomorphic theory, there exists a universal entity composed of the universal form and universal primary matter. This entity exists independently only in potency and receives actuality by the addition of the more specific, complementary forms. Avicebron argued that matter as potentiality need not necessarily be corporeal; it becomes corporeal (namely, quantitative) only through union with the form he called *corporeitas*. However, the formula *forma corporeitatis* does not appear in Avicebron and its origin was probably in Avicenna (c. 980–1037) (Gilson 1955).

Both Avicenna and Averroes (1126–98) went against Aristotle's assertion that prime matter was pure potentiality; they argued that it was indeed a kind of substance. As a substance, prime matter could now be shown to possess a form. They noted that all bodies possess certain constant properties, such as bulk, continuity and extension and identified the corporeal form with these properties. All bodies thus, in their view, were composed of first matter and corporeal form. This form rendered matter in itself apt for understanding and description. Both thinkers agreed that among the various properties common to all bodies, extension or dimensionality was primary (Hyman 1965).

For Avicenna, the *forma corporeitatis* is the first and most universal of all physical forms and is that which makes matter to be a body. The extension of bodies results from the form of corporeity, which is a substantial form that endures through all change (Gilson 1955; Pasnau 2010). Through the form of corporeity prime matter is taken out of its state of indetermination and disposed for the reception of a specific form. However, Avicenna refused to identify the form of corporeity with the dimensions themselves and instead determined that it is a form having a predisposition for receiving the three dimensions. In this way he preserved the form of corporeity as a substantial form. In Averroes, the corporeal form was identical with a body's extension, providing the body with an indeterminate extension, an accidental form which inheres in prime matter (Hyman 1965). Both these views were influential on the later Latin tradition, and competed against the view—associated with Aquinas—that extension (or quantity) is posterior to the substantial form and derives from it, just as other accidents are (Hyman 1965; Pasnau 2010).[7]

The concept of form of corporeity before Grosseteste did not account for activity in nature, nor was it quantitative or accessible to the senses. It was conceived within the frame of form-matter analysis in order to stress the one feature that is common

[7] In spite of the Thomistic attack, the notion of corporeal form received support from later Christian thinkers, such as John Duns Scotus (c. 1266–1308), who claimed that the body of Christ in the tomb must have possessed a form of corporeity since the body does not dissolve immediately and must possess the form which makes the body a body. After the soul departs, the body is still there and needs a proper form of its own. See, Copleston (1966).

to all matter, namely, extension. The concept of the form of corporeity demonstrates a shift in the Aristotelian idea of matter: instead of a pure potentiality that cannot be grasped or discussed apart from its link to a specific form, matter could now be turned into an object of study. The Aristotelian analysis of form-matter relations as a series of gradual changes was being slowly replaced by a reductionist analysis, comprising a search for an ultimate substance to which all material bodies can be reduced (Randall 1960). The shift received a significant enhancement by Grosseteste's identification of the corporeal form with light. In his *De luce*, Grosseteste wrote, 'The first corporeal form which they name corporeity, I consider to be light. For by its nature light spreads itself in every direction' (Grosseteste, trans. Lewis 2013).

Instead of a passive capacity to receive dimensions, Grosseteste rendered the corporeal form an active agent, which through its inherent, imperative and unceasing diffusion introduces dimensions into matter. Matter in itself, Grosseteste proclaimed, is a substance with no dimensions (Grosseteste 2013). It is light's capacity to radiate and expand that makes it the first corporeal form. In fact, Grosseteste made three important modifications in the conception of the form of corporeity: (1) he identified it with light, (2) endowed it with activity, and (3) described this activity as uniform and necessary. I will discuss each modification separately.

9.3 The Identification of the First Corporeal Form with Light

The identification of the first form with light rendered it a unique status. The *New Testament*, the writings of the church fathers, as well as those of the Greek philosophers, are filled with light imagery: God is light, truth is light, beauty is light. Grosseteste reviewed this position:

> Wise thinkers consider the first corporeal form to be more exalted than all subsequent forms and to have a more excellent and nobler essence that is more like the forms that are separate [from matter]. But light has a more exalted, excellent and nobler essence than all corporeal things, and more than any of them is like the forms that are separate [from matter], which are the intelligences. Therefore, light is the first corporeal form (Grosseteste 2013).

Indeed, the form of corporeity became a central theme in Grosseteste's philosophy, a place it did not have before. This new role is worthy of exploration.

Since the function of the form of corporeity is to add quantitative dimensions to matter, then by proxy both quantity and matter receive some of light's nobility and thus elevated to a new status. This is apparent in Grosseteste's assertion that matter, rather than being a mere potency, is in itself a substance (Grosseteste 2013).[8] Since

[8] 'Both corporeity and matter are in themselves simple substances.'

the form of corporeity precedes the specific substantial form, the object's dimensions become a part of its essence. In this way, dimensions can form a part of the explanation of the characteristic properties of a thing. As a consequence, quantitative properties can now be considered causes and used in *propter quid* explanations. Thus, considerations of lines, angles and geometrical properties turn to be the main line of argument in accounting for natural phenomena (Grosseteste, trans. Lindberg 1974a, b).[9] For Grosseteste, this was not an empty methodological declaration; he applied these principles in several cases, such as the rainbow, climatic regions and the tide. In all these cases, it is not the specific forms of the elements that play the central role, but the general form, common to all matter—the form of corporeity or light.

Scholars are divided in their evaluation of the place that Grosseteste's light metaphysics had in the project of the mathematization of nature. Crombie grants it a considerable role, and argues that Grosseteste and Bacon 'had a conception of physical nature in which the essence or 'form' itself is mathematically determined, and a conception of the immediate objective of inquiry as mathematical and predictive laws' (Crombie 1959). Weisheipl disagrees with this position, claiming that Grosseteste did not consider that mathematics gave the real causes of natural phenomena (1984). Wallace claims that Grosseteste did think that geometry provides *propter quid* knowledge of the physical aspects of natural phenomena. However, when dealing with complex phenomena such as thunder and lightning, he could not supply convincing geometrical explanations (1972). Lindberg credits Grosseteste with an enlargement of the common ground between mathematics and physics by subordinating a large part of physical science to mathematics (Lindberg 1982). The discussion on the part played by Grosseteste in the emergence of an early form of mathematical physics is indeed intriguing, but goes far beyond the scope of this paper, which does not explore the place of mathematics in the investigation of nature, but rather the reduction of specific phenomena to universal processes.

The ability to provide explanations based upon a form present throughout nature enables a reduction to a minimal number of principles and renders those explanations universal: the same few principles can be used in different ways, in accounts of different phenomena. Accordingly, Grosseteste wrote in *De lineis*, that, 'The utility of considering lines, angles and figures is very great, since it is impossible to understand natural philosophy without them. They are useful in relation to the universe as a whole and its individual parts' (Grosseteste 2013).

This is a version of the early principle of parsimony: the same principle can be used to account for both the universe as a whole and its parts. The geometrical properties can unite our scientific accounts because they are present throughout nature.

[9] '...all causes of natural effects must be expressed by means of lines, angles and figures, for otherwise it is impossible to grasp their explanation.'

9.4 Endowing Activity

Grosseteste defined the form of corporeity from the outset as 'multiplying itself', 'instantaneously spreading itself', and 'extending matter' (Ibid.). In *De operationibus solis* Grosseteste added the attribute 'active potency of three-dimensionality' (*corporeitas est potentia activa triplicis dimensionis*) (Grosseteste, ed. McEvoy 1974a, b). In other words, the form of corporeity is inherently active. It extends in all directions and draws matter with it. This means that the corporeal form is a driving force, a causal principle and motor in nature. Indeed, in Grosseteste's later writings (such as the *Hexaemeron*), the corporeal form became, as McEvoy notes, the source of all causal action (McEvoy 1982).

9.5 Uniform and Necessary Activity

As the source of all natural causation, the multiplication of the form of corporeity is made equally in all directions. Grosseteste stressed that in the creation of the universe, light's multiplication has spread 'uniformly (*equaliter facta*) in every direction' (Grosseteste 2013) and that light 'extends matter uniformly (*undique equaliter*) on all sides into a spherical form'(Ibid.). This uniformity is linked to the beauty and nobility of the corporeal form. Light in itself is beautiful, and therefore the spherical shape it creates by dragging matter with it is also beautiful, simple and the most united among corporeal things. All beauty, Grosseteste declared, can be reduced to the equality of proportions (*ad aequalitatem proportionis*). The movement of the heavens is the most uniform (*uniformitatem*), and it is *lux* which is the beauty of heavens (Grosseteste 1974a, b).

The uniformity of the activity of light is to be found not only in its spatial aspects; this activity is of one kind only, namely, multiplication and it is the same whatever it encounters. Light in Grosseteste's scientific writings always acts in the same way; it is invariant and the different effects are in the recipient and therefore accidental to light. Eastwood calls this 'the principle of uniformity in nature', according to which the agent considered by itself will always act in the same way, but the effects may vary with change in external conditions (Eastwood 1967). In *De motu corporali et luce* Grosseteste remarked that 'lux generates itself in one way' (*lux secundum unam viam se generat*) (Grosseteste 1912a, b), and in *De lineis* he stated that, '...the agent sends the same power into sense and into matter... For it does not act by deliberation and choice, and therefore it acts in a single manner whatever it encounters' (Grosseteste 1974a, b).

Moreover, it was made clear by Grosseteste (perhaps under the influence of the necessary emanation of the Plotinian *One*) that this typical activity of light, namely, multiplication, is its essential attribute, which cannot cease without light changing its nature. It is therefore necessary and incessant: 'Light, then... must, when infinitely multiplied, extends matter [*necesse est extendere*]' (Grosseteste 2013).

After reaching the uttermost parts of the cosmos, light 'of necessity [*de necessitate*] is spread from the first body into the center of the whole' (Ibid.). This inherent uniform activity adds two features to the form of corporeity: (1) it endows the universe with unity, and (2) it renders its description and study possible.

Grosseteste was not the first in the history of philosophy to stress the unity of nature. Plato and Plotinus had done that before, and several studies have stressed the Plotinian origins of this 'metaphysics of light', which have reached Grosseteste through Arab and Jewish mediators, such as al-Kindi (c. 800–870), Avicenna and Avicebron.[10] Certainly, both the idea of identifying corporeity with light and the necessary activity of corporeity derive from these mediators. What Grosseteste did was to turn Plotinus' metaphysical unity into a physical one.

The corporeal form as defined by Avicenna and Averroes could not have been examined by observation; it was an abstract, metaphysical concept, not accessible to description or observation. But by the identification of the corporeal form with light and its definition as active, one could describe and even measure its activity. With the study of light's actions within the material world, one could find out the details of light's behavior. And those details are the features of the basic causal mechanism of nature.

The most primordial action of the corporeal form according to Grosseteste, however, is the creation of the universe. The universe, both the heavenly spheres and the elements on Earth, was created in one act. The same actor, the same driving force, the same principle of uniform activity now unites both regions. In *De luce*, Grosseteste stressed the function of unity that light as form bestowed on the universe, by being contained in all corporeal things: 'all things are one on account of the perfection of the one light' (Grosseteste 2013). Light unites the world, but—at the same time—it is also the principle of distinction and multiplicity: 'things that exist are many on account of the diverse multiplication of light' (Ibid.). The difference among things becomes thus a difference of size and magnitude, since this is the distinction that the form of corporeity can make.

Scholars indeed noted this function of unity of the activity of light. McEvoy remarks that 'the lux or first corporeal form, made essentially one physical system out of what for Aristotle had been two separate ones' (McEvoy 1982). And Lindberg concurs, 'One of the most striking features of this cosmogenical scheme is its firm statement of the unity of the cosmos... underlying the differences is a fundamental unity, based on light as first corporeal form' (Lindberg 1986).[11]

---

[10] Such studies include, for example, Lindberg (1976), who argues that the origin of the idea that light is involved in the creation of the material universe as well as light's self-diffusion is the Neoplatonic doctrine of emanation. In another study Lindberg (1986) expands on the Plotinian origin and its significance. See also McEvoy (1982). For a study on the influence of Plotinus on the idea of the unity of nature in Grosseteste, see Raizman-Kedar (2006).

[11] The idea of the unity of earthly and celestial matter is manifest in Bacon as well. In *De multiplicatione specierum* 1.5, he wrote: 'it is evident that lower things can be influenced by higher things, since they share the same matter... the purpose of this conformity is that the more the parts of the universe are like one another, the greater are their well-being and utility.'

9.6 Bacon's 'Species'

Thus far I have claimed that the terminology of 'laws of nature' appears in Grosseteste's writings, and that it is pronounced more prominently in the writings of Bacon. I have also argued that what induced this discourse was Grosseteste's insertion of an active form of corporeity—identified with light—into the innermost constitution of natural processes. But here my thesis seems to encounter a difficulty, namely, Bacon did not follow Grosseteste in positing a form of corporeity or light as the ultimate constituent of matter. Instead he devoted much attention to the description and analysis of another entity, species, which 'causes every action in the world [*haec species facit omnem operationem hujus mundi*)]' (Bacon 1964 4.2.1). I therefore explore the relation between the two conceptions—corporeal form and species—first in Grosseteste and then in Bacon. My aim is to demonstrate that Bacon's theory of the multiplication of species is a direct derivative and in fact an elaboration on Grosseteste's concept of corporeal form.

In a famous paragraph in *De lineis*, Grosseteste set up a principle, which will be later repeated by Bacon almost verbatim:

> A natural agent multiplies its power [*virtutem suam*] from itself to the recipient, whether it acts on sense or on matter. This power is sometimes called species, sometimes a likeness, and it is the same thing whatever it may be called; and the agent sends the same power into sense and into matter...But the effects [*effectus*] are diversified by the diversity of the recipients (Ibid.).

This sounds strange, why did Grosseteste posit another term, another entity or mechanism of natural action in addition to light, and how does it relate to light as an active force? A partial answer is given in another text, *De operationibus solis*, where Grosseteste defined corporeity as an active potency, which is the form and species of the first body, the firmament: 'And this form and species of the first heaven is light [*lux*], which is perhaps the first form which by itself extends the first corporeal matter into huge dimensions' (Grosseteste 1974a, b). In the same text a few pages later, the same identification appears: 'the corporeal species of heavens, which is light [*lux*], would have their reasoning in the Divine mind' (Grosseteste 1974a, b). However, after turning a few more pages, the identification looks less clear. Grosseteste remarked:

> The Sun can be said to announce forth, because its light [*lumen*], primarily among corporeal powers, bring out and move forth the potential to act of the figures and forms and corporeal species in plants and animals (Ibid.).

There seems to be a distinction here between the Sun's light and the corporeal species in plants and animals, upon which it acts. It could be that the corporeal species in this citation are the light within the elements. However, a reasonable interpretation would be that the sun's light as *lumen* is not corporeal, while the light within the elements is *lux*, namely, corporeal light. Light as *lumen* is posited in the medium, while light as *lux* is posited in things, such as plants and animals, which do not emit visible light. Light as *lumen* is considered the species of *lux* (Bacon 1983 1.1),[12] and both have the same mode of action, namely, multiplication.

In order to explore the relation between the concepts of corporeal form and species in Grosseteste, a clarification is in place of his use of the two Latin words for light, namely *lux* and *lumen*. I have addressed this issue extensively elsewhere, so my discussion here will be concise (Raizman-Kedar 2006).

Lux in Grosseteste's cosmogeny is both the matter and form of the firmament (Grosseteste 1996 5.4.1); it is the form of corporeity, 'a substantial form inseparable from matter' (Grosseteste 2013); it is a bodily substance, inherently active and self-diffusive; it is the generator of all motions and transmutations in nature. In relation to *lumen*, *lux* is a source and origin and precedes it ontologically; *lumen* is created out of *lux*'s self-diffusion. It is thus the offspring, copy or image of *lux*. Grosseteste stated that 'one point of *lux* can fill a whole hemisphere with *lumen*' (Grosseteste 1996 2.10.1). Being bound with matter, *lux* can express or reveal itself outwardly only through *lumen*, which is not a body or in a body, but is a 'spiritual body' or a 'corporeal spirit' (Grosseteste 2013), and thus can travel in different mediums. While *lux* is a substance, *lumen* is an accidental quality (Raizman-Kedar 2006). This is why Grosseteste ascribed visibility to *lumen* alone: 'the air is being lit up only as long as *lumen* is present, and when lumen is gone—it goes back to darkness' (Grosseteste 1996). It is light as *lumen* which is perceived by sense and reveals that which otherwise remains hidden: 'the sun is in the view of the seeing eye through the strength of its *lumen*. . . and [the sun's] *lumen*, being reflected from the moon and the stars, reveals, and by revealing it announces that which is under the darkness of the night' (Grosseteste 1974a, b). Light as *lux* serves as a principle of activity, causality and unity, as a substance and the essence of corporeality in Grosseteste, while light as *lumen* is the visible light: an accidental quality, perceived by the sense.

So it seems as though the identification of the form of corporeity with light is in fact made with regard to *lux* as a substantial form; *lumen*, as an accidental form, can be identified only with the visible expression of the activity of the corporeal form. In this way one can understand the citation establishing a universal multiplication of power or species as an expression of the activity of the corporeal form, that is, as its external manifestation in the natural, physical, sphere.

Let us turn now to Bacon. The term which is at the center of the present paper—the form of corporeity—is almost entirely absent from his works. I found one place where it is mentioned, in the *Questions on Metaphysics* (an early work), which seems to be a direct reference to *De luce* by Grosseteste:

> . . .but the corporeal form is the principle of being in the heavens, while at present we presume to be light (*lux*), or something else of this kind, therefore corporeal form shall be the principle of operation and so of movement (Bacon 1926, 12).[13]

[12] '. . .the *lumen* of the sun in the air is the species of the solar *lux* in the body of the sun'.

[13] ". . .set forma corporalis est principium essendi celum, que est lux ad presens supponitur, vel aliquid hujusmodi, ergo forma corporalis erit principium operandi, et ita rnovendi."

However, as his thought developed, Bacon rejected the metaphysics of common corporeity, and denounced the use of *lux* to denote substantial forms. His criticism can be viewed as directed toward the purportedly metaphorical function that light assumed in the writings of Grosseteste. Bacon declared that metaphors of this kind (or 'extensions of meaning' as he put it) led to much confusion and called for a clearer use of language:

> Not all the effects of colours or of the elements arise from this light [*lux*], but [some arise] from substantial powers [*virtutibus*], as appears elsewhere; and the name of light [*lux*] is transferred to designate these powers, because these are unknown to us, while *lux* is most evident to us (Bacon 1926, 12).[14]

This is the point at which Bacon adopted the terminology of species and developed a causal theory of his own, which has common features with the one proposed by Grosseteste, but also some original motifs. In Bacon's theory, it is not light (as *lux* or as *lumen*) which plays the central role; instead he posited species at the heart of his system.

The term 'species' had a long history of philosophical and theological use. The Latin origin of the term meant 'aspect', 'form', 'exterior appearance' and 'beauty', and it was the translation given by Cicero (106–43 BCE) to the Platonic *eidos* (Spruit 1994). In Christian theology, species was often used to denote the beauty of the highest beings, and most of all the earthly expressions of this beauty. Since the Son was considered the most sensible expression of divinity as incarnated in flesh and blood, species came to be applied to him as a proper name.[15] In the eleventh century, Hugh of St. Victor (1096–1141) characterized the species of the world as 'the natural pictures of God'. Hugh thought that each visible species bore in itself a certain resemblance to the 'invisible demonstration' (Ritter 1971).

In Victorinus (fourth century CE), a Neoplatonist whose conversion to Christianity had considerable influence on Augustine, the Father and Son can be called by various pairs of names, among which are found both 'God and "form or image"' and "substance and species"' (Victorinus 1981 1.41). In the thirteenth century, Bonaventure (1221–74) suggested that by observing the process of the generation of species in the medium we are eventually led to the origin or model that this process imitates—that is, to the duality of the Father and Son; he described the Son as the 'first Species, in which there is the utmost proportion to and equality with the

[14] 'Non omnes effectus colorum nec etiam elementorum fiunt a luce hac, set a virtutibus substantialibus, ut patet alibi, ad quas virtutes substantiales designandas transumunt nomen lucis eo quod sunt nobis ignote, et lux est nobis manifestissima.'

[15] In analysing the meaning of Christ as the image of God, Marius Victorinus distinguished the sensible image, which is 'a sort of shadow' and nothing by itself, from the image 'up there', which is 'living and life giving and the seed of all existents'. The image realizes what is potential and expresses it outwardly; by doing so it renders the general concrete. At this juncture Victorinus introduced the term *species*: 'For every being has an inseparable species, or rather, the species itself is the substance itself, not that the species is prior to "to be", but because the species defines "to be"... and for this reason "to be" is the Father, the species is the Son' (Victorinus 1981 1.19).

One generating' (Bonaventure 1993 2.8). In the Aristotelian translations made by William of Moerbeke (c. 1215–86),[16] species were the forms representing external realities within the senses and intellect.

The traditional load of the concept of species fits well with the function and meaning it received in Bacon's system. He took up species as the earthly and sensible aspect, which represents and expresses the hidden order and beauty of the universe. Accordingly, he defined species as physical and natural. In the *Perspectiva* he wrote '. . .of necessity such species have corporeal being. And if they have corporeal being, then they [also] have material being. And therefore they must obey the laws of material and corporeal thing' (Bacon 1996 1.6.4). And in *De multiplicatione specierum* he added: 'I therefore state unconditionally that the species of a corporeal thing is truly corporeal and has truly corporeal being' (Bacon 1983 3.2).

Bacon thought that *species* were produced by every active nature, and certainly not by light and colour alone. An 'active nature', according to Bacon, is a class comprising both substances and proper sensibles (Ibid.). He wrote:

> . . . indeed, every active nature seems to be doing so, by reason of its substantial form or of a certain accidental form, or both, which form ought not be light [*lux*]. For in truth coldness and dryness are such active natures through which the multiplication of virtues and species can occur (Bacon 1926).[17]

It is not only *lux* which is self-generated now, but every active nature sends out species. *Lux* or the corporeal form ceases to be a universal force and is replaced by the more general relation of agent and species.

An inseparable part of the depreciation of light is the bringing down of the explanatory principle from the metaphysical level into the physical one. Bacon's species are therefore material, physical entities operating in space and time. Bacon gave up the notion of the substantial corporeal form and was left with its accidental visible representative—species—rendering it the central concept in his scientific outlook. In fact, he abandoned the search for essences altogether. In the beginning of *De multiplicatione specierum* Bacon argued that essence, nature, virtue and force signify the same thing and differ only in relation. 'Essence' is used with respect to itself, while 'virtue' and 'force' in reference to eliciting an action. The term 'nature' means, according to Bacon, an aptitude for acting, and 'things of similar essence have similar operations' (Bacon 1983 1.1). Thus, one need not search for a thing's essence in order to achieve knowledge; it is enough to inquire about its operations.

Instead of using light's prestige to elevate the status of matter, Bacon saw the relation of agent and species as an imitation and a reminder of the Trinitarian relation of the Father and Son. Every natural action is an expression of this relation.

---

[16] Moerbeke's was the third translation of the *De anima*. James of Venice (fl. 1125–50) made the first translation before 1150.

[17] '. . .omnis enim natura activa sic videtur facere, aut ratione forme substantialis aut alicujus forme accidentalis aut utriusque, que forma non oportet quod sit lux. Frigiditas enim et siccitas vere sunt nature active per quas potest fieri multiplicatio virtutis et speciei.'

Nature as a whole is now worth exploring; not because of its inherent light, but rather because of the order and beauty endowed by the imitation of the Trinity. Accordingly, one finds in Bacon a similar elevation of the status of matter as was found in Grosseteste: Bacon claimed that 'matter is not nothing, but true nature and essence' (Bacon 1988 2.3.80).

The laws of nature of which Bacon spoke of were in fact the laws of the activity of species. In most cases he had used the term *lex* or *leges*, the context was one or another attribute of species. Here are a few examples.

The law of refraction, which according to Grosseteste describes the bending of light-rays (light being one with the corporeal form) upon passing from one medium to another, was used by Bacon to account for the difference between the true location of the stars and the location in which they appear to us. He explained this difference by the angle the star's species, formed as it passes from aether to air, and concluded that: 'therefore the species of all stars must be refracted' (Bacon 1983). In Bacon the law of refraction applies to the radiation of species, light rays being a mere example of such radiation.

In discussing generation and corruption, Bacon stipulated that the nobler is the agent the more it is active and, therefore, better able to complete its species in the elements and natural matter (Ibid.). A species becoming 'complete' means that as a vehicle of change it has done its part and rendered the patient similar to the agent in a certain respect. But if the species of spiritual and celestial substances would be completed in natural matter, they would overcome the terrestrial corporeal things and render them similar to themselves, namely, they would make them spiritual or celestial. Such a consequence is undesirable if the order of nature is to be preserved. Therefore, 'by divine ordination and a universal law of nature... the power is withheld and the actuality excluded, so that spiritual and celestial substances can do no more than produce their specie', but not complete them (Ibid.). Here, too, the 'law' spoken of applies to the activity of species, and their ability to transform natural matter into celestial or spiritual matter is excluded.

Bacon made a similar claim when he argued against the view that because species have spiritual existence in the medium, the species of colours do not mix. He claimed that if that view was correct, it would entail that species do not obey the 'laws of material forms'. This, in his view, is a good enough reason to reject the species' spiritual existence and to declare that 'they must obey the laws of material and corporeal things' (Bacon 1996 1.6.3–4).

In order to support my contention that Bacon's position is in fact an elaboration on Grosseteste's form of corporeity, I briefly discuss five points of agreement between Grosseteste corporeal form and Bacon's conception of species. Grosseteste based his view of an ordered, unified universe on the concept of form of corporeity; Bacon used the concept of species for the same purpose.

1. In Grosseteste's scheme, *lumen* was not the corporeal form itself, but the first product of its activity. Bacon took up this first effect and entitled it species, since he wanted to stress that it is produced in all cases of activity in nature, and that such activity is not special to light alone. Bacon defined species as 'the first

effect of any natural agent' (Bacon 1983 1.1) and the first example he provided for such an effect is the *lumen* of the sun in the air, which is the species of the solar *lux* in the body of the sun (Ibid.). Bacon replaced *lumen* with species because according to Grosseteste *lumen* was not strictly corporeal, but a 'bodily spirit'. Bacon, however, was determined to show that every natural activity was physical and corporeal.

2. Both the form of corporeity and species necessarily diffuse and multiply in all directions. It was Bacon who dwelled on the details of that multiplication, in which a species is brought forth out of the potentiality of the matter of the recipient. The agent does not deposit a species in the recipient but elicits a species out of the recipient's matter. The effect is then produced 'out of the active potentiality' of the recipient matter (Ibid.). It is noteworthy that 'active potency of three-dimensionality' was the definition of the form of corporeity given by Grosseteste in *De opertaionibus solis*. In another place, Bacon clarified that, '. . .there is no motion, but a generation multiplied through the different parts of the medium; nor is it body that is generated there, but corporeal form that does not have dimensions of itself' (Bacon 1996 1.9.4).

   This is a straightforward identification of species with the notion of corporeal form. The definition of species as both form and corporeity stands in a sharp contrast to other contemporaneous thinkers who applied the term species to 'immaterial' or 'intentional' kinds of beings (Normore 2007). However, it is a different corporeal form than the first form spoken of by Grosseteste, since it is not substantial but accidental.

3. Grossetetste reduced all natural causality to the radiation of light; Bacon, in a similar manner, identified species with a physical force responsible for all efficient causality in the universe. In the *Communia naturalium* he declared:

> I hold that two things drive to the production of things, that is, the efficient cause and matter. Now it ought to be proceeded about the efficient cause as much as required, because Metaphysics has to fully certify concerning the influence of the agent upon the patient, that which all operations in sense and intellect and the matter of the world are made by influences of this kind, namely, the aforementioned [entities] which are called species (Bacon, *Communia naturalium* 1.1.2.1).[18]

In *De multiplicatione specierum* he added that 'all judge that through species [all] other effects are produced (Bacon 1983 1.1).

4. As efficient causes, the operation of both the form of corporeity and species is uniform and universal. Bacon accepted Grosseteste's declaration that 'the agent brings about the same [effect] in whatever it acts on' for 'this is a purely natural

[18] 'Habito quod duo exigantur ad rerum produccionem, scilicet, efficiens et materia, nunc procedendum est circa efficiens quantum hie requiritur, quia Metaphisica habet certificare ad plenum de influencia agencium in paciencia, eo quod omnis operacio in sensum et intellectum et materiam mundi fit per hujusmodi influencias, scilicet, predictas que vocantur species'.

action and therefore cannot occur except in one way' (Bacon, *Communia naturalium* 1.1.2.2).[19]

5. Species, just like the form of corporeity, is the feature common to all things, whether earthly or celestial, and is the link binding the two regions together. Bacon opted for a two-ways mutual influence:

> ... although terrestrial things cannot resemble the heavens in their complete natures, they agree at least to the point [of being linked through the reception of species]... when it is objected that celestial nature is not generable or corruptible, and therefore that it will not be generated in elemental matter, we reply that this is true as regards complete being; however, as regards the being of species it is not unsuitable, but necessary [for celestial nature to be generated in elemental matter] (Bacon 1983 1.5).

Thus, Bacon replaced corporeity with species as the source of unity in nature. He shifted attention from internal aspects of essences and substantial powers, to their external manifestations. Bacon preserved however the distinctive features of Grosseteste's form of corporeity: a species necessarily diffuses and multiplies; it is a vehicle of efficient causality; it acts uniformly and universally, and it binds celestial and earthly matter together.

9.7 Conclusion: From 'Corporeity' to 'Species': The Foundation of the Nomological Image of Nature

I began by inquiring into the origin of the nomological image of nature. As I elaborated at the outset, a nomological image of nature consists of uniform and universal principles, governing an ordered nature, unified by these universal principles. What I tried to show is not only that these features are displayed in the philosophies of nature of Grosseteste and of Bacon, but that there is a common conception of the unity of matter which both thinkers posit at the center of their doctrines.

The unity of matter is ensured by Grosseteste in assuming the concept of 'corporeal form' or 'form of corporeity'. Grosseteste thought of the form of corporeity as inherent in every material substance, accounting for the corporeal character of matter and for its extension. It thus served as a universal element which binds the material world together, providing it with common features that can be described, measured, and reduced to quantitative properties. With the identification of the form of corporeity with light, Grosseteste turned the corporeal form into a principle which is not only of extension, but also of activity. In this way, he was

[19] See also Bacon (1983) *De multiplication specierum* 1.1: '... an agent naturally produces the same first effect [that is, species] in whatever it acts upon, because for its part it acts uniformly; for only an agent that possesses free will and acts by deliberation can, for its part, act difformly. But a natural agent possesses neither will nor the ability to deliberate, and therefore it acts uniformly ... since nature and natural mode have the same mode [of action].'

able to describe natural processes in general statements about the behavior of light, rather than by the definition of specific natures and the distinction between substantial and accidental forms. The important point is that light in Grosseteste's conception of nature is active, causing generation and corruption. The form of corporeity thus becomes a pivotal explanatory concept, by which an account of a living universe can be given. This is the moment at which Grosseteste departed from both the Platonic and Aristotelian conceptions of nature, for he used the Platonic mathematical principles to account for change, and shifted the Aristotelian attention away from individual objects and specific causes.

I have argued that a nomological image of nature prescribes that the explanatory terms of natural phenomena are universal, necessary and impersonal; that it finds general factors, linking discrete phenomena into a whole; that it stresses nature's unity and order and their quantitative description. I demonstrated how Grosseteste's conception of nature corresponds to these exact criteria. According to Grosseteste, given the corporal form, nature is a unified system, with necessary mutual connections. This image receives its definitive representation in *De Cessatione legalium*, where Grosseteste remarked that 'in this way all things are linked together in the most orderly way by natural connections' (Grosseteste 1982).

Bacon followed Grosseteste's conception, replacing the notion of corporeal form with that of species. In Bacon's philosophy of nature, species is the glue which binds the material universe together. Just like Grosseteste's form of corporeity, Bacon's species are the basic explanatory units behind all natural phenomena. Species, however, do not endow matter with extension or represent the measurable aspects of their material source. While Grosseteste portrayed a natural world that share a common feature, for Bacon the goal was to determine the behavior and activity of that common feature. What unites matter according to Bacon was the mode of activity, prevailing throughout the universe, which is always conveyed through the production and mediation of species. *Lux* in Grosseteste's cosmogeny is a metaphysical principle, displaying some physical facets; its multiplication is instantaneous and diffusion—infinite. Species, as the principle of conduct of natural things, is distinctively physical: its action is temporal, (Bacon 1983 4.3)[20] and its multiplication finite. Every physical entity is finite, and a finite nature can only multiply itself finitely (Bacon 1926, 23). It is also subject to corruption just as other physical things are. It is this activity, perceived by the senses, which can be measured and formulated as laws.

Bacon set aside the search for essences; he postulated that every active nature, whether substantial or accidental, produces species. What was crucial for Bacon was to describe the modes of the production and propagation of species, in regard to which all sources act the same. The distinction, therefore, between substantial and

[20] 'Therefore the motion of a species according to prior and posterior [parts] of space entails priority and posteriority in duration, and thus in time.' For an elaborated discussion concerning *species* as produced in time, see Bacon (1996) 1.9.3.

accidental features became redundant. Bacon sided with Averroes in considering species (or corporeal forms) accidental, concentrating his attention on their study.

Both form of corporeity and species are the common factors linking different phenomena, and both operate in the same manner; the description of their activity is the laws which Grosseteste and Bacon prescribed. The laws formulated by Grosseteste were the laws of the radiation of light; the laws which Bacon drew up were the laws of the propagation of species. But in both cases, the formulation was subsequent to an assumption that the material world was unified by common, active, forms, the very features of a nomological image of nature.

Acknowledgements This research was supported by the Israel Science Foundation (grant no. 1622/13): 'Roger Bacon (1214–94) and the Making of the Concept of Law of Nature,' with Giora Hon (Philosophy, University of Haifa) as the Principal Investigator.

References

Armstrong, D. (1983). *What is a law of nature?* Cambridge: Cambridge University Press.

Bacon. (1926). Quaestiones supra undecimum prime philosophie Aristotelis (Metaphysica 12 primae et secundae). In R. Steel & F. Delorme (Eds.), *Opera hactenus inedita Rogeri Baconi* (fasc. 7). Oxford: Clarendon Press.

Bacon, R. (1964). *Opus majus* (2 vols, J. H. Bridges, Ed.). Frankfurt/Main: Minerva.

Bacon, R. (1983). *De multiplicatione specierum* (D. C. Lindberg, Ed. & Trans.). Oxford: Clarendon Press.

Bacon, R. (1988). *Compendium of the study of theology* (T. S. Maloney, Ed. & Trans.). Leiden: Brill.

Bacon, R. (1996). *Perspectiva* (*Opus majus*, part 5) (D. C. Lindberg, Ed. & Trans.). Oxford: Oxford University Press.

Bonaventure. (1993). *The journey of the mind to god* (S. F. Brown & P. Boehner, Ed. & Trans.). Indianapolis, IN: Hackett Publishing.

Cartwright, N. (1992). Aristotelian natures and modern empirical method. In J. Earman (Ed.), *Inference, explanation and other frustrations: Essays in the philosophy of science* (pp. 44–70). Berkeley, CA: University of California Press.

Cartwright, N. (1999). *The dappled World: A study of the boundaries of science*. Cambridge: Cambridge University Press.

Copleston, F. (1966). *A history of philosophy*. London: Burns and Oates.

Crombie, A. (1959). *The significance of medieval discussions of scientific method for the scientific revolution*. In M. Claget (Ed.), Critical problems in the history of science—Proceedings of the Institute for the History of Science at the University of Wisconsin, September 1–11, 1957 (pp. 79–101). Madison, WI: The University of Wisconsin Press.

Eastwood, B. S. (1967). Grosseteste's 'Quantitative' law of refraction: A chapter in the history of non-experimental science. *Journal of the History of Ideas, 28*, 403–414.

Foster, M. B. (1934). The Christian doctrine of creation and the rise of modern natural science. *Mind, 43*, 446–468.

Giere, R. N. (1999). *Science without laws*. Chicago: University of Chicago Press.

Gilson, É. (1955). *History of Christian philosophy in the middle ages*. New York: Random House.

Grosseteste, R. (1912a). De statu causarum. In L. Baur (Ed.), *Die Philosophischen Werke des Robert Grosseteste, Bischofs von Lincoln. Beitrage zur Geschichte der Philosophie des Mittelalters, Texte und Untersuchungen 9* (pp. 120–126). Munster: Aschendorff.

Grosseteste, R. (1912). De motu corporali et luce. In Baur (Ed.), *Die Philosophischen Werke des Robert Grosseteste*.
Grosseteste, R. (1974a). Concerning lines, angles and figures. In *A source book in medieval science* (E. Grant & D. C. Lindberg, Ed. & Trans., pp. 385–391). Cambridge, MA: Harvard University Press.
Grosseteste, R. (1974b). De operationibus solis. In J. McEvoy (Ed.), The Sun as *res* and *signum*: Grosseteste's Commentary on Ecclesiasticus ch. 43, vv. 1–5, *Recherches de Theologie ancienne et medieval, 41*, 38–91.
Grosseteste, R. (1982). *De cessation legalibus*. In McEvoy (Ed.), *The Philosophy of Robert Grosseteste*.
Grosseteste, R. (1996). *On the six days of creation—A translation of the Hexaemeron* (C. F. J. Martin, Trans.). *Auctores Britannici Medii Aevi VI(2)*. Oxford: Oxford University Press.
Grosseteste, R. (2013). On light (N. Lewis, Trans.). In J. Flood, J. R. Ginther, & J. W. Goering (Eds.), *Robert Grosseteste and his intellectual milieu* (pp. 239–247). Toronto, ON: Pontifical Institute of Mediaeval Studies.
Henry, J. (2004). Metaphysics and the origins of modern science: Descartes and the importance of laws of nature. *Early Science and Medicine, 9*, 73–114.
Hyman, A. (1965). Aristotle's 'first matter' and Avecenna's and Averroes' 'corporeal form'. In S. Lieberman (Ed.), *Harry Austin Wolfson—Jubilee volume, on the occasion of his seventy-fifth birthday* (Vol. 1, pp. 385–406). Jerusalem: American Academy for Jewish Research.
Klaaren, E. M. (1985). *Religious origins of modern science*. Lahman: University Press of America.
Lang, M. (2009). *Laws and lawmakers*. New York: Oxford University Press.
Lewis, D. (1994). Humean supervenience debugged. *Mind, 103*, 473–490.
Lindberg, D. C. (1976). *Theories of vision from al—Kindi to Kepler*. Chicago: The University of Chicago Press.
Lindberg, D. C. (1982). On the applicability of mathematics to nature: Roger Bacon and his predecessors. *The British Journal for the History of Science, 15*, 3–25.
Lindberg, D. C. (1986). The genesis of Kepler's theory of light: Light metaphysics from Plotinus to Kepler. *Osiris, 2*(4–42), 12–17.
Maudlin, T. (2007). *The metaphysics within physics*. Oxford: Oxford University Press.
McEvoy, J. (1982). *The philosophy of Robert Grosseteste*. Oxford: Clarendon.
McEvoy, J. (1983). The chronology of Robert Grosseteste's writings on nature and natural philosophy. *Speculum, 58*, 614–655.
Milton, J. R. (1981). The origin and development of the concept of the 'Laws of Nature'. *Archives Europeennes de Sociologie, 22*, 173–195.
Mumford, S. (2004). *Laws in nature*. London: Routledge.
Needham, J. (1956). *Science and civilization in China* (Vol. 2). Cambridge: Cambridge University Press.
Normore, C. (2007). The matter of thought. In H. Lagerlund (Ed.), *Representation and objects of thought in medieval philosophy* (pp. 117–134). Aldershot: Ashgate.
Oakley, F. (1961). Christian theology and the Newtonian science: The rise of the concept of the laws of nature. *Church History, 30*, 433–457.
Pasnau, R. (2010). Form and matter. In R. Pasnau (Ed.), *The Cambridge history of medieval philosophy* (pp. 635–646). Cambridge: Cambridge University Press.
Raizman-Kedar, Y. (2006). Plotinus's conception of unity and multiplicity as the root to the medieval distinction between *lux* and *lumen*. *Studies in History and Philosophy of Science, 37*, 379–397.
Randall, J. H. (1960). *Aristotle*. New York: Columbia.
Ritter, J. (1971). *Historisches Wörterbuch der Philosophie* (Species, Vol. 9). Basel: Schwabe.
Ruby, J. A. (1986). The origins of scientific 'law'. *Journal of the History of Ideas, 47*, 341–359.
Schramm, M. (1981). Roger Bacon's Begriff vom Naturgesetz. In P. Weimar (Ed.), *Die Renaissance der Wissenschaften im 12. Jahrhundert* (pp. 197–209). Zurich: Artemis.

Spruit, L. (1994). *Species intelligibilis from perception to knowledge* (Classical roots and medieval discussions, Vol. 1). Leiden: Brill.

Van Frassen, B. C. (1989). *Laws and symmetry*. Oxford: Clarendon.

Victorinus, M. (1981). *Theological treatises on the trinity* (M. T. Clark, Trans.). Washington DC: The Catholic University of America Press.

Wallace, W. A. (1972). *Causality and scientific explanations* (Vol. 1). Ann Arbor, MI: The University of Michigan Press.

Ward, B. (2002). Humeanism without humean supervenience: A projectivist account of laws and possibilities. *Philosophical Studies, 107*, 191–218.

Weisheipl, J. A. (1984). Science in the thirteenth century. In J. I. Catto (Ed.), *The history of the University of Oxford* (The early Oxford Schools, Vol. 1, pp. 435–470). Oxford: Clarendon.

Wilson, C. (2008). From limits to laws: The construction of the nomological image of nature in early modern philosophy. In L. Daston & M. Stolleis (Eds.), *Natural law and laws of nature in early modern Europe—jurisprudence, theology, moral and natural philosophy* (pp. 13–28). Burlington, VT: Ashgate.

Wolfson, H. A. (1929). *Crescas' critique of Aristotle: Problems of Aristotle's physics in Jewish and Arabic philosophy*. Cambridge, MA: Harvard University Press.

Zilsel, E. (1942). The genesis of the concept of physical law. *The Philosophical Review, 51*, 245–279.

Part IV
Infinities and Transcendentals

Chapter 10
Robert Grosseteste on Transcendentals

Gioacchino Curiello

10.1 Grosseteste and the Doctrine of Transcendentals

From 1996 the historiography began to deal with a new paradigm, namely to interpret medieval philosophy as a transcendental thought. I am referring to Jan Aertsen's ground-breaking book *Medieval Philosophy and the Transcendentals: the Case of Thomas Aquinas*, and his further researches (Aertsen 1996, 2012). Aertsen's thesis is that at the core of medieval philosophy is the doctrine of transcendentals—among which 'being', 'one', 'good' and 'truth' are reckoned—at least from the *Summa de Bono* (c. 1225) by Philip the Chancellor. It is not a doctrine, says Aertsen, alongside many others because of two reasons: first, because every time a medieval theologian writes on it they write in first person (and those 'ego-statement' are relatively rare): second, because it concerns the foundation of thought (Aertsen 1996). Even though I do not accept completely Aertsen's thesis, as it will be clear in this article, his study is very impressive for completeness of information and depth of insight into texts from late Antiquity to the late Renaissance. Accepted or not, Aertsen's provocative position—that the true spirit of medieval thought is the transcendental philosophy—poses new questions and comes to envision new horizons that historians of philosophy have to deal with.[1]

The purpose of my article is to find a place, if any, for Grosseteste in this history. The questions that will lead my research are: is there a doctrine of transcendentals in Grosseteste and is it important in his thought? In his large work, Aertsen

[1] Aertsen's work was praised for its breadth of knowledge of medieval texts, but his historiographical thesis was criticized by eminent scholars, cf. (Gracia 1997; Bázan 2000).

G. Curiello (✉)
Università di Salerno, Salerno, Italy

Université Catholique Louvain, Louvain, Belgium
e-mail: Pius83@libero.it

J.P. Cunningham, M. Hocknull (eds.), *Robert Grosseteste and the pursuit of Religious and Scientific Learning in the Middle Ages*, Studies in the History of Philosophy of Mind 18, DOI 10.1007/978-3-319-33468-4_10

mentions Grosseteste once, referring to his Augustinian position on truth, i.e. what he calls 'adequation formula', namely that the Son of God is the Truth because He is the highest degree of similarity with his principle, God the Father (Aertsen 2012). I will show that Grosseteste has much more to say about the transcendentals, in particular in his *Commentary on the Divine Names* (henceforth *CDN*), one of the four treatises that make up the *Corpus dionysiacum*. As Aertsen himself recognizes, Pseudo-Dionysius—in particular his treatise the *Divine Names*—had a formative influence on the doctrine of transcendentals (Ibid.). Unfortunately Grosseteste's *CDN* is unedited and that is why scholars have neglected this aspect of his thought, and my aim is to fill this gap.[2] I think, in fact, that Grosseteste's Commentary on the *Corpus dionysiacum* in general, and on the *Divine names* in particular, is one of the most suitable places to search for Grosseteste dealing with the tradition of the transcendental thought.

I will divide my article into three chapters. In the first chapter I will give a definition of transcendentals discussing Aertsen's position and softening it by taking into account Klaus Jacobi and Luisa Valente's studies (Jacobi 2003; Valente 2005, 2006; 2007a). In the second chapter, I will present some elements of the transcendental doctrine before and during Grosseteste's time: in particular I will deal with the elements of transcendental doctrine in the twelfth century and then with the first treatises on transcendentals. I will conclude this part by drawing five conclusions which are five fundamental features of the transcendental thought up until Grosseteste. The third chapter will be devoted to Grosseteste's view on transcendentals in relation to the five conclusions of the second chapter.

At the end of this article, I hope to demonstrate that Grosseteste, even though did not develop a systematic account of transcendentals, did however uphold the core ideas of it thanks to his Neoplatonic sources. Grosseteste is in a middle position (not chronological, but theoretical) between the logicians of the twelfth century and the masters of theology of the first half of the thirteenth.

10.2 Transcendental: A Definition

We need to know what a transcendental is, before presenting Grosseteste's sources and position. First of all, we must remember the term 'transcendental' is a modern one; medieval authors, in fact, used words like '*transcendentia*', '*communissima*' or '*prima*'. That said, I will try to define transcendentals and though it is not my intention to report all the nuances that those words assumed in the Middle Ages as a general introduction we may quote Aertsen:

[2] The transcription of *CDN* is, unless stated otherwise, from Oxford, Merton College, MS 86. The text has been corrected against Paris, Marazine MS 787 and Paris, Biblioteque Nationale MS Lat.1620. English translation is mine.

> Some modern scholars define transcendentality in an extensional sense. Take for instance: 'A transcendental predicate runs through all the categories and extends beyond to their first cause' [Owens 1963]. Other scholars stress the necessity of an intensional account, which leaves more room for the diversity of medieval views on the meaning of the transcendental. A purely extensional definition can perhaps account for Aquinas' understanding of the transcendental as that which runs through the categories because of its commonness, but it explains neither Scotus' understanding of the transcendental as that which is not determined to a genus, nor the early Scotists' conception of degrees of transcendentality (Aertsen 2013).

This passage calls for some clarifications. It has to be noted that the extensional definition is not simply something held by modern scholars (as it seems from the passage or the whole article), but it is explicitly a definition from the twelfth-century logic (De Rijk 1967). I agree with Aertsen when he suggests that this kind of definition needs to be completed by an intensional one but I think that it is part of the doctrine of transcendentals and it is arbitrary to start the history of this idea from the 1225, when the authors reason more about the differences among transcendentals than to establish their commonality.

Aertsen retraces the great variety of meanings of '*transcendentia*' or '*communissima*' in detail and one of the definitions is 'that which can be said of everything'. This means that transcendentals are coextensive, that is whatever is said to be 'transcendental X' is at the same time 'transcendental Y'. The original group of names which have the property to be said of everything includes 'being', 'one', 'thing', 'something'. In the thirteenth century the names 'good', 'true' and 'beautiful' were added.[3] This semantic meaning of transcendence (De Rijk 2003)—found in the logical treatises of the twelfth century—is one of the most important and maybe the first to definition to appear, as Aertsen himself admits (Aertsen 2012). I agree with him when he states that what is entirely missing in those treatises is a systematic analysis of the relations of the transcendentals to one another and the metaphysical dimension, which is constitutive for the transcendental theory in the thirteenth century (Ibid.). Therefore the difference between the twelfth and the thirteenth century in the conception of transcendentals is not so much in the definition (the intensional meaning does not replace the extensional one, but rather complements it) but in the capacity of building a comprehensive theory, a capacity that began to develop with the arrival of Aristotle's new works. We will see, in fact, that in the twelfth century, the transcendentals are considered mostly in their semantic aspect—as *communissima*—but with the arrival of *Metaphysics*, *Posterior Analytics*, *Nicomachean Ethics* and the Arab commentaries on Aristotle, the transcendentals will assume an epistemological status—as *prima*, the first conceptions of the intellect—and an ontological status, as *transcendentia*, that which transcends the categories of being (Ibid.).

[3] On the transcendentality of beautiful there is not unanimity among scholars. Aertsen (1991) discusses at length his position against Eco (1988)). My position, that I cannot develop here for reasons of space, is that 'beauty' belongs to the transcendentals.

10.3 Sources of Grosseteste's Thought on Transcendentals

Grosseteste was intellectually active during both twelfth and thirteenth centuries, which gave him access to two traditions about transcendentals. The first of these traditions refers to logical treatises (based on *logica vetus*), and theological treatises (in particular those about theological language); these belong principally to the twelfth century. The second tradition refers to the first treatises on transcendentals, written during the first quarter of the thirteenth century. Notably, Poullion held that the *Summa de Bono* written by Philip the Chancellor was the first treatise on the transcendentals (Pouillon 1939). This thesis was confirmed by Aertsen, however not everyone agrees. In fact, Alessandro Ghislaberti (1990) and Jack Zupko (2003) believe that we can find traces of the doctrine of transcendentals—or, as Pouillon suggested, an 'embryonic' doctrine—some years earlier in the *Summa Aurea* by William of Auxerre and in the *Magisterium divinale et sapientiale* by William of Auvergne. We will see what those treatises add to the accounts given in the twelfth century and what is the novelty of Philip's treatise.

Grosseteste's Commentary on the *Corpus dionysiacum*, which is my reference work for this study, was written in the 1240s. I will demonstrate that he was aware of those traditions. He does not elaborate a keenly developed theory like Philip the Chancellor, but nor does he propose again the twelfth-century account. Neither does he achieve a synthesis of those traditions, and nor was he searching for it. The fact that the Commentary was written during the period of his episcopacy is the key to understanding why he arrived at his particular consideration of the transcendentals and how he dealt with them.

10.3.1 The Twelfth Century: Boethian transcendentals

Luisa Valente and Klaus Jacobi hold that in the twelfth century there was some anticipation or proto-theories of the doctrine of transcendentals. Valente has rightly pointed out that those theories came from two kinds of genre: on one side the logical tradition dealing with Boethius' logical treatises where he poses the convertibility of '*ens*' et '*unum*' (Boethius 1906, 2000). On the other side the theological tradition dealing with the divine names and their equivocity or univocity (Valente 2005). Now I will list the fundamental points of these 'proto-theories' as found in the studies of those scholars.

(1) In the logical tradition of the twelfth century, transcendentals were names that can be said of everything. The lists of *nomina transcendentia* are longer and different from what we usually think (i.e. '*ens*', '*bonum*', '*unum*' and '*verum*'), they are often incomplete and usually end with the expression '*et similia*'.[4]

[4] An example may be found in (De Rijk 1967).

From this kind of lists miss other terms that will be decisive in the next century, namely 'true', 'good' and 'beautiful'.

(2) The theological tradition was more involved in questioning about the over categorial terms as divine names, and trying to qualify them as equivocal or univocal. The question was: Can transcendentals like '*ens*' and '*unum*' be properly predicated of God? According to Valente, there is an evolution of this debate in three stages: the first one, led by Gilbert of Poitiers' followers, was to consider divine names equivocal names and therefore they were not properly attributed to God.[5] As a reaction to this position, thinkers like Prepositino of Cremone held the univocity of those names; the last development was an attempt to reach for a mediation between the previous positions. This last position is particularly important for our subject because it manifests the conscience of the particular status of some terms that will be recognized as transcendentals. Words like '*res*', '*ens*', '*unum*', '*aliquid*' are considered a particular group with its own characteristic (to be predicated univocally of both God and creatures) and are different from all other divine names (which are said equivocally). Those names are called *superpredicamentalia* or, using a circumlocution, 'those that transcend the most general genres' (*illa quae trascendunt generalissima*) (Valente 2006).

The commonness of the notion 'good', absent in the logical treatises on the over categorial term, is an important element that appears in this metaphysical/theological tradition. Valente rightly completes her studies on what we may call 'transcendentals before a doctrine on transcendentals', adding the contribution of Boethius' *De hebdomadibus* to the culture of the twelfth century (Ibid.). From the identity of 'being', 'being one' and 'being good' in God there derives (*fluere*) the identity of these three features in the creatures. This thesis will become more and more decisive in the following century.

(3) There was not a developed theory caring about distinctions among the transcendentals themselves. Using the terminology of Aertsen, in the twelfth-century thinkers formulated an extensive definition of transcendentals (what can be said of everything), but not an intensional definition.

10.3.2 The Thirteenth Century: The First Treatise(s)

The twelfth century's reflection about over categorial terms was characterized in particular by Boethius' thesis that *ens* and *unum* are convertible; at the beginning of the thirteenth century, theologians became more and more influenced by another Boethian work, the *De hebdomadibus*, which concerned with the coextensiveness of good and being. It was not just a coincidence that in the first decades of the

[5] Among those authors we have Simon of Tournai and Alan of Lille (Valente 2007a, b).

twelfth century some of the most important works were devoted to this theme.[6] As was pointed out in the introduction of this chapter, some scholars believed it is possible to find some traces of an 'embryonic' doctrine of the transcendentals in William of Auxerre and William of Auvergne, who wrote a decade before Philip the Chancellor. It is not easy to take a position in a debate based on an unclear word like 'embryonic'. It is undeniable that the level of orderliness reached by Philip is far from the sketched lines drawn up by the two Williams. In my opinion, the works of William of Auveregne and that of William of Auxerre have to be traced back to the attempts made by Gilbert of Poitiers and his school to make the *De hebdomadibus* the source for the account of the transcendentality of goodness. If we still want to use metaphors, I think that more than say 'embryonic', which presupposes an organism already ordered but not well developed, the two Williams added another important tile in the mosaic of a doctrine of transcendentals.

What is important for my argument is to highlight what is common to all three authors and what Philip adds to the account. The first point in common is that they share the preoccupation with the notion of the good, and their interest is motivated by the challenge of Neo-Manichaeism and its dualism (Teske 1993). Philip is particularly explicit in his prologue in connecting the foundations of thought (which are the transcendentals) with Manichaeism, which ignores them. The second point in common is the recovery of Boethian distinction between *bonum in essentia* and *bonum per participationem*: only God is *essentialiter* Good and all other creatures are good since their being has flowed (*fluxit*) from the first good (Boethius 2000). This means that Good can be predicated, in different ways, of everything just like 'being', and therefore it is a transcendental.

The originality in Philip the Chancellor's account consists in the fact that good is not only the most common principle, but it is clearly linked with being, one and true. Not only the most general features of reality are coextensive and convertible (*convertuntur*) but they are also intensionally distinct as MacDonald rightly points out (MacDonald 1992). It is from Philip's mind that there arises the famous formula '*idem in re sed differunt ratione*', which means that transcendentals are the same in the subject but they are conceptually or logically different.[7]

10.3.3 Conclusions

The great part of Grosseteste's works, which I will take in consideration in the next chapter, were written during or after Philip's *Summa*, therefore it is important to list

[6] Besides the authors mentioned in this paragraph, I want to recall Albert the Great's *De natura boni* (1236/7) and *De bono* (in the '40) and the collective work of the first Franciscans at Paris, the *Summa Halensis* (1240–1256).

[7] To be precise, the formula comes from Aristotle *Physics* III, 3 (202b7-202b22), but its application to the transcendentals is Philip's originality.

the principal characteristics of the previous period in order to understand how he dealt with his sources, showing his debt and his originality.

We can draw the following conclusions from the first part.

(1) In the twelfth century transcendentals are names which have the particularity to be predicated of everything. Logical treatises list various and several names but their number is not established clearly.
(2) 'Truth' is not present in the lists of transcendentals either in the logical or in theological traditions.
(3) In the debate about theological discourse, some leading figures like Simon of Tournai and Alan of Lille held that transcendentals are equivocal terms, consequently they are not properly said of everything; relying on the authority of Pseudo-Dionysius they stressed that there are no distinctions among God's names and we cannot say properly anything about God.
(4) During the first half of the thirteenth century, in the context of a Neo-Manichaen controversy, the notion of goodness became a central issue for many authors. There is a clear shift from a semantic treatment of transcendentals to a more metaphysical conception of them: transcendentals are not only words, but properties of being.
(5) It is in Philip the Chancellor's *Summa* that the investigation into the distinctions and order among the transcendentals began.

In the second part of the article, I will present Grosseteste's position on each point, showing how he dealt with his sources.

10.4 Grosseteste on Transcendentals

10.4.1 List of Transcendentals

As we saw in the second chapter, in the twelfth century a lot of terms were considered over categorial, namely that which had the capacity of being said of everything because they did not belong to any Aristotelian category in particular, however their number was not established. Grosseteste never makes a list of transcategorial terms but we can deduce that he also considered some terms to be predicable of everything. I would like to present two passages from two different texts: *Commentary on the Mystical theology* (henceforth *CMT*) and the *Commentary on the Posterior Analytics*.

10.4.1.1 Commentary on the Mystical Theology

There is at least one passage, taken from the *CMT*, the last treatise of the *Corpus dionysiacum*, that allows us to think that Grosseteste considered as transcendentals a long list of names. This passage is found in the last chapter of the work where he describes the last steps towards the mystical union with God. In order to reach it, according to a reading of Grosseteste, man has to remove and go beyond sensible things (first step) and then intelligible things (second step). The further move (third step) is to go beyond what is common to both sensible and intelligible things, namely things that are both perceptible by the senses and apprehensible by the intellect. What are those things? Grosseteste answers: 'These, however, are number, which follows upon everything as soon as it is there; and rest and motion, one of which follows necessarily upon the existence of each thing, insofar as every agent is said to be moved and every non-acting thing is said to rest; and natural power, which similarly belongs to everything; and the measure of being in which everything necessarily shares; and the love of good which is naturally within everything' (*CMT* 113) This is Grosseteste's original method of making sense of Dionysius' list of things to be removed from God, and he complies with this order even at risk of forcing Dionysius' text. For Dionysius God 'is not number or order, nor greatness or smallness, nor equality; nor similarity or dissimilarity; it neither stays at rest nor is it moved; nor does it bring about silence or have substance, or everlastingness, or time; it cannot be grasped by the understanding; it is not knowledge or truth; nor kingship or wisdom; it is neither one nor unity'. Grosseteste classifies those names in the categories mentioned before. For example, in the category 'number', there are 'number, order, greatness, smallness, equality' and in the category 'measure of being' we have 'everlastingness and time'. The strangest category is the 'love of good', which does not actually occur in Dionysius' text and in which fall 'understanding, knowledge, truth, kingship, wisdom, unity' that should pertain to the intellect alone. According to Grosseteste, those names represent 'the appetite of good that knows, and does not err in the act of knowledge, that commands the actions and does not err in acting' (*CMT* 117). This explanation reveals a leitmotif of Grosseteste's thought, namely the inextricable unity of *aspectus* and *affectus*, intellect and appetite, wisdom and love (Callus 1955; McEvoy 1982).

It has to be noted that Grosseteste's list is very different from those of other writers in the twelfth century. First of all here it is not just a question of terms, as in the logical treatises of the previous century. Dionysius—and Grosseteste as commentator—is speaking of concepts and realities: concept to be removed from human minds when it wants to go towards God; realities to be removed from God Himself because He is above everything. It is possible to consider Grosseteste's list a list of transcendentals? In a very loose sense yes, because Grosseteste states that all things are marked by those properties, therefore are *communia*. It is highly questionable if they are convertible with one another, but Grosseteste is not interested in this kind of questions, just like the logicians of the twelfth century.

An element that may surprise the reader is that the next passage (fourth step) of the ascent towards God, according to Grosseteste, is the removal of divine names given to the whole Trinity ('deity' and 'goodness') and, finally (fifth step) the removal of divine names concerning each Person of the Trinity ('Father', 'Son', 'Spirit'). This seems to imply that the names said in common of sensible and intelligible beings (third step) are not suitable to God. The strangeness of this part consists in the fact that a lot of names said in common of sensible and intelligible beings are in fact considered as divine names in the *CDN*. I am referring to names such as 'one', 'wisdom', 'great and small' etc… that are considered divine in the *CDN* and simply common to intelligible and sensible things in the *CMT*. Grosseteste, as commentator on two different treatises—the *Mystical theology* and the *Divine names*—does not seem to be coherent. After a deeper reading of the texts we realize that some special names that are common to sensible and intelligible beings are also considered divine names and in the mystical ascent must be denied of God not because He lacks something but because He exceeds the way in which the property signified by that name is realized among the creatures (*CMT* 117). These tensions in the *CMT* are almost inevitable when someone tries to make sense of the long list of things to be removed from God, made of repetitions and without an explicit given order. Grosseteste arranges a complex division of the text in which some particular names are said of every created being and contextually of God, but in order to ascend towards the union with God, they had to be denied. What is important for my argument is that those names are said of everything (God included), as the transcendentals, even though Grosseteste does not evidently employ the notion.

10.4.1.2 Commentary on the *Posterior Analytics*

We can look at another passage, in another context, for the explicit presence of the term '*transcendens*'. It occurs in Grosseteste's *Commentary on Posterior Analytics*, written probably in the late twenties. Aristotle is criticizing the method of definition by division because it could lead him to commit such mistakes such as introducing accidents of the subject which are not part of its essence, or fail to state the final difference of the subject and passing over an intermediate difference (Aristotle 1957, 639; Lib. II Cap V, 91b12-27).

Grosseteste comments that the method of division leads to vague definitions that say nothing about the essence, where is possible to exclude necessary elements or, on the contrary, include unnecessary items (*superflua*):

> Furthermore nothing prevents us from taking, through this division, keeping unnecessary elements in the definition, as if 'animal' were divided according to substantial and accidental differences; or from removing something necessary to the definition, as if 'animal' though divided not according to the proximate differences; or even from taking transcendental things (*res transcendentes*) which are not appropriate to a definition, but have to be set aside, as if someone began to divide 'man' into 'being' or 'non being' and then takes 'being' (Grosseteste 1981).

Among the unnecessary elements that should not be put in a definition, Grosseteste says, there are *res transcendentes* and 'being' is one of them. Since Grosseteste used the plural we can assume that there are other transcendentals but he does not add anything else. Being is beyond the substantial or accidental differences, because it is something that belongs to everything, which is why it is redundant to put it in a definition. With the last remark, it is worth noting that we are in a logical context like that of the twelfth century works analyzed in the first part of this article.

10.4.2 Truth

We saw that Grosseteste does not give a list of transcendentals nor he is clear on their number, like the vague lists of the twelfth century. Grosseteste, however is seen to be different from them and to be closer to the mature treatises on transcendentals, like that of Philip the Chancellor, because he considers 'truth' a transcendental. Grosseteste tackles the concept of 'true' as transcendental in his philosophical treatise *On Truth* (c. 1225), where he employs a reasoning very similar to Philip.

In this treatise Grosseteste answers the question of uniqueness or multiplicity of truth. The interesting part of the treatises for the subject at hand is where Grosseteste proposes several definitions of truth in supporting the uniqueness or the multiplicity of truth. Among the first group there is the following one, which Grosseteste seems to prefer, drawn from Augustine's *Soliloquia*: 'Truth is that which is. Therefore the being of each thing is the truth of it' (Grosseteste 1912).[8] Two other definitions, which fall in the second group, deserve our attention: 'Truth is that which shows what which is; its truth, therefore, reveals the being of each thing' (Ibid.); and another similar to it, 'The supreme truth reveals all goods which are true' (Ibid.).[9] It is not my intention to study the doctrine of truth by Grosseteste, but to show that, according to him, not only being, truth and goodness are correlative and coextensive, (Ibid.) but also truth adds something to being and goodness, namely truth is what shows, manifests being and goodness.

Although Grosseteste is not explicit on this point, a comparison with Philip the Chancellor, who also used Augustine's definition of 'truth', may help to clarify Grosseteste's point of view. Philip holds that the identity formula taken from Augustine—'truth is what it is'—is a definition that reveals what truth is substantially; however he is afraid to fall into tautology by identifying truth and being *sic et simplitciter*. That is why Philip prefers his own definition—'truth is the indivision of being and what it is'—and introduces the element of indivision to distinguish the two transcendentals.[10] I think that Grosseteste does not fall into tautology because, notwithstanding his preference for the identity formula, he also introduces an element of distinction, taken from Augustine, namely the capacity of truth to reveal being and goodness.

[8] Grosseteste reaffirms the identity of being and truth in *CDN* 252vb.

[9] Grossetste refers to Augustine's *De libero arbitrio* II 13, n. 36.

[10] For a full exposition of Philip's argument see (Aertsen 2012).

Grosseteste aims in this treatise is to harmonize the authorities, therefore the definitions of both groups are considered, approved and employed in the discussion. Those definitions are not redundant but each of them adds something to the others. In fact, some paragraphs later Grosseteste, trying to find a solution, recovers the idea that truth reveals being, and he tells us: 'Created truth too, therefore, shows that which is, but not in its own illumination [*lumen*], but in the light [*lux*] of the supreme truth, as colour shows body, but only in the light spread upon it' (Ibid.). It means that just as seeing a body as coloured is impossible without the sun shining upon the body, so also seeing a created thing as true is impossible without a higher light shining upon the thing.[11]

This passage sheds light on another aspect of truth and being considered as transcendentals. I am referring to the fact that transcendentals are not only *communia* to all things, but also *prima*, namely they are the first known by the intellect. Everyone consciously, as the pure in heart, or unconsciously, as the impure men, needs the supreme truth in order to reach any other certain knowledge (Ibid.). Grosseteste goes further and holds that just as the supreme truth is the condition for knowing every other truth, so also divine Being is the condition for the existence of every other being which, consequently, depends on it just as water needs a container to be shaped, otherwise it flows away, so also creatures without God would slip back into nothing. At the same time this example demonstrates that knowing the shape of water means to know firstly—consciously or not—the shape of the container (Ibid.). I want to stress that Grosseteste refers to divine being and not to common being (*ens commune*), which comes from the Avicennian tradition. Even though Grosseteste knew the Arab philosopher, his account of the first impression of the soul is different precisely on this point; both hold that there are some preconditions for all further conceptual knowledge, but according to Grosseteste it is the divine being and truth, which are above and prior to anything else, while for Avicenna these preconditions are common being, thing and one.[12]

Grosseteste's treatise *On Truth* is the source, together with Gilbert of Tournai, of Bonaventure's theory of illumination, as Servus Gieben and Camille Bérubé demonstrated (Bérubé and Gieben 1974). This aspect has been studied by Aertsen, but unfortunately he focused solely on Gilbert's contribution, neglecting the fact that Grosseteste is the source of Gilbert too (Aertsen 2012). This is a first step to the acknowledgement of Grosseteste in the long chain of sources that led to the doctrine of transcendentals.

---

[11] For an extensive comment on Grosseteste's analogy see (Cooper 2012).

[12] For a brief discussion of the Avicennian contribution to the doctrine of the transcendentals see (Aertsen 2012).

10.4.3 Analogical Names

Concerning the equivocity or univocity of transcendental terms, Grosseteste developed his considerations in two different works: the *Commentary on the Physics* and the *CDN*. Aertsen holds that the ontological discourse in the twelfth century was dominated by Porphyry's *Isagoge* and he concludes that: 'Both his (viz. Porphyry) claims—that of the primordial diversity of things and that of the equivocity of 'being'—reveal a horizon different from transcendental perspective. They are in fact incompatible with fundamental presupposition concerning the predictability and conceptual unity of the transcendentals [. . .] Another supposition is the recognition that the concept of being, although not univocal, nevertheless possesses a certain unity' (Ibid.). I will demonstrate that Grosseteste uses the concept of analogy implicitly and explicitly dealing with transcendentals terms.

10.4.4 Commentary on the Physics

Grosseteste does not show much interest in Aristotle's *Metaphysics* which he sporadically quotes, but we have another place that has been neglected by most of the scholars but that says a lot about being and its univocity, equivocity or analogy, namely Aristotle's *Physics*.[13] The work of Silvia Donati revealed the richness of medieval commentaries on the *Physics* in the Parisian and Oxford traditions (Donati 2003). Unfortunately she started her analysis from the commentaries of the late fifties of the twelfth century and overlooked the previous works. In what follows, I will argue for the analogy of being in Grosseteste's *Commentary*.

Donati stresses that British authors clearly distinguished between the logical and metaphysical level of discussion about being (Ibid.). She refers to authors like Geoffrey of Aspall whose *Commentary on the Physics* was written in the fifties 1250s. He held that in logic the sufficient condition of equivocity for a term is to be predicated according different notions; whereas in metaphysics and physics there are stronger conditions. A term is equivocal not only if it is predicated according to different notions but also the things predicated are not ontologically or gnoseologically dependent on each other. The consequence is that 'being', which is predicated according to different notions to the substance and to accidents, is equivocal according to a logical point of view, whereas it is analogical according to a metaphysical and physical point of view because of the dependence of accidents on substance.

Grosseteste's position is not as developed as that of Geoffrey, however there are interesting elements common to the two authors. Grosseteste comments on

[13] The doctrine of the analogy of 'ens' in Grosseteste's commentary of the *Physics* has been studied by Neil Lewis (2009) to which I refer the reader for further details. In this paragraph I have omitted all the information that can be found in that study.

Aristotle's criticism of Eleatic monism and connects the question of unity of being with the presence of one superior Being, namely God, saying that: 'If being were said univocally about everything, then everything would be "being" substantially, for being is said substantially of the First Being; but it is not possible, because what is being substantially is necessarily being in virtue of itself: but what is being in virtue of itself can be absolutely only one eternal being; therefore if it were univocal, all things would turn out to be only one single thing' (Grosseteste 1963). But Grosseteste is not content with affirming that being is said equivocally, but he also specifies the kind of equivocity, that is equivocity due to a dependence of one thing on another. Grosseteste says: 'Since being is said substantially of only one thing, and of other things in respect of a dependence on that one thing in respect of prior and posterior, it is clear that it is said equivocally' (Ibid.). God is the only one to be called 'being' substantially, while all other things are said to be 'being' after Him. There is an ontological dependence just like the one evoked by Geoffrey, but Grosseteste is more interested in the dependence of creatural being on God, than the dependence of accident on the substance. We already realized this dependence reading the treatise *On Truth* in the previous paragraph. On this point it is clear that the difference between this kind of equivocity, that is de facto analogy, with the kind of equivocity stated by Boethius and the twelfth-century logicians, namely the equivocity from one thing (*ab uno*). Boethius considers the dependence of two things on another one, like instruments and potions are said 'medical', because they descend from the art of medicine; Grosseteste, instead, is stating a dependence of one thing on another superior to it, and there is no room for a third element shared by both.

Moreover, it is worth noting that the expression 'according to prior and posterior sense' (*per prius et posterius*) stands for analogy in a lot of authors of the first half of the twelfth century (Ashworth 2013). Jennifer Ashworth noted that Grosseteste (and many others after him) connects analogy with the word 'ambiguous' (*ambiguum*) in his *Commentary on the Posterior Analytics*, saying that 'Aristotle's use of analogy to find a common term produces ambiguous names said according to a prior and a posterior sense, and he uses the phrase "*ambiguum analogum*"' (Ibid.).

A passage that is more useful for my argument, because it involves two transcendentals, being and truth, is at the very end of the treatise *On Truth*. There Grosseteste realizes that the concepts of truth and being are analogical because they are predicated of all things and at the same time they are just one, and their uniqueness is realized, as I showed in the previous paragraph, in God. Grosseteste concludes his treatise holding that 'the intention of truth, as the intention of being, is ambiguous: from one part it is one in all truth and, nevertheless, by appropriation it is diversified in the particulars' (Grosseteste 1929). Grosseteste then, following a Boethian interpretation of the *Isagoge*, tries to maintain a certain unity of the transcendentals using the concept of analogy: implicitly in the *Commentary on the Physics* and then explicitly in other works.

10.4.4.1 Commentary on the Divine Names

The account on the analogy of transcendentals would not be completed without a discussion about another kind of analogy, different from that treated above. In Grosseteste's Commentaries on the *Corpus dionysiacum* the vocabulary changes: here the word '*equivocatio*' simply means misunderstanding, *quid pro quo* (*CDN*, 183va). In the Commentary Grosseteste, following Dionysius, uses the concept of analogy to say that God participates Himself to all creatures according to their receptiveness. It is not the Aristotelian analogy, a comparison of proportions, but another type of analogy, often used by theologians, which appealed to a relation of likeness between God and creatures. Creatures are (called) good or just because their goodness or justice imitates or reflects the goodness or justice of God. This type of analogy was called the analogy of imitation or participation (Ashworth 2013). There are some scattered remarks about Grosseteste's use of analogy in the *Commentary on the Celestial hierarchy*, in the study of James McEvoy (1982). I will linger, instead, on the pervading use of analogy in the *CDN*. In Book IV of the *Divine names* for example, Dionysius holds that the Highest Good surpasses the sun like the archetype surpasses its dark images. This means that, as the sun radiates on everything, Goodness does so even more, for the very fact that it exists, it reaches out to everything, which participates in Him according to a proportion, i.e. analogy (*CDN* 198rb). As we will see in the next paragraph, everything is good even non-being, because it comes from God who calls non-being to existence. This capacity of reaching out to everything is not a prerogative of goodness alone. Recalling a Dionysian adage, it is not just goodness that is self-diffusive (*diffusivum sui*), but being too. In fact 'Being', namely divine being, extends itself to everything (*CDN* 234va), and Grosseteste uses the verb 'extend' (*extendere*) both for being and for goodness (*CDN* 202va).

Grosseteste overcomes the controversy of the twelfth century on the equivocity or univocity of transcendentals, thanks to the notion of analogy. He, relying on the authority of Dionysius, comes to the opposite conclusion of some twelfth-century thinkers, like Simon of Tournai and Alan of Lille, who denied any possibility of a meaningful theological language; according to Grosseteste, it is possible to say something of God, because everything participates in His perfections: rational being as His image (*imago*), all other creatures as simple similitude (*similitudo*), as he states in his *Hexaëmeron* commenting on the biblical passages about God's creation of man (Grosseteste 1996).

10.4.5 Primacy of Good Against the Neo-Manichaeism

As set out above Grosseteste maintained the idea of the transcendentality of 'truth', an element in common with Philip the Chancellor and the later treatises. Another element that brings Grosseteste near to the 'spirit' of his contemporaries is the topic of 'goodness' in the context of a Neo-Manichaeism controversy. Goodness becomes the first concern and the first divine name in Grosseteste's thought thanks to his commenting on the *Divine names*.

There are several passages in the IV Book, dedicated to the name Good, where Grosseteste states that good is the first divine name because it is a procession, namely something that proceed from God to creatures, that comprehends every other procession. We can say that goodness is not simply a procession, but the act of God proceeding. Goodness precedes being because goodness extends itself to non-being too, calling it to existence (*CDN* 198ra). Grosseteste confirms his view in the next Commentary, that on the *Mystical theology* where he states: '[...] whatever can be said concerning him positively can be comprehended in one single, positive word, which is to say the name of good, as is clear from what has been said concerning the *Divine names*' (*CMT* 77).

Those passages reveal another element Grosseteste held in common with contemporary masters such as Philip the Chancellor, William of Auxerre and William of Auveregne, beside the commonness of good. All these authors believe that God is the source of good, which flows from Him to creatures according to priority or posteriority or, more explicitly, according to analogy (Aertsen 2012). It is interesting to note Aertsen's remark on Philip, who employs the expression '*per prius et posterius*' even though he knew the notion of analogy. We have already seen something similar in Grosseteste: before his commentaries on Dionysius' works the occurrences of analogy are sporadic. Grosseteste, drawing from another source, namely the *CDN*, instead of Boethius' *De hebdomadibus*, comprehended the coextensiveness of good and being and their proportional (analogical) reception by creatures.

There is also in Grosseteste an explicit reference to the Manichaeism, a heresy that he refutes throughout his career, from the *De libero arbitrio* till the Commentaries on the *Corpus Dionysiacum*, passing through the *Hexaemeron* (Grosseteste 1912).[14] Among the several occurrences of 'Manichaeism' in Grosseteste's work, I have chosen one passage that could be more suitable to the discussion about transcendentals. It occurs in the XI Book of *CDN* dedicated to the name 'peace'. Dionysius says that peace means to have unity and steadiness. But someone could object that everything is in motion and is multiple, various, not one. Dionysius replies that there is nothing lacking completely of unity or steadiness, otherwise it does not exist at all. Grosseteste, commenting on this passage, states that those Dionysian words are against the Manichaeans, according to whom the nature is unstable and with no limits which is impossible, due to the premise (*CDN* 268va). This passage confirms also the transcendentality of '*unum*', one. This name will be treated properly in the last chapter of the *Divine names*, where Grosseteste reaffirms

[14] In the Hexaemeron (Grosseteste 1996, 81f) Grosseteste argues against the manichean interpretation of Genesis 1, 2, (about the 'darkness over the abyss') which is the source of that heresy. Manicheans are mentioned also in the *Commentary on the Ecclesiastical Hierarchy* in a passage on Christology and several heretics are mentioned and confuted by the authority of Dionysius see (Grosseteste 1991).

Dionysius' idea that everything that exists is one, from God to the lowest creature. Once again, drawing from the *CDN*, instead of Boethius, Grosseteste reached the same conclusion as the logicians of the twelfth century, namely that being and one are coextensive and, again, are participated in creatures according to a prior and posterior sense (*CDN* 274ra).

The several mentions of Manichaeism, along with other heresies, in the commentary on the *Corpus dionysiacum* lead us to reflect upon the time when Grosseteste commented on the *Corpus*. In the 1240s, when he was already a bishop and he was very much involved in the life of his diocese he was so occupied in pastoral activity that he had a reputation for sanctity among the faithful (McEvoy 2000). The commenary was not an intellectual tantrum, but it represented the study of an authoritative masterpiece—along with his beloved Augustine—which gave him the tools to answer to the pastoral needs that was the struggle against heresy. This pastoral use of Dionysius is already known for issues like hierarchy (Hogan 1996). I think that we can add the problem of goodness as a common feature of everything as another Dionysian legacy upon Grosseteste in general and his pastoral activity in particular.

10.4.6 Differences Among the Transcendentals

We saw already that Grosseteste was closer to Philip the Chancellor than to the twelfth-century tradition about truth considered as transcendental. Truth is not only coextensive with being, but it also adds to being light and intelligibility. Unfortunately Grosseteste did not develop his thought, but he limited himself to quoting Augustine. There is another passage, in the *CDN*, where the distinction between transcendentals is clear and technical; a passage already known to scholars, that demonstrates that Grosseteste was aware of the most recent development on the transcendental doctrine (Pouillon 1946; Eco 1988). It occurs in the IV Book of *CDN* devoted to the divine names 'good' and 'beautiful'. Grosseteste realizes that, according to Dionysius, those names signify the same thing and nevertheless they are not identical. Grosseteste wonders what the difference is between them and says:

> Good and beautiful are the same, because everything tends to good and beautiful with respect to all causes, as efficient, final and formal cause [. . .], and there is no being which does not participate in beautiful and good, otherwise the natural appetite of everything would be useless and vain. [. . .]. But someone could say that good and beautiful are the same according to the thing (*secundum rem*) but, since divine names signify His beneficent processions into creatures, good and beautiful are different according to the reason (*diversa ratione*). God is called good because He confers being and well-being on everything and He increases and perfects and preserves. But He is called beautiful in that makes all things, both in themselves and reciprocally, concord in their identity with Him. But these differences of intentions are contained both in themselves and reciprocally, so that neither being nor appetite could be without the other (*CDN* 204vb. Dioysius' words are in italics).

This passage is very dense and there are a lot of points to be elucidated. First, it is a matter of fact, that everything participates in goodness and beauty and therefore good and beautiful are the same. Wherever there is beauty there is goodness and vice-versa. Everything participates in good and beautiful because otherwise its natural appetite, or tendency, would be in vain. This point is confirmed in the passage from *CDN* (113) mentioned above. The tendency to goodness (*appetitus boni*) is naturally inscribed in everything, from the stones to the angels.

Grosseteste then considers good and beautiful not as divine processions into creatures, but as the names signifying those processions. In this case Grosseteste refers to the opinion of someone who says that they have the same reference, namely they are coextensive, but they are conceptually different. Goodness is attributed to God in so far as He is the source of existence and perfects it; beauty is ascribed to Him in so far as He is the principle of concordance and harmony. This expression '*idem in re sed diversa ratione sunt*' is a technical expression of the transcendental theory as I have shown above. To whom is Grosseteste referring? Grosseteste's source is probably the *Summa Halensis*, written almost in the same years, which is the first to use this expression to discriminate beauty from good (Alexander of Hales 1924).[15] The *Summa* also considers the relationship of good and beauty to the causes, but it is more accurate than Grosseteste who does not distinguish between them; the *Summa* connects the formal cause with beauty and the final cause with goodness (Ibid.).

Pouillon and Eco believe that Grosseteste accepts the view of the *Summa*, but a closer reading of the text allows us to conclude the opposite. Pouillon transcribed Grosseteste's text until the point quoted above, but it goes on to reveal something different (Pouillon 1946). Grosseteste refers to the conceptual differences between good and beautiful in order to reject it, because he does not want to imply any kind of multiplicity in God, nor in His processions into creatures. A few lines later, in fact, Grosseteste–after having stressed that good and beautiful are so connected that they cannot be separated, concludes that even if goodness and beauty were considered as processions and not as God's properties in themselves, they would be identical in their intentions (*in illis rationibus realem habent ydemptitatem*), because the appetite for good and beautiful is one and the same (*CDN* 204vb). This means that Grosseteste is aware of a tradition that considers the transcendentals as conceptually different but he rejects it. What is important for Grosseteste is to determine the identity of the object of desire, namely God considered as beautiful and good and for this reason the word 'one' (*unum*) occurs twice, at the beginning and at the end of his consideration. This passage clearly shows the distance between Grosseteste and other writes of the thirteenth century reflection upon the transcendentals.

[15] '*Pulchrum at bonum sunt idem in substantia, sicut habetur a Dionysio [...], sed aliqua est differentia rationis*'. The difference between good and beautiful in these terms will be recovered by Aquinas (*Summa Theologiae* I, q.5, a. 4, resp.) almost with the same words.

10.5 Conclusion

In conclusion I want to answer to question posed at the beginning of this work: does Grosseteste belong to the tradition of transcendental thought? We can answer this in two ways: general or particular. On the one hand we can consider the doctrine of the transcendentals in general, like John Marenbon who dealt with Abelard and the transcendentality of one (*unum*). Marenbon says: 'The Doctrine of the Transcendentals, is a particular manner of talking about a problem, which any thinker at any time might consider in some form. What might be called the "Problem of Transcendentals" is the question whether there are certain non-trivial ways in which absolutely anything can be characterized, and, if so, what they are?' (Marenbon 1992) We can say that if this is the problem of transcendentals, then the answer is positive, Grosseteste was a thinker who, following Dionysius, wondered if anything was absolutely characterized, and he concluded that being, goodness, truth, beauty and unity are the most general condition of everything.

On the other hand, if we consider the doctrine of the transcendentals as a particular theory, as envisioned by Aertsen (2012), with some specific features—like the explicit presence of Aristotle's *Metaphysics*, the analogy of being, the transcendentals as condition of intellectual knowledge, the acknowledgement of the limits of the categorial order—then we can say that Grosseteste followed his own path. He had access to sources that were not known in the twelfth century and studying them he came to some conclusions shared by other authors of his time such as the transcendentality of truth, the primacy of goodness and the analogy of being. However he was not interested in analysing and developing a series of distinction among transcendentals, nor in building up a theory. His scattered remarks on the coextensiveness of being, good/beautiful and truth manifest his lack of interest in this kind of comprehensive thought. Almost every passage that I quoted from Grosseteste's writings has a theological reference: it is God that benevolently offers his gifts (processions) to creatures, he calls them to existence and to a good existence (*in esse et bene esse*) and knowing them means to know (even unconsciously) Him as their source.

Therefore it is not so strange that someone attributed a treatise entitled *Transcendentia* to Grosseteste, even though it is neither Grosseteste's nor about transcendentals.[16] I think that it confirms the fact that Grosseteste's thought was perceived in the wake of the transcendental tradition of the first half of the thirteenth century. Grosseteste's idea about transcendentals is another aspect of his positive, aesthetic view of the universe, ordered by God, alongside its metaphysics, physics and aesthetic of light. But this topic could be a starting point for another piece.

[16] I am referring to Salamanca, Bibliotheca Universitaria MS. 1986. In this manuscript are contained authentic and spurios Grosseteste's works. For further details see (Beaujouan 1962; Bermon and Rothschild 2004).

References

Aertsen, J. (1991). Beauty in the middle ages. A forgotten transcendental? *Medieval Philosophy and Theology, 1*, 68–97.

Aertsen, J. (1996). *Medieval philosophy and the transcendentals: The case of Thomas Aquinas.* Leiden: Brill.

Aertsen, J. (2012). *Medieval philosophy as transcendental thought: From Philip the Chancellor (Ca. 1225) to Francisco Suarez.* Leiden: Brill.

Aertsen, J. (2013). Medieval theories of transcendentals. In *Stanford encyclopedia of philosophy.* http://plato.stanford.edu/entries/transcendentals-medieval

Alexander of Hales. (1924–1948). *Summa theological* (4 vols). Rome: Quaracchi.

Aristotle. (1957). *Prior and posterior analytics* (D. W. Ross, Ed.). Oxford: Clarendon.

Ashworth, E. J. (2013). Medieval theories of analogy. In *Stanford encyclopedia of philosophy.* http://plato.stanford.edu/entries/analogy-medieval

Bázan, C. (2000). Thomas d'Aquin et les transcendantaux. Retour sur un livre de Jan A. Aertsen. *Revue des Sciences Philosophiques et Théologiques, 84*(1), 93–104.

Beaujouan, G. (1962). *Manuscrits scientifiques médiévaux de l'Université de Salamanque et de ses 'Colegios mayors'.* Bordeaux: Féret.

Bermon, P., & Rothschild, J. P. (2004). *Bibliographie annuelle du Moyen Age tardif 14.* Turnhout: Brepols.

Bérubé, C. & Gieben S. (1974). *Guibert de Tournai et Robert Grosseteste sources inconnues de Saint Bonaventure, suivi de l'éditíon critique de trois chapitres du* Rudimentum Doctrinae *de Guibert de Tournai.* In J. G. Bougerol (Ed.), *S. Bonaventura 1274-1974* (Vol. II, pp. 627–654). Roma: Grottaferrata.

Boethius. (1906). *In Isagoge Porphyrii commenta.* (G. Schepss & S. Brandt, Eds.). Vienna-Leipzig: Corpus Scriptorum Ecclesiaticorum Latinorum.

Boethius. (2000). *De consolatione philosophie; opuscola theologica* (C. Moreschini, Ed.). Munich-Leipzig: De Deutsche Bibliothek.

Callus, D. A. (1955). Robert Grossetste as Scholar. In D. A. Callus (Ed.), *Robert Grosseteste. Scholar and Bishop* (pp. 1–69). Oxford: Clarendon.

Cooper, T. J. (2012). *One truth or many truths. Two medieval account of truths: Anselm of Canterbury and Robert Grosseteste.* PhD Dissertation, The Catholic University of America.

De Rijk, L. M. (1967). *Logica Modernorum* (2 vols). Utrecht-Assen: Van Gorcum.

De Rijk, L. M. (2003). *The aristotelian background of medieval transcendentia: A semantic approach.* In Pickavé (Ed.), *Die Logik* (pp. 3–22).

Donati, S. (2003). La discussione sull'unità del concetto di ente nella tradizione di commento della 'Fisica': commenti parigini degli anni 1270-1315 ca. In Pickavé (Ed.), *Die Logik* (pp. 60–139).

Eco, U. (1988). *Art and beauty in the middle ages* (H. Bedin, Trans). New Haven, CT: Yale University Press.

Ghisalberti, A. (1990). *Medioevo teologico. Categorie della teologia razionale nel Medioevo.* Roma-Bari: Laterza.

Gracia, J. J. E. (1997). Critical study. Medieval philosophy and the transcendentals: Aertsen's characterization of medieval thought and thomistic metaphysics. *Recherches de Théologie et Philosophie médiévales, 64*(2), 455–463.

Grosseteste. (1912). *Die Philosophischen Werke des Robert Grosseteste, Bischofs von Lincoln.* (L. Baur, Ed.) Münster: Aschendorff.

Grosseteste. (1929). On truth. In R. McKeon (Ed.), *Selection from medieval philosophers* (pp. 263–281). New York: Scribners.

Grosseteste. (1963). *Roberti Grosseteste episcopi Lincolniensis commentarius in viii libros Physicorum Aristotelis* (R. C. Dales, Ed.). Boulder, CO: University of Colorado Press.

Grosseteste. (1981). *Commentarius in Posteriorum Analyticorum Libros* (P. Rossi, Ed.). Florence: Leo S. Olschki.

Grosseteste. (1991). *Commentary on De ecclesiastica hierarchia* (2 vols, C. T. Hogan, Ed.). PhD dissertation, Cornell University.
Grosseteste. (1996). *On the six days of creation: A translation of the Hexaemeron* (C. F. J. Martin, Trans.). Oxford: Oxford University Press.
Hogan, C. T. (1996). Pseudo-Dionysius and the ecclesiology of Robert Grosseteste: A fruitful symbiosis. In J. McEvoy (Ed.), *Robert Grosseteste: New perspectives on his thought and scholarship* (pp. 189–213). Turnhout: Brepols.
Jacobi, K. (2003). Untersuchungen von Logikern des 12. Jahrhunderts über transkategoriale Terme. In Pickavé (Ed.), *Die Logik* (pp. 23–36).
Lewis, N. (2009). Grosseteste on being. *The Modern Schoolman, 86*, 25–47.
Macdonald, S. (1992). Goodness as transcendental: The early thirteenth-century recovery of an Aristotelian idea. *Topoi, 11*(2), 173–186.
Marenbon, J. (1992). Abelard, ens and unity. *Topoi, 11*(2), 149–158.
McEvoy, J. (1982). *The philosophy of Robert Grosseteste*. Oxford: Clarendon.
McEvoy, J. (2000). *Robert Grosseteste*. Oxford: Oxford University Press.
Pickavé, M. (2003). *Die Logik des Transzendentalen. Festschrift für Jan A. Aertsen zum 65. Geburtstag*. Berlin: Walter de Gruyter.
Pouillon, H. (1939). Le premier traité des proprietés transcendentales. La Summa de bono du Chancelier Philippe. *Revue néo-scholastique de philosophie, 42*, 40–77.
Pouillon, H. (1946). La Beauté, propriété transcendentale chez les scholastique. *Archives d' Histoire doctrinale et littéraire du Moyen-Age, 21*, 263–329.
Teske, R. J. (1993). William of Auvergne and the Manichees. *Traditio, 48*, 63–75.
Valente, L. (2005). 'Illa quae transcendunt generalissima': elementi per una storia dei termini trascendentali nella teologia latina del XII secolo. *Quaestio, 5*, 217–239.
Valente, L. (2006). 'Ens, unum, bonum': elementi per una storia dei trascendentali in Boezio e nella tradizione boeziana del XII secolo. In S. Caroti, R. Imbach, Z. Kaluza, G. Stabile, & L. Sturlese (Eds.), *Ad Ingenii Acuitionem. Studies in honour of Alfonso Maierù* (pp. 483–545). Louvain-la-Neuve: Brepols.
Valente, L. (2007a). Names that can be said of everything: Porphyrian tradition and the 'transcendental' terms in the twelfth-century logic. *Vivarium, 45*(2–3), 168–180.
Valente, L. (2007b). *Logique et Théologie. Les écoles parisiennes entre 1150-1220*. Paris: Vrin.
Zupko, J. (2003). William of Auxerre. In J. J. E. Gracia & T. B. Noone (Eds.), *A companion to philosophy in the middle ages* (pp. 688–689). Oxford: Wiley.

Manuscripts

Grosseteste. *Commentary on the divine names*. Oxford, Merton College MS 86.
Paris, Marazine MS 787; Paris, Bibliotèque Nationale MS Lat.1620.

Chapter 11
A Theoretical Fulcrum: Robert Grosseteste on (Divine) Infinitude

Victor Salas

11.1 Divine Infinitude

In addition to his numerous scientific contributions, Robert Grosseteste, as some recent literature has suggested, is rightly regarded as a 'master of the sacred page' (Ginther 2004). Yet, as Étienne Gilson has noted with some bewilderment, among the various descriptions Scripture affords theologians about God, *sacra pagina* is relatively mute with respect to the issue of divine infinitude, for no text directly and unambiguously asserts that 'God is infinite' (1954). As Gilson points out, however, there can be no denial of the fact that, for medieval theologians, infinity, 'becomes one of the primary characteristics of the Christian God, and the one which, after Being, most clearly distinguishes Him from all other conceptions of God'(1991; cf. Burns 1998, 57). Of course few Christian theologians have ever been stymied by what Holy Writ has failed to say or even by what it has said, and Grosseteste is certainly no exception. I shall argue, however, that the Bishop of Lincoln is unique among his contemporaries and important for succeeding generations of Christian theologians in that he generated a theoretical framework that could substantiate and develop further religious belief in an intrinsic and entitively perfective concept of divine infinitude which was a central tenet of much medieval metaphysical speculation at the end of the thirteenth century and beyond. The backdrop of Grosseteste's thinking are ancient philosophies of nature, especially that of Aristotle, which find infinitude to be an incomprehensible morass of indeterminacy. Motivated equally by (1) his theological commitment to Patristic thinking about the nature of God as infinite and (2) his cosmogonical accounts of the universe (as found most especially in his *De luce*) that are as much mathematical (if not

V. Salas (✉)
Sacred Heart Major Seminary, Detroit, MI, USA
e-mail: salas.victor@shms.edu

J.P. Cunningham, M. Hocknull (eds.), *Robert Grosseteste and the pursuit of Religious and Scientific Learning in the Middle Ages*, Studies in the History of Philosophy of Mind 18, DOI 10.1007/978-3-319-33468-4_11

more so) as they are physical, Grosseteste reformulates and deploys the concept of ‘infinitude’ in a positive direction that, I hope to show, has decisive significance for succeeding generations of medieval thinkers, especially Richard Fishacre, but also Thomas Aquinas, Bonaventure, and Duns Scotus who are committed to a positive understanding of God’s entitive infinitude.

11.2 The Context

Here, let me begin by pointing out that Grosseteste is not the first Christian theologian nor is he even the first thinker to maintain the infinitude of the first principle of all being, for even the pre-Socratics were content to identify that principle as the ἄπειρον, i.e., the unbounded or undetermined, from which all particular determinate and defined things emerge. Yet, between the ἄπειρον of antiquity and the divine infinitude of the Middle Ages stretches a gaping abyss that could only filled perhaps by Damascene’s ‘*pelagus substantiae infinitum et indeterminatum.*’ Joseph Owens puts it well when he writes, ‘Perfect Being for the Greeks meant limitation and finitude; for the Christians, the perfect Being is infinite. Limitation for the Christians denotes imperfection; while for the Greeks, imperfection was implied by infinity’ (Owens 1978). Medieval theologians were themselves keenly aware of the difference in the stance they maintained over and against their Greek antecedents. Thomas Aquinas, for example, raises the question in his *Summa theologiae* whether or not God is infinite and notes that while ‘all antique philosophers’ attributed infinitude to the first principle, they did so in the line of imperfection insofar as they understood infinitude to follow from a material first principle’ (*Summa theologiae* I, q. 7, a. 1). God, however, is not a material principle—David of Dinant notwithstanding (*Summa contra gentiles* I, c. 17)—and so if the divine being is to be regarded as ‘infinite,’ it will have to be along entirely different lines than what the ancients had held. The question here is: what exactly marks the distinction between the two conceptions of infinitude such that the logic of both positions could be maintained simultaneously without contradiction?

Those familiar with the life-long research of Leo Sweeney will be aware of his observation that Christian attitudes towards God’s infinitude largely took two different directions: one that regarded God’s infinitude as an extrinsic attribute and another that maintained infinitude to be an intrinsic entitive property (Sweeney 1957). Patristic and early medieval teachings on divine infinitude, argues Sweeney, principally unfolded their understanding of God’s infinitude in relationship to creation. That is, God is called infinite in power, wisdom, or goodness not so much because of what the divine being is in itself but because of what God can effect, namely, an infinite number of things. Never was it the case that ‘infinitude’ was attributed to the divine being itself as an intrinsic entitive perfection (Ibid.). Yet, Sweeney points out that beginning around 1250 a shift occurred in medieval thinking such that infinitude came to be considered as pertaining to the divine essence in itself. In addition to Aquinas, who held that God is infinite because the divine *esse* is not contracted by a distinct essence (*Summa theologiae* I, q. 7, a. 1), one may also consider Bonaventure who, after noting those opinions that accord infinitude to God only on account of some extrinsic relation, argues that God's being is in itself infinite. There are those, complains the Seraphic Doctor, 'who wish to say that the divine essence under the aspect of essence [*sub ratione essentiae*] is finite, but under the aspect of power [*sub ratione potentiae*] is infinite' (Bonaventure, *In Sent.*, I, d. 43). Here, it is worth noting that the reason given for the opposing position that Bonaventure reports is that 'essence' names God as He is in Himself; and thus God must be finite because the divine being is both perfect and capable of being comprehended (*comprehenditur*) by the blessed who are themselves finite (Ibid.). This latter point, namely, God's knowability, will be of crucial importance for both Fishacre and Grosseteste, and, as we shall see, directly contributed to their understandings of God as intrinsically infinite *per essentiam*.

We need not presently concern ourselves with the details of Bonaventure's response to these objections, and it will be sufficient to point out that Bonaventure, together with a number of other medieval doctors at the time including Aquinas, and later Henry of Ghent (*Summa quaestionum ordinariarum*, a. 44, q.1.), and most especially Duns Scotus (*Ordinatio* I, d. 3, pars 1, q. 2.) all held the common view that God is infinite with respect to His very being (Sweeney, 'Divine Infinity'). What is worthwhile to take away from the Franciscan master, however, are the data of the problem that any medieval thinker hoping to identify God as intrinsically infinite would have to deal with: namely, (1) a sense in which 'infinity' could stand for a perfection and (2) how an infinite being could serve as an intelligible object for a finite intellectual creature. It was the successful solution of these two difficulties that allowed a shift in thinking with respect to divine infinitude and, historically, the solution came just before the 1250s with Richard Fishacre, or so Sweeney argues at least. Why then is Fishacre important for a study on Robert Grosseteste? The simple answer is that Grosseteste's fingerprints are all over Fishacre's thinking on divine infinity. As James Ginther points out, Fishacre's argument for God's intrinsic infinitude is a recapitulation of Grosseteste's *Dictum* 60 (2004). While that is correct, I think Fishacre draws more from Grosseteste than just the *Dictum* in question. What is more, several ideas associated with infinity as spelled out in *Dictum* 60 have parallels to a number of Grosseteste's other works, most especially the *De luce*. In short, while Fishacre may have played a significant role in the shift of the medieval understanding of divine infinitude, as Sweeney contends, Grosseteste's contribution to that shift cannot be ignored and should not be downplayed. For it may very well be the case that Grosseteste helped establish the conceptual framework in which Fishacre could conceive of a positive and perfective notion of infinitude and efficaciously draw the conclusion that God is intrinsically infinite. Let us briefly consider Fishacre's teaching on divine infinitude, then, so as to appreciate all the better the Bishop of Lincoln's contribution to the subject.

11.3 The Begetting of an Idea

According to Sweeney, Fishacre composed his Commentary on the *Sentences* shortly before 1245 (Sweeney and Ermatinger 1958). In the second distinction Fishacre has occasion to raise the issue of divine infinitude, but the context, it is worth noting, concerns one of the issues surrounding divine infinitude noted above, namely, God as an object of a finite creature's knowledge (Ibid.). Just a few years prior, on 13 January 1241 the theology faculty at the University of Paris issued a condemnation of ten articles that were contrary to 'theological truth.' The first of these ten was the claim that 'the divine substance in itself will be seen by neither men nor angels' (*Chartularium universitatis parisiensis* 1889). In opposition, the theology faculty 'firmly believe[s] and affirm[s]' that 'God's essence as it is in itself will be seen by angels, all the saints, and is seen by glorified souls' (Ibid.). The question remaining for theologians to figure out is: how was such a vision possible?

As late as the 1260s Aquinas was protesting the same opinion condemned by the Parisian theology faculty (*Summa theologiae* I, q. 12, a. 1), thus it should hardly come as a surprise that Fishacre would likewise address himself to the matter only a few years after the condemnation was issued. In his commentary Fishacre raises four questions on the matter: (1) whether God is infinite; (2) in how many ways one is able to say that God is infinite; (3) in what manner infinity or numerousness (*numerositas*) is compatible with divine simplicity; and, finally, (4) in what manner is a rational creature, with a finite power, capable of attaining that which is infinite (Fishacre, *In Sent.*, I, d. 2, cap. 1). The order of these questions suggests that their progressive answers are intended to setup a solution to the overarching question: how can a finite creature 'see' the essence of the infinite God? That this question is a real one and not simply ill-framed is guaranteed by the fact that God truly is infinite, which Fishacre first establishes.

Fishacre argues for God's infinitude on the basis of the divine power, wisdom, and goodness. To establish God's infinitude first with respect to power (*potentia*), Fishacre employs what Sweeney calls the '*distantia* argument' (Sweeney and Ermatinger 1958). 'So much as what is made and that by which it is made stand apart [*distant*], so great is the power of the maker' (Fishacre, *In Sent.*, I, d. 2, cap. 1, q. 1), writes Fishacre. But the distance between prime matter and that from which it is made, viz., nothing (*de nihilo*), is an infinite distance, such as what is found between something (*aliquid*) and nothing. Therefore, the power to bring it about must be equally infinite (Ibid.). That God is also infinite in wisdom, Fishacre thinks is equally clear. The figures and dispositions of artifacts indicate the wisdom of the artisan. But in the smallest mote of dust (*atomus*)[1] are infinite figures. Therefore, the Dominican concludes, the wisdom of the artisan, God, by whom these things are made, is itself infinite (Fishacre, *In Sent.*, I, d. 2, cap. 1, q. 1). Finally, with respect to

[1] I have chosen to translate '*atomus*' as 'dust mote' to avoid any connotations that the term 'atom' may convey to the contemporary reader. Fishacre, I believe it safe to say, was not a proponent of contemporary quantum theory.

the infinitude of divine goodness, Fishacre explains that it is the nature of goodness to clothe the naked and supply what is needed to the destitute. But prime matter is 'naked' and 'needy' in the sense that it 'desires to clothe itself in infinite forms.' Therefore the goodness of God is infinite inasmuch as He 'clothes' matter with an infinite number of forms (Fishacre 1958).

In arguing for God's infinitude on the basis of the divine power, wisdom, and goodness relative to creation, it seems that Fishacre has done little more than what previous theologians had before him who argued for God's infinitude on the basis of some extrinsic relation. There is more to the story and Fishacre goes on to draw the conclusion, albeit perhaps somewhat understated, that God's being is infinite in itself. But before considering that aspect of his argument, we are presently in a position to examine Grosseteste's role in the Dominican's speculation about divine infinitude. Indeed, as noted above, the arguments we have just identified have their origin entirely in Grosseteste's *Dictum* 60,[2] and if Fishacre concludes to the fact of God's infinitude from them, then Grosseteste's own thinking is at the heart of the medieval transition to a positive perfective consideration of infinitude.

In *Dictum* 60 Grosseteste explores how each creature is a 'mirror' or similitude of the creator with respect not only to divine unity but also with respect to God's Trinitarian character (*Dictum* 60). While Grosseteste does advert to the common Augustinian intelligence-memory-will[3] device to account for a rational creature's similitude to the Trinitarian God—the greatest kind of similitude which is that of an 'image' (*imago*) (*Dictum* 60)—the Bishop of Lincoln thinks that every creature—not just rational ones but also the lowliest—bears some similitude to the Trinitarian character of God. To explain his thinking, Grosseteste offers a kind of thought experiment. 'Let us posit only two creatures,' he says, 'one rational and one corporeal' (Ibid.). Likewise, with respect to the corporeal creature, Grosseteste imagines the most insignificant, smallest, and least useful body possible: 'a dust mote [*atomus*] wafting about in the sun [light]' (*Dictum* 60). Even in such a lowly being, Grosseteste thinks, is the rational creature able to discover the Trinity 'as if by beholding God through a mirror' (Ibid.). If one considers the nature of that speck of dust as composite, mutable, material, and diffused throughout space, one can conclude something about its creator since that bit of dust, he notes, was made from nothing. One understands, then, that the power (*potentia*) of the creator of such a lowly being must not be finite but immense and infinite (Ibid.). The reason Grosseteste gives is virtually the same as Fishacre, and, though the Bishop of Lincoln does not use the term '*distantia*,' he argues similarly that every power is measured by the proportion of what is made to that by which it is made. But everything, however small and insignificant, infinitely exceeds nothing. Since a

[2] Thomson places the date of composition of the majority of Grosseteste's *Dicta* after 1220 and most likely between 1229 and 1232 during Grossetest's archdiaconal period. See (Harrison 1940). It is fairly safe to say, then, that Fishacre's Commentary on the *Sentences*, written around 1245, did not antedate the *Dictum* in question.

[3] Grosseteste mentions 'love' (*amor*) instead of the more common 'will'; cf. *Dictum* 60, 156.

speck of dust is something from nothing (*aliquid ex nihilo*), it can be seen, concludes Grosseteste, that that particle of dust comes into being by a power that is no less than infinite (Ibid.).

Here, it is important to note that Grosseteste seems to think that if the particle of dust 'exceeds' nothing infinitely, it is because the dust itself is in some fashion infinite. In fact, according to Grosseteste, it is on account of that infinitude which the particle of dust possesses that it resembles or possesses some similitude to the infinite power of its efficient cause: God (Ibid.). But if that infinitude which the dust possesses, for example, is that on account of which even the smallest and most insignificant creature is like God, the creature's infinitude cannot represent an imperfection since it is a mirror reflection, so to speak, of the divine infinitude, which itself certainly cannot be imperfect.

Grosseteste adds more to his discourse which makes his understanding on divine infinitude clearer. In much the same manner as Fishacre, the Bishop of Lincoln argues that God is infinite in wisdom and goodness. Still considering the particle of dust, Grosseteste points out that one discovers in it three lines that intersect at three right angles, in which one is able to inscribe a sphere (presumably on equidistant points along the intersecting x-y-z axes) within the dust particle. Then, within that sphere it is possible to inscribe infinite circles and then within those circles infinite figures (Ibid.). But any one of these inscribed infinite figures can give rise to a demonstrative science, and it is discovered that there is an infinite science—not only of magnitude but also of numbers—in each particle of dust. But there cannot be an infinite science so inscribed in each bit of dust unless there be an infinite wisdom and power through which that dust and its corresponding science is brought into being (Ibid.).

In this step of his proof, Grosseteste has concentrated not so much on a physical notion of infinitude, such as Aristotle would either regard as actually impossible and only potentially feasible, but a mathematical or numeric infinitude. Once again, inscribed within every creature is not only the infinitude of matter,[4] which Aristotle would link with imperfection, but the positive and perfective infinitude of number, which Grosseteste will work out in terms of proportions in what turns out to be a kind of proto-set theory put forth in his *De luce*. But more about that later, in the meantime, Grosseteste draws yet another conclusion: that there must be an infinite goodness, which corresponds to the final conclusion that Fishacre himself drew in his *Sentences*. The infinite science inscribed in each bit of dust has a positive and perfective value for the created intellect, as Grosseteste sees it, and thus the creator of that dust has created something greatly useful (*valde utile*) for the mind, and this, the Bishop of Lincoln observes, without any merit on the part of the created intellect itself. This beneficence of the creator manifests its goodness (*Dictum* 60). To this conclusion Grosseteste adds that as good or useful as the science contained in a

[4] That Grosseteste is a universal hylomorphist seems clear from his *De motu corporali et luce* in which he describes magnitude as a consequence of first form and first matter. There he states that there is nothing common among all bodies except prime matter, prime form, and magnitude.

particle of dust is for one intellect, so is it useful (inasmuch as it is in itself) for an infinite number of intellects should they be created. Therefore, the goodness of a particle of dust, just inasmuch as it is in itself, is infinite, which, Grosseteste reasons, means that it was created by a power that is not only (infinitely) wise but also infinitely good (Ibid.).

With these three attributes of God unfolded in terms of their infinite character, Grosseteste returns to his principle point about how creation manifests the Trinitarian character of God, writing, 'behold how reason is able to see in a bit of dust the infinite power, infinite wisdom, and infinite goodness of the creator. And to see this is to see the creator Trinity' (Ibid.). To be sure, Grosseteste is not the first to describe God as infinite in terms of the divine power, wisdom, and goodness, as Sweeney's researches have shown. But previous efforts had always seemed to yield an extrinsic attribute relative to some creature. While Grosseteste identifies and unfolds the divine attributes in question in relationship to creation, the point the Bishop of Lincoln is attempting to make is that creation is manifesting or reflecting to the theorizing intellect something about the divine reality itself, a reality that is infinite. Here, Grosseteste's argument is innovative insofar as it locates an infinitude with the creature itself as well. James McEvoy is correct, then, when he writes, 'An infinite mind was at work in the production of the world; Grosseteste was ... the first figure in the Judeo-Christian tradition to find a real corresponding infinity in the world itself' (McEvoy 1982). In other words, the point here is not so much that Grosseteste has established God's infinitude, but that he has established it on the basis of an infinitude that resides within creation, an infinitude that denotes a perfection rather than, as the Greeks had thought, an imperfection. And, since *nemo dat quod non habet*, a perfect creaturely infinitude can only point to a corresponding and perfect divine infinitude. There is within creation an actual infinitude within a perfective order. But, again, for Grosseteste, that created infinitude points to a corresponding infinite reality within the divine being. God, one might say, is the grand *numerator* in whose wisdom is contained exemplaristically the creaturely infinitudes that image the divine infinitude itself.[5] In his commentary on the *Physics* Grosseteste writes, 'For indeed there are the *rationes* of infinite things and infinite wisdom in the divine mind, which wisdom is the principle of all effects' (*Commentarius in VIII Libros Physicorum Aristotelis*).

Grosseteste then discusses further what he has in mind by the creaturely infinitudes that mirror the infinitude of God in his *De luce*. The *De luce* has been amply celebrated for its originality and philosophical richness, and as legions of commentaries and analyses have been written on it (McEvoy 1982; Miccoli 2001), we need not presently concern ourselves with a detailed exposition of its principle argument here. It will be enough to identify the main problem addressed in the work and the role that infinitude plays in its solution. In short, Grosseteste is concerned with giving an account of how the world-machine (*machina mundi*) comes into being, an account which is complicated by the fact that, while the world is three dimensional,

[5] For God as divine *numerator* see (McEvoy 1982).

its basic principles—first form and prime matter—are themselves non-dimensional. In other words, how is corporeity, which is three dimensional, generated from that which lacks magnitude entirely? (Ginther 2004) Grosseteste's answer is, as we know, light (*lux*): for light is the form of corporeity (*De luce* 1912a, b). But light can only attain three dimensions if it 'infinitely multiplies itself.' This latter point is what concerns our immediate interests.

Light, says Grosseteste, by its very nature 'diffuses itself in every direction,' such that 'from a single point of light a sphere of light however big is immediately generated….' (Ibid.). But since light is the form of corporeity, in diffusing itself, light, 'cannot leave matter behind since it is not separable from matter'(Ibid.). Light brings matter along with itself, so to speak, and in diffusing itself, 'extends' matter into tri-dimensionality (Ibid.). Grosseteste explains, 'Light therefore, which is first formed in created prime matter, *infinitely* multiplies itself through itself everywhere and pours itself out equally in all directions, drawing matter along with itself, which it is not able to leave behind [*relinquere*], into such a great mass—as great as the world-machine—extending [it] from the first moment of time' (Ibid.). This passage is the first mention of 'infinitude' in the *De luce* and it pertains to the generation of tri-dimensional mass. But, as Grosseteste will go on to say, it is only the *infinite* self-multiplication of light that can produce a finite *quantum*, that is, extended body, (Ibid.) 'because the product of something multiplied infinitely, infinitely exceeds that from whose multiplication it is produced' (Ibid.). In other words, if two simple things be considered, neither exceeds the other, and certainly one will not exceed the other infinitely. But if a finite thing be considered, it exceeds a simple thing infinitely, just as the number of points composing a line segment are infinite and thus infinitely exceed the single point itself (Ibid.).

Be that as it may, if tri-dimensionality supposedly arises from the infinite self-multiplication of light, one might put to Grosseteste the following question: how can there ever be more than one body or even diverse kinds of bodies if everything proceeds from the apparent homogeneity of a single point of light? After all, as Avicebron taught in his *Fons vitae*—and which would be a particular challenge for metaphysicians of light such as Albertus Magnus[6]—'*ab uno non nisi unum*' (Avicebron 1892). Once again, Grosseteste's thinking on infinitude serves as an answer. While light must diffuse itself infinitely to generate a single *quantum*, there are diverse proportions or sets of infinity that give rise to different *quanta* or finite bodies. Here, Grosseteste has in mind various proportions or relations that can obtain between different sets of infinite numbers. He mentions four in total[7]: (1) the relation that obtains between an infinite sum or numbers can be in terms either of numeric or non-numeric proportions (*De luce* 1912a, b) (2); the sum of all numbers

[6] For Albertus Magnus's treatment of this problem see his *De causis et processu universitatis*, 1.4.8, vol. 17.2.

[7] James McEvoy helpfully enumerated and described these four (1982).

is greater than the sum of only even numbers (Ibid.); (3) the sum of numbers doubled or tripled continuously to infinitude is greater than the sum of the corresponding halves or thirds of those numbers also infinitely doubled or tripled (Ibid.); (4) and, finally, the proportion that obtains between two infinite sets of numbers that are related 2:1 in which a finite number has been subtracted from the latter set resulting in a non-numeric or irrational proportion (Ibid.; McEvoy 1982; *Libros Physicorum Aristotelis* 1963).

From these considerations of various relations and proportions between different sets of infinite numbers Grosseteste concludes, 'This [set of proportions] therefore holding, it is manifest that light, by its infinite self-multiplication, extends matter into lesser finite dimensions and greater finite dimensions according to any proportion they may have to one another, namely, numeric or non-numeric [relations]' (Ibid.). The product of light's self-multiplication, then, as Grosseteste sees it, is the various celestial spheres. Light extends itself, and along with itself matter, in all directions spherically and becomes more and more rarified until it forms the 'first body,' which is the outermost sphere of the fixed stars. This outermost sphere in turn radiates its own light towards its center, condensing as its progresses. The inward diffusion of light continues until the earth, composed of its four elements, and gives rise to thirteen spheres in total: the nine celestial and four elemental that constitute our terrestrial experience (Ibid.).

In short, the world in its very corporeality is constituted by infinity, more precisely, light's self-multiplication according to various proportions of infinitude to one another since, for Grosseteste, 'one infinite number is able to relate to another infinite number in every numeric and non-numeric proportion' (*Libros Physicorum Aristotelis* 1963). This notion of infinitude, we recognize, is far from the indeterminate potentiality that goes hand-in-hand with prime matter as Aristotle understood it. Rather, the infinitude that Grosseteste develops corresponds to the infinite wisdom that a creator-God exercises over the creation He has made precisely according to that infinite wisdom. What seems infinite to us is in itself only really finite, but what is truly infinite in itself—to God—is as if finite (Ibid.; McEvoy 1982). After all, God has—as Wisdom 11:21 claims and as the Bishop of Lincoln observes—'created everything according to number, weight, and measure' (*Libros Physicorum Aristotelis* 1963). McEvoy puts it well, 'Granted an infinite mind… an element of actual infinity in the creation becomes a thinkable possibility, and something which Aristotle would have rejected on axiomatic grounds becomes for the first time plausible, even in a certain way congruent, as a fuller expression of unbounded creative wisdom and power' (McEvoy 1982). God is a wise and provident creator, who creates not through blind impulse or natural necessity, but freely according to reason and design. If infinitude exists within creation, it is because it corresponds to eternal divine ideas that are themselves infinite and exemplar principles of all (infinite) created things (*Libros Physicorum Aristotelis* 1963).

11.4 In the Wake of Infinitude

Let us return now to Fishacre's Commentary on the *Sentences* to see how the Dominican concludes that God is infinite *per essentiam* and consider whether Grosseteste's own account of infinitude—as just described—may have been the catalyst for Fishacre's thinking. So far, we have noted that—together with Grossesteste and adducing arguments similar to those of the Bishop of Lincoln—Fishacre holds that God is infinite with respect to power, wisdom, and goodness.

Fishacre, while making his argument for God's infinitude on the basis of the divine power, wisdom, and goodness, draws the conclusion that God is, in His very being, infinite. His argument, as Sweeney points out, follows upon the notion of 'infinite distance,' which, as we have seen, has its origin in Grosseteste's *Dictum* (Sweeney and Ermatinger 1958). As we have also seen, for both the Dominican and Grosseteste, God's power is infinite because it is capable of traversing the infinite 'distance' between nothing and something. Grosseteste unfolds the notion of God's infinitude more fully in terms of God's being a 'divine numerator,' one that holds and marks all sets and relations of infinite sums in His divine mind. If God holds an infinite number of divine ideas and knows the infinitude of created numeric and non-numeric relations, could this be because God is Himself intrinsically infinite? Grosseteste is, admittedly, not explicit on this point, but the inference, I would suggest, does not run contrary to his teachings. For his own part, Fishacre is much more forthcoming. If one poses to him the question 'What is the ground of God's infinite power?' Fishacre seems to have a ready answer:

> Since God is simple in Himself and lacking composition with another, as if part of a composite, it is clear that [God] is infinite virtually, not according to additions of power made [to God], but rather because God is elongated [*elongatus est*] from impediments and matter, since [God] is entirely a separate substance (*In Sent.*, I, d. 2, cap. 1).

In other words, for the Dominican, God is infinite in power because the divine being is free from the limitations that matter imposes. Although Fishacre does not say so explicitly in the passage cited above, his meaning is clear: if God is free from matter and 'entirely a separate substance,' then it must follow that God is a pure form. Dare one suggest, as Grosseteste did in a letter to his former student Adam Rufus, that '*Deus est forma omnium*?' (*De unica forma omnium*). Admittedly, it may be too strong a claim to make that Grosseteste's *De unica forma omnium* is the immediate source of Fishacre's thinking on this particular point regarding God's so-called 'elongation' from matter, but certainly the Dominican was himself concerned with unraveling the Augustinian claim that 'God is the form of all things' (Sweeney and Ermatinger 1958). The same Augustinian explanation of God's functioning as the exemplar cause of all creatures—albeit flavored by the (condemned) twist Eriugena gave it in the *Periphyseon* (Ginther 2004)—took center stage in Grosseteste's own account of the proposition.

Now infinitude of power that the infinite divine essence generates is crucial for Fishacre's solution to the question of the beatific vision and, not insignificantly, for Grosseteste too. In his *De cessatione legalium* Grosseteste addresses the issue of rational creatures' happiness. Ever inspired by Augustinian wisdom, the Bishop of Lincoln holds that rational creatures have been created for happiness. But the rational creature cannot attain happiness through its own power, as is obvious if one considers that no one can raise himself from the dead, which it seems would be a minimal requirement for happiness (*De cessatione legalium*, 1.4.1. 1986). What is more, happiness consists in an 'infinite reality,' that is, in cognition of infinite uncreated ideas, knowing infinite numbers and figures, in short, to contemplate God and see in the divine being the archetype of the infinitude discovered within creation (Ibid.). But no creature, says Grosseteste, has the *power* on its own to gaze upon such an infinite reality. And so if the rational creature is to attain happiness, it can only be through God who has the *power* to bestow upon the creature an 'infinite act of happiness' (Ibid.).

Fishacre offers a similar answer to the question whether a finite creature can attain the beatific vision. The Dominican notes first of all that the soul has been created with the possibility of attaining an apprehension of infinity, and, like Grosseteste, he points to Augustine, who tells us famously in his *Confessions* that our hearts are restless until they rest in God (*In Sent.*, I, d. 2, cap. 1). Nevertheless, our souls cannot attain to the infinite God without divine assistance, that is, the divine power, which elevates the soul to attain the beatific vision, in the same way that God is able to elevate (corporeal) fire, which by nature has no affect upon spiritual beings, so as to punish evil souls in hell. How much more, asks Fishacre, can God raise the soul in act above its finite nature so as to attain the infinite beatific vision? (Ibid.).

The point in this discussion is that, for both Fishacre and Grosseteste, the solution to the question of how a finite mind can attain to the vision of an infinite God resides in the infinitude and beneficence of the divine power. There can only be so many coincidences in thought before the scales tip and a definite influence of one philosopher upon another should be acknowledged. It is not unreasonable to suggest that Grosseteste put in place throughout his various works—from the *De luce* to his *Dicta* and beyond—the pieces from which Fishacre would draw his own conclusions concerning divine infinitude. If Fishacre could see the truth of the divine infinitude *per essentiam*, it is only because he was standing on Grosseteste's shoulders, so to speak, and able to see better the contours of the metaphysical horizon towards which Grosseteste's thinking was progressing. But this is no small matter, for if Fishacre is, as Sweeney has suggested, a pivotal figure in the Western tradition's coming to regard divine infinitude as an intrinsic entitive perfection, then Grosseteste, I have argued, can rightly be regarded as laying the foundation for the tradition that would follow suit. It is remarkable to observe, then, with what relative speed Western metaphysics transitioned from Grosseteste's somewhat understated claims about divine infinity in the 1230s to Duns Scotus' basic metaphysical insight, made at the end of the thirteenth century, that God is properly understood as *ens infinitum*. In fact, given that Scotus' notion of *ens infinitum* represents a truly positive, as opposed to merely negative understanding of infinity such as what one finds with Thomas and a host of other thinkers, one might find in the Subtle Doctor the consummation or apotheosis of the Bishop of Lincoln's germinal insight regarding the positive and perfective value of infinitude, both creaturely and divine.

References

Avicebron. (1892). *Fons vitae*. In C. Baeumker & G. von Hertling (Eds.), *Beiträge zur Geschicte der Philosophe des Mittelatter* (Vol. 1, pp. 2–4). Munster: Aschendorff.

Burns, R. (1998). Divine infinity in Thomas Aquinas: I. Philosophico-theological background. *The Heythrop Journal, 39*(2), 123–139.

Chartularium universitatis parisiensis. (1889). (H. Denifle & É. Chatelain, Eds.). Paris: Delalain.

Fishacre, R. (1958). *Commentarium in librum 1 Sententiarum* (C. J. Ermatinger, Ed.). *The Modern Schoolman, 35*(3), 213–235.

Gilson, É. (1991). *The spirit of mediaeval philosophy*. Notre Dame, IN: University of Notre Dame Press.

Gilson, É. (1954). L'infinité divin chez saint Augustin. In *Augustinus Magister: congrès international augustinien* (Vol. I, pp. 569–574). Paris: Études Augustiniennes.

Ginther, J. (2004). *Master of the sacred page: A study of the theology of Robert Grosseteste, ca. 1229/30-1235*. Burlington, VT: Ashgate.

Grosseteste. (1912a). *De luce* (L. Baur, Ed.). Münster: Aschendorffsche Verlagsbuchhandlung.

Grosseteste. (1912b). *De unica forma omnium* (L. Baur, Ed.). Münster: Aschendorffsche Verlagsbuchhandlung.

Grosseteste. (1963). *Commentarius in VIII Libros Physicorum Aristotelis* (R. C. Dales, Ed.). Boulder, CO: University of Colorado Press.

Harrison Thomson, S. (1940). *The writings of Robert Grosseteste, Bishop of Lincoln 1235-1253*. Cambridge: Cambridge University Press.

Albertus Magnus (1993). *De causis et processu universitatis* (W. Fauser, Ed.) Cologne.

McEvoy, J. (1982). *The philosophy of Robert Grosseteste*. Oxford: Claredon Press.

Miccoli, L. (2001). Two thirteenth-century theories of light: Robert Grosseteste and St. Bonavenure. *Semiotica, 136*(1), 69–84.

Owens, J. (1978). *The doctrine of being in the Aristotelian metaphysics*. Toronto, ON: Pontifical Institute of Mediaeval Studies.

Sweeney, L. (1957). Divine infinity: 1150-1250. *The Modern Schoolman, 35*, 38–51.

Sweeney, L., & Ermatinger, C. (1958). Divine infinity according to Richard Fishacre. *The Modern Schoolman, 35*, 213–235.

Chapter 12
The Fulfillment of Science: Nature, Creation and Man in the *Hexaemeron* of Robert Grosseteste

Giles E.M. Gasper

> *I should like to dedicate this article to Professor Joseph W. Goering on the occasion of his retirement, as a small contribution to the larger body of work which celebrates his role in, and guidance of, the history of the Middle Ages, and in particular that of Grosseteste. For my own part I am extremely grateful for the generosity of Joe's scholarship, the precision of his editing and writing, and the sureness and soundness of his advice for all walks of life.*

12.1 The *Hexaemeron* and Grosseteste's Scientific Learning

The breadth of subjects which engaged Robert Grosseteste's interest and the range of genres in which he explored them are considerable. His writings cover pastoral care (including his allegorical Anglo-Norman poem *Le Chasteau d'amour*), speculative theology, biblical exegesis, philosophy, scientific commentary on Aristotelian texts, and shorter scientific works focused on particular natural phenomena. He was, in later life, an important translator of Greek philosophy and theology, of Aristotle, the Pseudo-Dionysian corpus and of John Damascene (McEvoy 2000). How Grosseteste's works relate to each other is a central question for consideration of the evolution of his intellectual interests, from issues connected to the sources with which he was familiar to the subjects to which he devoted himself and themes to which he returned. The relative paucity of historical evidence for his life makes the matter more complex, and, in this way, leads to an historiographical debate of long-standing. Different interpretations of the relationship between Grosseteste's

G.E.M. Gasper (✉)
Department of History, Durham University, Durham, UK
e-mail: g.e.m.gasper@durham.ac.uk

J.P. Cunningham, M. Hocknull (eds.), *Robert Grosseteste and the pursuit of Religious and Scientific Learning in the Middle Ages*, Studies in the History of Philosophy of Mind 18, DOI 10.1007/978-3-319-33468-4_12

works formed one of the major avenues for scholarly disputation throughout the last century. The resolution of these disputes, although unlikely to be more than tentative, remains fundamental to how Grosseteste's intellectual priorities and achievements are to be judged. Close to the heart of this problem lies the relationship between Grosseteste's scientific and theological thought. Here, the *Hexaemeron*, the *Commentary on the Six Days of Creation* holds a crucial, if under-explored, place. Emerging from Grosseteste's Genesis commentary, produced at the same time as his later scientific work, is the *Hexaemeron* which, it will be argued, fulfills his scientific learning.

Whatever the position taken on where and how Grosseteste acquired his scientific and theological learning, it is clear that for a period of years, as part of teaching and as part of a subsequent writing, the two areas of study overlapped. By quite how much is an intriguing and challenging question. This is the case not only because of the difficulties in establishing any convincing and coherent chronology in the absence of modern critical editions for many of these works, but also because these works become milestones in themselves for suggesting a chronology and framework for Grosseteste's career and intellectual development. It is well established that Grosseteste began his studies focusing on the liberal arts and quadrivial subjects, moving to consideration of Aristotle's scientific works and their medieval Arabic commentary in Latin translation. His scientific corpus is taken here to include the *Commentaries* on Aristotle's *Posterior Analytics* and *Physics* and the thirteen *Opuscula*, shorter treatises on specific phenomena including the *Computus correctorius*, and the *De operationibus solibus*. The precise dating of the composition of the last of these works is debatable, but a general consensus on the broad chronology would place it between 1230 and 1232.[1]

When Grosseteste began his theological career is a vexed question. Grosseteste's career, as he himself outlined it at the end of his life, consisted of three phases: cleric, master of theology and priest, and then bishop (Ginther 2004).[2] The second phase which ended with his appointment as bishop of Lincoln in 1235, included the period in which he served the recently established Franciscan community in Oxford, starting in 1229–30. When he began his regent mastership in theology cannot be established with any degree of certainty: dates as divergent as 1214 and

[1] The scientific works have occasioned most efforts at relative chronologies notably Dales (1961) McEvoy (1983) and Southern (1992). Panti has provided their suggestions in tabular form, together with her own tentative chronology (2013). The major dissenting voice to this broad chronology is Southern's, who projected a longer period of continuous theological and scientific speculation through the 1230s and 1240s. The test-case for Southern was the treatise *De luce* which is assigned to 1235–1240 on the basis that the work contained Grosseteste's 'final view of the role of Light in the universe' (Southern 1992). As Southern pointed out, this was only suggestive. The current weight of scholarly opinion places the *De luce* firmly in the mid-1220s (Panti 2013).

[2] *'fui clericus, deinde magister in theologia et presbiter; et tandem episcopus'*. The Latin is taken from *Sermo* 31; the sermons are unedited, here Ginther uses: London, British Library MS Royal 7. E.ii, fol 344rb.

1225 have been suggested (Callus 1955; McEvoy 2000; Southern 1992),[3] although neither commands total confidence. As Ginther has more recently underlined, the documentary evidence, as it stands, indicates a regency that began in or around 1229, and ending in 1235 (Ginther 2004). This would preclude either an earlier date, or Grosseteste's theological literacy in earlier years. Nevertheless, the bulk of Grosseteste's theological works can be dated to the period of the regency, with the caveat that their interrelationship is even more difficult to establish than the scientific texts (Ginther 2004).[4] The theological works are taken here as constituting the *Commentary on Psalms*, the extracts of glosses on the Pauline Epistles and comments on Galatians, the *Hexaemeron* and the *De cessatione legalium* (preserving the lectures on Genesis, Daniel and Isaiah), the records of disputation, *De dotibus*, *De veritate*, *De ordine* and *De libero arbitrio*, a number of sermons from the *Dicta*, and the pastoral works *De decem mandatis* and the *Speculum confessionis* as well as the first ten letters of his collection (Ginther 2004).

While Grosseteste's theological and scientific works were chronologically closely related, the point of real significance is the overflowing of ideas and themes from one field of study to the other. In this, the *Hexaemeron* holds the most important place in Grosseteste's corpus. It occupies a central place in Grosseteste's theological vision in its formation and in its development (Grosseteste 1982; Dales and Gieben 1968; Sharp 1930; Phelan 1943; Muckle 1944, 1945, 1951).[5] Grosseteste's self-appointed task is to explain why the Bible begins with Genesis, with Creation, and to lead the reader through the account of Creation, following the opinions of authoritative guides. The canvas on which he paints, however, is vast. The subject matter allows, and indeed compels him to use his scientific learning, including elements of his treatise *De luce* 'On Light', to present a theology of the cosmos, and of the relation between Creator and Creation, revolutionary in its implications, and multi-layered in its presentation. The goodness, the completeness and the unity of creation, and the centrality of man within that creation lie at the heart of his theological claims. The *Hexaemeron* reveals the range of Grosseteste's intellectual interests and virtuosity in a startling manner, and allows exploration of the context and interplay of his various bodies of knowledge, in particular his

[3] The issues turn on, first, the question of whether Grosseteste could have learnt and taught theology as a deacon rather than as a priest: Southern insists on the priesthood, and a later date (1225) (Southern 1992); McEvoy and Eastwood dispute whether this was necessary (McEvoy 2000; Eastwood 1988). A masterly summary is provided by Joseph Goering (1995). The second question concerns the Chancellorship of the University at Oxford and regent master in theology, from 1214, or as Southern suggests, later in the 1220s (Callus 1955; Southern 1992; McEvoy 2000).

[4] A new edition of the shorter theological works is under preparation under the care of Pietro B. Rossi.

[5] Grosseteste (1982), is the modern critical edition. The history of earlier efforts to make a critical edition is recorded in Dales and Gieben (1968). Some extracts were printed by Dorothea Sharp before Gerald B. Phelan began a full critical edition in 1934 at the Pontifical Institute of Mediaeval Studies in Toronto. The task passed to J. T. Muckle, at PIMS, who made a number of preliminary studies, before abandoning the project because of lack of funding.

religious and scientific learning. In important ways it brings together not only aspects of his earlier scientific work, but also his other scriptural commentaries, on Psalms and on Isaiah, the latter being incorporated in considerable part into the treatise *De cessatione legalium* 'On the Cessation of the Laws', on the relationship between the Old and New Testaments, the old covenant between the Jewish faith and God, and that promised in the Messiah, Jesus Christ. Indeed, Creation commentary, and the themes of light, beauty, order, and unity, are explored in many contexts other than the *Hexaemeron* by Grosseteste.

The *Hexaemeron* is a long work, full and varied in its content, and replete with imagery drawn from a wide range of reading, and from its author's capacious and inventive mind. Grosseteste's decision to explore Creation and the first creation story reveals a wide range of questions and topics. At a heuristic level, he reflects on how knowledge is to be expressed and classified, finding a specific manifestation in the extent to which theology may be adequately described as a science. The role of imagination in learning is another key theme, pointing to the relationship between the perception and insight of the individual investigator and the way in which authorities are to be balanced and explored. The text reveals the ways in which Grosseteste respects authorities, allowing contradictory positions to remain in tension, while finding space for his own judgement. His disagreements with more contemporary discussion erupt at various points; identifying their particular targets is an interesting task. The range of sources deployed by Grosseteste, show both continuity and development from earlier works, and a radical shift in western theological attitudes. Running through the *Hexaemeron* are other themes, ubiquitous in discussion of methodology and of content: Unity and Diversity, Order and Beauty, and Light. These themes and their variations will be explored in what follows, by way of highlighting and illuminating the ways in which they are deployed by Grosseteste in the service of his exegetical task.

Reading the *Hexaemeron* against and alongside the *De luce* and the *De cessatione legalium* in particular, offers sharper insight into Grosseteste's cosmological thought, and provides a clearer framework for the place and purpose of scientific learning in his religious context. The *De luce*, probably written in 1225, offers homage to ancient cosmology and produces a mathematically based evocation of the universe of the spheres, presenting broadly speaking Aristotelian and Platonic views overlain if not reconciled. Grosseteste's discussion is so detailed that it can be modelled mathematically in two and three dimensions, as shown by recent research conducted by the Ordered Universe Project. Richard Bower, Tom McLeish and Brian Tanner, along with others, stress, in this research, the unifying principles that govern Grosseteste's cosmological thought (Bower et al. 2014). The *Hexaemeron* not only fills some of the imaginative gaps, but most importantly provides a fuller context for what its author conceived as the point and purpose of human learning.

12.2 Exegesis

In turning to scriptural commentary, Grosseteste was well aware of the complexities that he would encounter. In what has become known as the *Proemium* to the *Hexaemeron*, Grosseteste comments on two of Jerome's letters which were frequently prefixed to the Bible in the medieval period (Dales and Gieben 1968; Southern 1992).[6] Observing through Jerome the importance of the 'living voice' of the teacher in theological study, Grosseteste emphasises the special care to be taken over the interpretation of Scripture:

> And since someone might say that theology could be learnt without an instructor, since it is an easy study, especially for one who is trained in secular literature, he proves that this is not so: that Scripture has a hidden and sealed-up sense, which it is hard to reach (Grosseteste 1982/1996, Proemium, 3).[7]

All other arts require a teacher, the liberal arts and the mechanical, so how much more so is one needed for theology 'which is the most inclusive of all arts, and the one whose understanding lies deepest. But this is the only art which we often find people presuming to teach without having studied'. Theology should not be approached lightly or presumptuously, and Grosseteste hints at his views of how the subject fits into larger schema of human learning.

The *Hexaemeron* is fundamentally exegetical, and in this respect conforms to general patterns of high medieval thought and practice. Scripture is at once all-encompassing, it is all relevant, every phrase and every word carries meaning, and it is relevant to all human beings at all times. It is also secret, mysterious and difficult to interpret, as Grosseteste takes pains to point out:

> Therefore, Scripture contains everything that nature contains, since after the creation of the world, there are no new natures or species to be added. It also contains the whole of the supernatural, that is to say, our restoration and future glorification. It also contains the whole of morality and the whole of rational knowledge. This is because the archetypal world is the reason, the art, the rule, and the rational knowledge of every single thing. In it is every single cause of existence, every reason of understanding and every ordering of life.... And while anything valuable which is taught elsewhere can be found in Scripture, in yet more abundance are found things that are ever taught anywhere else, but which are learned

[6] Jerome's *Letter* 53, to Paulinus, dating from 394, was a defence of his translation of the Bible, and his letter to Desiderius, formed a preface to the Pentateuch (Jerome 1996). The *Proemium* is included in the extant manuscripts of the *Hexaemeron* containing the complete text, that is six of the seven, bar one. R. W. Southern suggested, however, that the *Proemium* was in fact a quite separate work for a separate audience, based on the content, in particular the more elementary nature of the grammatical questions addressed, concluding that, 'The combination of the lectures on Jerome's Introduction with the *Hexaemeron* appears to be a factitious union of two disparate works originally intended for different audiences'.

[7] '*Et quia posset quis dicere quod theologia esset sine instructore addiscibilis, tanquam scientia aliqua facilis, et precipue viro in secularibus literis excercitato, ostendit econtra, quomodo sacre Scripture sensus occultus sit et signatus; cuius non modicum difficilis sit auditus*'. Unless otherwise indicated, all subsequent translation from the *Hexaemeron* follows that by Martin, with citation of the Dales and Gieben edition.

> from Scripture alone, in marvellous sublimity and marvellous lowliness (Grosseteste 1982/1996, 1.IV.1).[8]

To compare this to the statement from 1109 by Anselm of Canterbury, with whose thought Grosseteste engaged extensively, is to see something of the shift in perspective and scope which had taken place over the twelfth century. Anselm is well known for his statement that in discussing the incarnation of Christ he wished to do so, without reference to Scripture, that is to show its necessity in reason. This did not mean, as he later clarified, that he was in any way promoting human reason above Scripture; in fact the opposite was true.

> For, indeed, in our preaching, nothing which Sacred Scripture—made fruitful by the miracle of the Holy Spirit—has not sent forth or does not contain is conducive to spiritual salvation. Now, if on the basis of rational considerations we sometimes make a statement which we cannot clearly exhibit in the words of Scriptures, or cannot prove by reference to these words, nonetheless in the following way we know by means of Scripture whether the statement ought to be accepted or rejected. If the statement is arrived at by clear reasoning and if Scripture in no way contradicts it, then (since even as Scripture opposes no truth, so it favours no falsity) by the very fact that Scripture does not deny that which is affirmed on the basis of rational considerations, this affirmation is supported by the authority of Scripture. But if Scripture unquestioningly opposes a view of ours, then even though our reasoning seems to us unassailable, this reasoning should not be believed to be supported by any truth. So, then, Sacred Scripture, in that it either clearly affirms them or else does not at all deny them, contains the authority for all rationally derived truths (Anselm 1946–1961c, 1976, 3.6).[9]

Anselm's statement has nothing in it with which Grosseteste would disagree, and the foundational role for Scripture in human learning is clear. Grosseteste's scriptural engagement is, however, both more specific and projected onto a larger scale, in the way in which he identifies nature and Scripture as part of the same process: to understand the place of man in Creation, but with the former placed in the embrace of the latter.

---

[8] *'Continet igitur in se hec scriptura totum quod continet natura, quia post mundi creacionem non est nove speciei seu nature adiectio. Continet eciam totum quod est supra naturam, quod videlicet est nostre reparacionis et future glorificacionis. Continet eciam totam moralitatem et totam scienciam racionalem. Ipse enim mundus archetipus est omnis rei racio et ars et regula et racionalis sciencia. In ipso est omnis causa subsistendi et racio intelligendi et ordo vivendi… Et cum in ista quisque invenerit omnia que utiliter alibi didicit, multo habundancius inveniet ea que nusquam omnino alibi sed in istius tantummodo scripture mirabili altitudine et mirabili humilitate discuntur'.*

[9] *'Siquidem nihil utiliter ad salutem spiritualem praedicimus, quod sacra scriptura spiritus sancti miraculo foecundata non protulerit, aut intra se non contineat. Nam si quid ratione dicimus aliquando quod in dictis eius aperte monstrare aut ex ipsis probare nequimus: hoc modo per illam cognoscimus, utrum sit accipiendum aut respuendum. Si enim aperta ratione colligitur, et illa ex nulla parte contradicit—quoniam ipsa sicut nulli adversatur veritati, ita nulli favet falsitati - hoc ipso quia non negat quod ratione dicitur, eius auctoritate suscipitur. At si ipsa nostro sensui indubitanter repugnant: quamvis nobis ratio nostra videatur inexpugnabilis, nulla tamen veritate fulciri credenda est. Sic itaque sacra scriptura omnis veritatis quam ratio colligit auctoritatem continent, cum illam aut aperte affirmat aut nullatenus negat'.*

Biblical interpretation matters, then, since it concerns the totality of human experience. It does so in complex and hidden ways, and Grosseteste follows the standard medieval paradigm of identifying different levels of interpretation: the literal (about the thing or deed itself), the allegorical (what should be believed), the tropological or moral (how people should behave), and the anagogical or eschatological (where creation is going to end up). The six days of creation can therefore be understood as days or modes of creation, but also as emblematic of the six ages of history, or the six natural ages of man, or the six ages of the new man, that is a schema of spiritual growth after the baptism. The creation of light on the first day may be understood allegorically as the creation of free will (Grosseteste 1982/1996, 8.XXX–XXXIV). Whales, he notes, are understood in their literal sense as creatures of the sea; allegorically as those engaged in substantial self-reflection, their greatness as exemplars drawn over the great size of the creature. Notable whales include Hezekiah, David, Paul, Augustine, Jerome, Gregory the Great as well as Plato, Aristotle and Pythagoras (Grosseteste 1982/1996, 6.VI.1 and 6.XV.1). Grosseteste pays attention to whales because they are specified in Genesis 1:21, as part of the creation of Day Five.[10]

This attention to the precise wording of Scripture pervades the whole of the *Hexaemeron*. An example is the separation of water and earth on Day Three, where he notes that the divine injunction is phrased as: 'Let the dry land appear' not 'Let there be an appearance of dry land', and in distinction to 'Let light be made' and 'Let there be a firmament'. The significance of this grammatical change, Grosseteste explains, is the revelation of a hierarchy in the creative process, and the proximity of parts of Creation to the true Existence. So, species of things of light more closely resemble the Word, that is, light and the firmament, whereas those species below, the species of things here, in this case on earth, resemble the Word in a lesser way. This framework is reflected in the change in phrasing. Grosseteste comments further that the fact of a grammatical change is in itself a foreshadowing of the more powerful 'Let us make', said in relation to the creation of man. In the direct reference to creation in this last example 'something even greater than these other two is hinted' (Grosseteste 1982/1996, 4.I.1).[11]

The intricacies of scriptural commentary flow from Grossetestes's use of interpretative norms, and his attention to the details of the text. He articulates the magnitude of the exegetical tasks, registering the sheer volume of commentary on the words of Scripture, in one case on precisely what is understood by heaven, earth, water, formlessness, voidity:

---

[10] 'And God created the great whales, and every living and moving creature, which the waters brought forth, according to their kinds, and every winged fowl according to its kind. And God saw that it was good [*Creavitque Deus cete grandia, et omnem animam viventem atque motabilem, quam produxerant aquae in species suas, et omne volatile secundum genus suum. Et vidit Deus quod esset bonum*].' All quotations from the Bible are from the Vulgate and Douay-Rheims translation.

[11] [*In homine vero faciendo*] *quasi hiis utrisque maius aliquid insinuatur...*'

> So let the reader of this science take note that until he can comment in this way on what has gone before and on what will follow, he is studying rather in the way that someone might look from a long way off, across a great intervening space, at a very fine piece of carving, nor distinguish the varied formed surface of the carving from the rough and unformed wood (Grosseteste 1982/1996, 1.XIX.2).[12]

The need for careful commentary is underlined by Grosseteste's reminder of the seriousness of wrong or mistaken interpretation. The most serious is heresy, and Grosseteste presents as an example that of the Manichees in their interpretation of dark and light as two competing divine powers. Grosseteste outlines both the errors of the Manichees, and how these might best be refuted (Grosseteste 1982/1996, 1. XXIII.1–2 and 1.XXIV.1–2). With the campaign against the Cathar heresy in the south of France running to its close, identified in the high medieval period with ancient Manicheism, this would have been a matter of no small contemporary resonance to Grosseteste and his audience (Marvin 2008; Pegg 2008; Power 2013).

In writing a hexaemeral commentary Grosseteste joined a long and distinguished series of commentaries and commentators (Robbins 1912; Gasper 2011). As a genre of Christian exegesis, the commentary on the opening of the Bible, and the six days of creation may find its roots in Paul's letter to the Colossians 1:15–17, and in the writings of Philo of Alexandria, who was the first author to invoke the phrase *hexaemeron* for the work of the six days (Wenham 1987). From the writings of Theophilus and Origen in the second and third centuries came the background for the magisterial series of homilies on the first Genesis creation story by Basil the Great (329/30–379). Basil's was the first extant work to be limited to the six days, and there are from this point on a number of other treatises which are so limited. However all Genesis commentary must include, as a matter of course, treatments of the six days of creation. Too sharp a distinction between the two genres would, for example, exclude any of the five occasions which Augustine devoted to the subject. Grosseteste himself does not confine his commentary to the six days, moving to the fall and expulsion from Paradise. The fact that he composed the *Hexaemeron* after delivering his lectures on Genesis should not be forgotten in this context.[13]

The Genesis creation story features strongly in Greek and Latin Patristic writing, and is one genre in which the medieval west inherited a significant part from the Greek commentaries, from the Patristic period to the mid-twelfth century (Freibergs 1981; Gasper 2011) Basil's work was translated into Latin at the end of the fourth century, as was the continuation on the making of man by his brother Gregory of Nyssa. Moreover, Ambrose's text drew heavily on Basil, and even Augustine shows his knowledge of the Greek tradition in this context. Isidore of Seville and Bede continued the western tradition, and form the basis for Carolingian

[12] '*Unde noverit lector huius sciencie quod, donec sic possit exponere tam predicta quam ea que sequuntur, speculatur velud a longe distans qui subtilem sculpturam magno interiecto loci spacio contuetur, nec signatas apprehendit sculpture protracciones, nec distinguit sculpture varietate formatum a lingo rudi et informi*'.

[13] A structural comparison could be made between Ambrose's *Exameron* and his related homiletic surveys on *Paradise* and *Cain and Abel* and the boundaries of Grosseteste's discussion.

commentary, with the exception of John Scottus Erigena, whose thought is particularly creative in its engagement with Greek thought. Creation commentary experienced a marked increase in popularity in the later eleventh and twelfth centuries, capturing the interest of leading scholars, from Anselm to Peter Abelard, and through to Peter Lombard, Hugh of Amiens, Anders Sunnensen and Alexander Neckham. A preliminary survey indicates over 90 commentaries on Genesis in the century around Grosseteste's own (Gasper 2014).

As James McEvoy colourfully expressed it: 'The desk at which he [Grosseteste] worked must have been huge, for he had the Vulgate, the Septuagint, and several Patristic commentaries on Genesis constantly open before him' (McEvoy 2000). This image aside, a significant number of the ancient and early medieval sources find their way into Grosseteste's treatment of the text: some 36 authors and 98 titles (Grosseteste 1982). Augustine, Ambrose, Jerome Gregory and Isidore are deployed extensively, the former dominantly. Basil, Gregory Nyssa, John Chrysostom and John Damasence are widely used too, and, at this point in Grosseteste's career mostly in existing Latin translation. Classical and Muslim authors also find their way into the text, in contexts to be explored later, some Aristotle, Plato, Ptolomey, Avicenna and Alpetragius. Modern authors are not used much, but this fits a genre where authorities are carefully balanced. Anselm's *Cur Deus homo* is never cited, but his influence, along with Hugh of St Victor, can be detected in Grosseteste's elaboration of a spiritual economy cosmic in its proportions and implications.[14]

Within the treatise Grosseteste exhibits four approaches towards his authoritative sources: fiercely condemning of points of view that damage the Christian interpretative framework, balancing authorities against each other even when contradictory, adopting a critical stance towards the subject under question, and offering of his own opinion, or marshalling of his own thoughts on the matter. The first emerges from the first part, and Grosseteste's commentary on 'in the beginning':

> The first word then, 'In the beginning' proclaims the start of time, and that the world was made at the beginning of time, and does not have an unlimited and infinite past. Hence, in the use of this single word, 'in the beginning', Moses overthrows the error of the philosophers who said that the world has no start in time. Aristotle said this, and tried to prove it, in the eighth book of the *Physics*, and Plato, likewise, in the *Timaeus*, brings in someone who claims that there have been an infinite number of Deluges (Grosseteste 1982/1996, 1.VIII.1).[15]

Grosseteste warns further against those of his contemporaries who claim that Aristotle did not think of a beginning-less universe, and 'while they make a Catholic of Aristotle, will make heretics of themselves' (Grosseteste 1982/1996, 1.VIII.4).[16]

---

[14] Anselm's influence on Robert Grosseteste has been explored recently in Cooper (2012).

[15] *'Primum itaque verbum, videlicet:* In principio, *resonat temporis inicium, et mundum a temporis principio esse factum, et non esse ex parte anteriori interminatum et infinitum. Unde in hoc unico verbo quod dicit:* In principio, *elidit errorem philosophorum qui dixerunt mundum non habuisse temporis inicium, quemadmodum dixit et probare nisus est Aristotiles in octavo* Physicorum*; similiter Plato in* Thimeo *inducit quondam qui infinitas inundaciones diluviorum asserit precesisse'.*

[16] *'...et Aristotilem catholicum constituendo, se ipsos hereticos faciant.'*

Patristic commentators are approached with respect and deference, although Grosseteste is not shy of pointing to their idiosyncrasies and disagreements. In a long discussion on the nature of goodness, he observes:

> That this feat is highly difficult and inexplicable, I will not try to disguise. Basil and Ambrose sweated away over explaining the natures of the things created on each of the six days, and to the best of their ability showed forth the goodness of their creator: but many have thought that their aim in writing all that was rather to show off how learned they were in the natures of things (Grosseteste 1982/1996, 2.VI.1).[17]

More often than not, Grosseteste stresses humility in front of the ancient authors. On the meaning of the terms 'heaven' and 'earth' as anagogical of the mind of God/created things, he uses the image of juvenility:

> This anagogical understanding of Basil, which takes us up from created things to the uncreated ideas in the mind of God, I pass over without interpretation, since I have no idea of how to interpret it. Indeed, even with the other interpretations I am as a child, and can only speak of them stammeringly (Grosseteste 1982/1996, 1.XII.4).[18]

One of the most significant areas where Grosseteste leaves questions between authorities unresolved concerns whether the Genesis account of creation indicated simultaneous or successive creation. Where Basil, Ambrose and Gregory the Great all stress the successive nature of creation, Augustine disagreed, preferring to take Genesis 2.4 as the controlling passage for the hexaemeron account: 'These are the generations of the heaven and the earth, when they were created. In the day that the Lord God made the heaven and the earth.' One day is implied, the seven 'days' a way of dividing different aspects of creation and their meaning. Bede sought to introduce greater clarity to the issue in noting that Augustine interpreted 'day' in Genesis 2:4 as synonymous with time, and therefore a different way of saying the six days. In his own account of the six days, Bede offers a straightforwardly successive interpretation. In the twelfth century these positions became more entrenched: where Anselm and Abelard followed Augustine, Peter Lombard in his theological case-book, of the 1150, the *Sentences*, did not: 'And he did not form them simultaneously, as it pleased some of the Fathers [to hold], but at intervals of time and in the course of six days, as it has seemed to others' (Lombard 1971–1981/2008, Dist. 12.1.2).[19] Grosseteste refers to the differences between his authorities,

---

[17] *'Quod quam difficile factum sit et inexplicabile, nullum reor latere. Basilius itaque et Ambrosius, qui in explicandis naturis rerum singulis sex diebus creatarum desudaverunt, pro modo facultatis sue creatorum bonitatem exposuerunt, licet multis videatur quod magis ad ostentacionem pericie sue in naturis rerum talia conscripserunt'.*

[18] *'Huius igitur anagogiam, que ex rebus creatis sursum ducit in raciones earum increatas eternas in mente divina, interpretari omitto quia interpretari nescio. Circa alias namque interpretaciones puer sum et non nisi balbuciendo loqui scio...'*

[19] *'quae non simul, ut quibusdam sanctorum patrum placuit, sed per interualla temporum ac sex uolumina dierum, ut aliis uisum est, formauit'*. Dist. 12.2 discusses the questions of simultaneous creation in more detail, with Augustine as a singular voice opposed to the view which is commended and preferred by Gregory, Jerome, Bede, and many others (Gasper 2014).

but deftly eschews final judgment while inclining towards Augustine (Grosseteste 1982/1996, 2.IV.1-3 and 2.V.6).[20]

Grosseteste is happy also to criticise the positions of his authorities especially on subjects where the limitations of human knowledge should inspire circumspection. On the nature of the heaven below the firmament and its spatial and conceptual extent, he notes the careful investigations by many authors, unnamed. However, he concludes that: 'I do not know whether any of them have found out the truth. Or if they perhaps found it, I do not know whether any of them have grasped that they have found the truth with any true or certain reasoning (Grosseteste 1982/1996, 3. VI.1).'[21] In this case he is happy enough to acknowledge openly his own ignorance about the subject (Grosseteste 1982/1996, 3.VIII.1). Following a similar logic Grosseteste will, in certain circumstances, put forward his own views, as his own:

> I want the reader to know that if I occasionally put in my writing words which are not from any authority, I am not putting them forward in an assertive manner, but am making them known to the audience as a sort of exercise for them, 'following the trail of the truth by conjectures and clues' (Grosseteste, 1982/1996, 4.I.4).[22]

The original context for the lectures on Genesis emerge here, with the threefold task of thirteenth century to the fore: *lectio* (reading), *disputatio* (disputation) and *praedicatio* (preaching) (Ginther 2004). Grosseteste is careful to circumscribe the occasions when he puts his own reasoning forward, and claims no greater authority than an intellectual exercise and discipline.

12.3 Knowledge: Faith and Reason

The question of the claim to knowledge, and the status of knower and known, form the central part of Grosseteste's *Hexaemeron*, and its most radical intellectual discussion. This discussion is grounded in the definition of faith and science, with which the commentary begins, and represents a decisive shift in the history of western theology, and in speculative thought more generally. Grosseteste addresses the complex question of what theology consists, and does so both within and beyond the context of Aristotelian science. The three requirements for an Aristotelian science, are (1) a defined and unified subject, (2) with conclusions resulting from syllogistic arguments, and (3) conclusions based on a set of premises

[20] Grosseteste raises but leaves unresolved the issue of successive days of creation. Where he favours Augustine it is with reference to the sight of angels rather than men.

[21] *'Sed nescio an aliqui veritatem invenerunt; aut si forte invenerunt, nescio an eorum aliqui se invenisse veritatem veraci et certa racione deprehenderint'.*

[22] *'Volo autem scire lectorem quod si qua non ex auctenticis verbis scribendo intersero, non enunciativo modo eadem profero, sed exercicii loco auditoribus intimo, "coniecturis quibusdam atque indiciis veritatis persequens vestigia".'* The text quoted is from *De opificio hominis* (Gregory of Nyssa 1855-1861, XVII.15), which would have been available to Grosseteste in the Latin translation by Dionysius Exiguus (c.470–c.544).

necessary and prior to the conclusions. Grosseteste, as Ginther has demonstrated, was the first to explore the first of these questions and unlike, for example, William of Auxerre (d.1231), he provides a delineation of what he understand theology to mean (Ginther 2004).[23] The subject on which theology turns, is, for Grosseteste, the whole Christ, the *Christus Integer*.

By making the incarnate God, or the God-man, the focal point of his theological vision, Grosseteste is able to develop the implications of Christ as the point of unity between the Creator and Creation. A powerful series of statements about the unities that Christ invokes follow; he is the union of divine and human natures, he unites with humanity through the incarnation, he reunites with the Church in the Eucharist and as Creator is the reason why all created reality belongs to the study of theology. Christ is the centre of theology, positioned above Scripture, and Grosseteste accordingly emphasises the encompassing nature of theological exposition. Theology includes all other sciences, and does so in an active way, allowing Grosseteste the theologian a creative and dynamic engagement with the secular sciences of the quadrivium: the mathematical arts. The created order does not consist of the natural world only; it is an element of salvation history and the way to seek God.

What theology is not, in itself, for Grosseteste, is a science. The *Posterior Analytics* provided him with a precise meaning for *scientia*: true knowledge is unchangeable and incorruptible, and descends from the archetypical world, through the created intelligences to the celestial bodies, and finally, in the sub-lunary, to the notion of universals. *Scientia* is located in the knowledge based on things as they generally are (natural philosophy), as they always are (mathematics) and in the immutable cause of their existence. Theology deals with far more than this, being focused on its object, the dual natured, unifying figure of Christ. It is therefore not so much a science as wisdom. Theology requires faith since it embraces both Creator and creation, it cannot be understood unless it is first believed, a formulation of Grosseteste close to that of Augustine and Anselm (Grosseteste 1982/1996, 1.II.1). Faith, like science, rises up, and in so doing does not encounter first the objects of faith but rather is a medium by which these objects are believed. The medium is that of authority, of which the highest is Scripture. Wisdom, Grosseteste states, following Job 28: 12–14, 18 and 21, is hidden, complex, and has to be drawn out; our knowledge is not God's, and it involves the whole created order, including Scripture.

> It was of this that Job spoke: 'But where is wisdom to be found, and where is the place of understanding? Man knoweth not the price thereof, neither is it found in the land of them that live in delights. The depth saith: It is not in me: and the sea saith: It is not with me. Wisdom is drawn out of secret places.' And: 'It is hid from the eyes of all living' (Grosseteste 1982/1996, 1.II.1).[24]

---

[23] This and the following paragraph draw heavily on Ginther's magisterial exposition.

[24] '*De quo dicit Iob:* Sapiencia vero ubi invenitur, et quis est locus intelligencie? nescit homo precium eius, nec invenitur, in terra suaviter vivencium. Abissus dicit: Non est in me, et mare loquitur: Non est mecum.. Trahitur autem sapiencia de occultis. Et: Abscondita est ab oculis omnium viventium'. Job, as Grosseteste and his listeners would have been fully aware, goes on to

12.4 Imagination

How this wisdom is revealed lies behind Grosseteste's original question pertaining to the *Hexaemeron*, namely why the bible begins with creation. The believability of Scripture, and the role of imagination are important elements in his answer:

> The species of this world, in so far as regards the way they are now governed, have the certainty of sense and of science. But in so far as regards the ordering in which they were created, they cannot be grasped at first except by faith. So the creation of the sensible world, on account of the way in which the world is imaginable and graspable by the external senses of the body, should be told in the opening part of Scripture. This is in order that anyone, even among the uneducated, may be able to grasp a story of this kind easily, through his imagination and through the images of corporeal things, and grow stronger in faith through the authority of the one who speaks (Grosseteste 1982/1996, 1.II.3).[25]

Grosseteste stresses there the authority, and hence the believability of Scripture, and the subordinate role of imagination in the articulation and realisation of faith. The role of the uneducated is noteworthy and might be taken particularly in the context of Grosseteste's capacity as lector to the Franciscans. How to preach effectively to all, educated and uneducated plays a central role in his pedagogic approach to his charges.

Imagination plays an important role in Grosseteste's writing, although it should be emphasised that his conception is quite different to modern understandings of imagination as fantasy. Grosseteste inherited the traditions of the twelfth and early thirteenth centuries, which, particularly under the influence of Avicenna, defined imagination as a faculty. According to this scheme, imagination is a fixing of sensory perception, prior to the application of reason (McEvoy 1983; Southern 1992). This process is illustrated in one of Grosseteste's most striking images, proposed in *Letter 1*, composed in the later 1220s, during his regency in theology and at about the same time as the Genesis commentaries. Running through the ramifications of the image of the object in the architect's mind as a metaphor for

state: 'God understandeth the way of it, and he knoweth the place thereof. For he beholdeth the ends of the world: and looketh on all things that are under heaven. Who made a weight for the winds and weighed the waters by measure. When he gave a law for the rain, and a way for the sounding storms. Then he saw it, and declared, and prepared, and searched it. And he said to man: Behold the fear of the Lord, that is wisdom: and to depart from evil, is understanding...' [Job 28: 23–8]. *'Deus intelligit viam ejus, et ipse novit locum illius. Ipse enim fines mundi intuetur, et omnia quae sub caelo sunt respicit. Qui fecit ventis pondus, et aquas appendit in mensura. Quando ponebat pluviis legem, et viam procellis sonantibus: Tunc vidit illam et enarravit, et praeparavit, et investigavit. Et dixit homini: Ecce timor Domini, ipsa est sapientia, et recedere a malo, intelligentia.'*

[25] *'Species autem huius mundi, secundum quod nunc gubernantur, habent sensus et scientie certitudinem. Secundum ordinem vero quo creabantur, non accipiuntur primo nisi per fidem. Mundi igitur sensibilis creacio, per modum quo mundus ymaginabilis est et per corporis exteriores sensus apprehensibilis, in primordio huius scripture debuit enarrari, ut quivis eciam rudis huiusmodi narracionem facillime possit per ymaginacionem et rerum corporalium ymagines apprehendere, et per dicentis auctoritatem in fide firmare'.*

creation, Grosseteste asks the recipient, Adam of Exeter, to imagine a house made of liquid, sustained only by the will of the architect (Grosseteste 1861/2010, Letter 1).[26]

However, imagination has to be employed correctly. False use of imagination led the ancients astray in their cosmological conception of a universe without beginning:

> This made them imagine before any given time, another time; just as the fantasy imagines a place outside any given place, and a space outside any given space, and so on to infinity. To cleanse oneself of this error, then, one can only cleanse the affection of one's mind of its love of temporal things, so that the glance of the mind, untouched by images, can go beyond time and grasp the simplicity of eternity, in which there is no extension of before and after, and from which all time and every before and every after proceed (Grosseteste 1982/1996, 1.VIII.5).[27]

---

[26] *'Imaginare itaque in mente artificis, artificii fiendi formam, utpote in mente architecti, formam et similitudinem domus fabricandae, ad quam formam et exemplar solummodo respicit, ut ad eius imitationem domum faciat. Et imaginare cum hoc per impossibile ipsius architecti volentis domum fabricare voluntatem ita potentem, quod se sola applicet, materiam formandam in domum formae in mente architecti, qua applicatione figuraretur in domum. Et imaginare cum his quod materia domus esset fluida, nec posset permanere in forma accepta in se, si esset separata a forma in mente architecti, sicut aqua figurata sigillo argenteo, separato sigillo, statim amitteret figuram receptam. Imaginare itaque voluntatem artificis applicantem materiam domus ad formam in mente architecti, non solum ut per hanc applicationem formetur in domum, sed etiam applicantem illam ei, quamdiu domus manet in esse, domus ut formata in esse servetur. Eo itaque modo quo forma huius, in mente huiusmodi architecti, esset forma domus, est ars, sive sapientia, sive verbum omnipotentis Dei, forma omnium creaturarum. Ipsa enim simul et exemplar est, et efficiens est, et formans est, et in forma data conservans est, dum ad ipsam applicantur et revocantur creaturae';* 'So, imagine in the mind of a craftsman the form of an object to be crafted, as, for example, in the architect's mind, the form and likeness of a house he is to build. It is on this form and archetype that he focuses exclusively so that he may build a house in imitation of it. And imagine along with this, despite the impossibility, the will of that architect who wants to build that house, a will so powerful that it could by itself apply to the form in his mind the material to be formed into the house, an action by which the material would be shaped into a house. And imagine along with these mental images that the house's building material were liquid and incapable of remaining in the form it had received if separated from the form in the architect's mind, just as water given a shape by a silver seal would, once the seal is taken away, immediately lose the shape it had received. So, imagine the will of the architect applying the building material of the house to the form in his mind not only so that by this action the material may be shaped into the house, but also applying it here as long as the house remains in existence as a house, so that the house thus formed may be kept in existence. In the same way, then, in which the form of this material in the mind of that architect would be the form of the house, the creative imagination or wisdom or the Word of the almighty God is the form of all creatures. For it is simultaneously creation's archetype, and that which brings it about and imparts its form, and that which conserves it in the form it has been given, when creatures are brought into contact with it and recalled to it'. Goering and Mantello make clear the relation between their English translation and the edition (Grosseteste 2010).

[27] *'. . .qua coacti sunt ymaginari ante omne tempus aliud, sicut ymaginatur fantasia extra omnem locum locum alium, et extra omne spacium spacium aliud, et hoc usque in infinitum. Unde et huius erroris purgacio non potest esse nisi per hoc quod mentis affectus purgetur ad amore temporalium, ut mentis aspectus immunis a fantasmatibus possit transcendere tempus et intelligere simplicem eternitatem, ubi nulla est extensio secundum prius et posterius, et a qua procedit omne tempus et prius et posterius'.*

What Grosseteste terms the *affectus mentis*, or the disposition towards receptivity and the *aspectus mentis*, the openness of the mind to perception, unlocks proper understanding. These are terms common to mystical discourse, but here lay emphasis on the caution with which sensory data in the changeable world should be taken; true knowledge is unchanging. All knowledge, of Scripture, of the natural world, of the created world requires the proper use of reason.

12.5 Light

Such proper use of reason, in the case of the *Hexaemeron*, involves and invokes the breadth of disciplines which Grosseteste deploys in the service of theology. Human sciences are kept firmly in their place: Grosseteste offers an allegorical reading of the firmament as Scripture, with the waters below as human sciences, inferior to Holy Scripture (Grosseteste 1982/1996, 3.XIV.8). Nevertheless, in pursuance of the literal interpretation of Scripture, scientific knowledge is used and elaborated upon.

While numerous examples could be adduced, this is particularly the case in his treatment of the subject of light. Light is especially associated with Grosseteste's science and theology, as is well known. It is striking within Part Two of the *Hexaemeron*, which deals with the creation of first light, to find a number of sections in which the *De luce* is recapitulated (taking the *De luce* as composed before the *Hexaemeron*).[28] Light's ability to constantly self-generate, and its self-manifesting quality are discussed. In the *Hexaemeron* however, Grosseteste makes use of other comparisons and sources. He turns to John Damascene for the observation that light has no hypostasis, and to Augustine for the description of light as the queen of colours (Grosseteste 1982/1996, 2.X.2). The beauty and harmony of light are called to mind, quoting Basil:

> Light is beautiful in itself, since 'its nature is simple and in every way homogenous': therefore it is united with itself to a very high degree, and most harmoniously proportioned to itself by its equality. Harmony in proportion is what beauty is: hence even without shapes of bodies light is beautiful, by its own harmonious proportion, and is most pleasing to the sight (Grosseteste 1982/1996, 2.X.4).[29]

De luce is used to contribute to the discussion of light's qualities and grounded in the commentary tradition. Moreover, Grosseteste uses only the first section of the *De luce* (Grosseteste 2013a, lines 1–31, pp. 226–228; Grosseteste 2013b, pp. 239–240). He does so in a condensed form; he does not enter the mathematical explanations of infinities or the creation of the spheres of the universe. These are

---

[28] See above, note 2.

[29] *'Hec per se pulchra est, quia eius "natura simplex est sibique per omnia similis;" quapropter maxime unita, et ad se per equalitatem concordissime proporcionata. Proporcionum autem concordia pulcritudo est; quapropter eciam sine corporearum figuram armonica proporcione ipsa lux pulcra est et visui iocundissima'.*

unnecessary to his exegetical purpose. More pointedly, he omits any mention of light, 'lux', as the first corporeal form: 'Formam primam corporalem' (Grosseteste 2013a, lines 1–2, p. 226; Grosseteste 2013b, p. 239). This might be read in the context of a theological debate revealed in his first letter, written probably between 1225 and 1229 to Master Adam Rufus, which opens with Grosseteste's acknowledgement of Adam's invitation to dwell on the words 'God is the first form and the form of all things': 'Deus est forma et forma omnium' (Grosseteste 1861/2010, Letter 1, p. 1/35). This is a phrase, as McEvoy suggests, which may derive from Eriugena's *Peryphiseon*, condemned in 1225, and Grosseteste's reply to Adam might be seen as a cautious and careful defence of Eriugena's position (Grosseteste 2010; McEvoy 1995). The *De luce* is not mentioned in Grosseteste's response, but the formulation raised by Adam also has similarities to the opening of the treatise. Grosseteste takes pains in his letter to emphasise God as the first form: '. . .in my view it is true that 'God is the first form and form of all things'. 'And since he is form, he is of necessity the first form, because before him there was nothing: he is the first and the last' (Grosseteste 1861/2010, Letter 1).[30] The theological debate over the question of first form may, perhaps, have been an additional factor in how the *De luce* was used in Grosseteste's later reflections in the *Hexaemeron*.

Grosseteste takes, and expands on, the motive power of light in his discussion of the firmament and the heavens, developing a detailed geometric argument for a terra-centric universe based on light and its life-giving power. The principle established in the *De luce*, that light expands omni-directionally from a single point is repeated, but in the *Hexaemeron* it becomes part of an argument of lines and angles to demonstrate that in a hollow spherical light-giving body, light will concentrate in the centre. Light from the heavenly bodies generates heat, and the growth of plants and animals, and, this being the case, the most appropriate place for the earth is at the centre of the universe. The action of light is transformed here by Grosseteste into one of the most important features of his theology of creation, the centrality of life, and in particular of man:

> . . ..all things are for the sake of human beings, that is, in order that the generation of the human race should be completed up to the bringing to completion of the body of Christ, the church. Hence the movement of the heavens will be solely for the sake of the generation of the human race and those things which are here below are of the service of human beings (Grosseteste 1982/1996, 1.XVII.1).[31]

Grosseteste's universe in the *Hexaemeron* is full and bursting with life: it is, in this respect, the *De luce* fulfilled.

> God is all things in all things: the life of living things, the form of things with forms, the species of things with species [Or, the beauty of beautiful things]: and human beings are in

[30] *'scilicet quod Deus est forma et forma omnium; et cum sit forma, necessario est forma prima, quia ante ipsum nihil; ipse enim est primus et novissimus'.*

[31] *'Cum enim omnia propter hominem sint, ut compleatur videlicet humana generacio usque ad complementum corporis Christi quod est ecclesia, motus celorum non erit nisi propter generacionem hominum et eorum que hic inferius ministrant homini'.*

> all things God's closest likeness in resemblance. For this reason human beings, in so far as they are the image of God, are also in some way all things (Grosseteste 1982/1996, 8.I.2).[32]

12.6 Beauty and Order

Light as beauty is a powerful image and heuristic device for Grosseteste; and one where scientific observation is fused with scriptural commentary. A passage on stars, worth quoting at length, exemplifies this use of science theologically.

> All the stars have a superb and very great beauty, not from the arrangement of their parts, for they have no parts, but because of the joyful and merry shining of their light. Hence they are more beautiful on dark nights than on moonlit nights. And the great stars, because of their size, are more beautiful than the small stars. Also, the stars that are separate and distinct, because they are divided and distinct, are more beautiful than the stars that are spread out and joined together; and more beautiful than the stars of the galaxy, just as separate candles are more beautiful than is a fire. But they cheat the human sight in estimating their resting-place because they are so very far away from us. We do not see the stars directly, but in a reflection, as is known in the science of perspective. And in the same way that something seen in water seems bigger than it is because of the reflection of the sight towards the deeper water, so the stars seen in heaven seem smaller because of the reflection of the sight in crossing the body of heaven towards the lesser depth of the body of heaven. Hence the stars seem small for two reasons: first, because they are far away from the eye, and secondly because of the reflection of the rays towards the less deep parts of heaven (Grosseteste 1982/1996, 5.XXIII.2).[33]

This passage is reminiscent of his treatise on the rainbow, the *De iride*, and Grosseteste's interest in optics and perspective in particular. The science of perspective, including refraction, reflection and magnification all feature in the *De iride*, and are applied here on a grander scale (Grosseteste 1912). Grosseteste not only pursues the line of thought that what is observable in nature on earth can apply equally to the heavens, but integrates this into his observation with biblical commentary. The contribution of the science of perspective to the intrinsic beauty of the

[32] *'Deus autem est omnia in omnibus, viventium vita, formosorum forma, speciosorum species; et homo in omnibus eius propinquissima similitudo imitatoria. Quapropter et homo, in hoc quod ipse est imago Dei, est quodammodo omnia'.*

[33] *'Omnibus autem stellis eximia et maxima est pulcritudo corporalis, non propter compaginacionem membrorum que nulla sunt eis, sed propter letum alacremque fulgorem luminis, pulcriores que sunt in noctibus obscuris, quam in noctibus a luna lustratis; et stelle magne propter magnitudinem pulcriores sunt stellis parvis; et stelle separate et distincte, propter divisionem et distinctionem, pulcriores sunt stellis extensis et coniunctis; et pulcriores stellis galaxie, quemadmodum candele distincte sunt pulcriores igne. Fallunt tamen humanos visus estimacione quietis propter magnitudinem elongacionis sue a nobis. Comprehenduntur autem stelle a visu non recte, sed reflexe, ut ostenditur in perspectiva. Et quemadmodum res visa in aqua apparet maior quam sit propter reflexionem visus ad profundius aque, sic stelle in celo vise apparent minores propter reflexionem visus in pertransitu corporis celi ad minus profundum in corpore celi. Unde stelle habent duas causas quare apparent parve: elongacionem videlicet a visu, et reflexionem radiorum ad minus profundum celi'.*

universe is not something that Grosseteste mentions in a treatise such as the *De iride*. However, in the *Hexaemeron* the service rendered by this science to the unfolding and exploration of divine creation is made manifest.

Beauty and order go hand in glove with the economy of creation. God does not create, as Grosseteste points out, randomly or pointlessly (Grosseteste 1982/1996, 3.XVI.3). Moreover, the goodness of creation is stressed, and all the more so in light of man's fall, and redemption.

> The goodness of a thing consists in the action on account of which the thing has been specially made, and in the usefulness of that action, in an ordering both towards that thing itself and towards others, and indeed towards the universe as a whole. Hence the goodness of each of the works of each day was to show forth the special natures of the special works of each day, and their natural actions and functions, and to bring out the beauty of their ordering towards the universe (Grosseteste 1982/1996, 2.VI.1).[34]

Order comes with consistency, scientifically and morally, and, importantly, derives from unity.

12.7 Unity

It is around unity that Grosseteste forms his most powerful arguments on the purpose of theology; and it is a theme explored in his other scriptural exegesis, occupying a prominent place in the *De luce*. From the opening of the *Hexaemeron*, unity is stressed: from the consistency of God's actions in creating small things as well as big things, whales and frogs and dolphins, shellfish and snails, to the overarching arguments of the *Christus Integer*, the whole Christ (Grosseteste 1982/1996, 6.I.4). The same emphasis is to be found in the *De cessatione legalium*, which derives from lectures on Isaiah and parts of the Genesis Commentary. Unity here is focused around Christ's incarnation, and Grosseteste's bold expansion of Anselm's argument that the God-man was necessary (man's sin was so great in the Adamic fall that only God could repay, while only man should repay it, hence the God-man). Instead, Grosseteste presses for the notion that the incarnation would have happened even without man's fall, to fulfil and unite creation. Human nature is both corporeal and rational. Grosseteste concludes with the chorus that:

> If, then, God should assume man in a personal unity, all creation has been led back to the fullness of unity; but if he should not assume man, all creation has not been drawn to the fullness of unity possible for it. If, therefore, we leave aside the fall of man it is nonetheless

[34] *'Bonitas autem rei consistit in accione propter quam res specialiter facta est et eiusdem accionis utilitate, et in ordine eiusdem rei ad se et ad alia queque in universitate. Quapropter, singulorum operum singularum dierum bonitates exponere esset operum specialium singularum dierum speciales naturas, et naturales acciones et utilitates, et ordinis sui pulcritudinem in universe pretractare'*. The argument here is proximate to Anselm's on right behavior and uprightness of will, as explored in his *De libertate arbitrii* and on the truth of things, as explored in his *De veritate* (Anselm 1946–1961a, b).

> fitting that God assume man into a personal unity, because he could do it and it would not be inappropriate for him to do it; but even more, it would be appropriate, because without this the created universe would lack unity. But if this were done all creation would have the fullest and the most fitting unity, and through this all natures would be led back into a circular fulfilment; because without God assuming man into a personal unity, one finds in the above-mentioned way a certain joining of angels and men (Grosseteste 1988/2012, 3. I.28).[35]

Although naturally lacking an exposition on the Unity and Incarnation, the same accent on unity suffuses the *De luce*, whether below the moon or part of the celestial spheres matter and light are fundamentally the same, they only act differently in, and because of, these different locations. A unified account of the ancient universe again finds fuller expression in the account of Genesis creation. Whether the experience of working through the cosmology of the *De luce* inspired parts of Grosseteste's theological vision is a different question, but the creative fusion of the scientific texts within biblical commentary remains the dominant feature.

12.8 Conclusion

The *Hexaemeron* presents Grosseteste as a capacious thinker at the height of his theological and scientific prowess, before he was to begin his sustained engagement with Greek thinking, theological and philosophical, in its original language. The richness of thought and the range of authorities leap from the pages, and make his commentary one of the largest and most significant within the genre. In turning his attention to Scripture and its exegesis, Grosseteste deploys his earlier explorations of natural phenomena, his emerging skills as translator and linguist, alongside his familiarity with, and sensitive response to the writings of the Christian Fathers and a very select number amongst their medieval successors to create a majestic, powerful and original theological vision. Amongst these authorities, science plays a central role. This is true of the exercise in defining theology to include Aristotelian investigation with which Grosseteste begins his discussion, and throughout the rest of commentary he makes extensive use of cosmological and meteorological works within the compass of his exegetical frame. In this sense, the *Hexaemeron* may be compared helpfully to the *Dicta*. These discourses incorporate nearly one-third of Grosseteste's commentary on Psalms, as well as other sermons and notes (Goering 2013). They were composed around 1230 and compiled later, and make frequent

[35] '*Si igitur assumat Deus hominem in unitatem persone, reducta est universitas ad unitatis complementum. Si vero non assumat, nec universitas ad unitatis complementum sibi possibile deducta est. Circumscripto igitur hominis lapsu, nichilominus convenit Deu assumere hominem in unitate persone, cum et hoc possit facere nec dedeceat ispum hoc facere; sed multo magis deceat, cum sine hoc careat universitas unitate. Hoc vero facto, habeat universitas plenissimam et decentissimam unitatem, redacteque sint per hoc omnes nature in complementum circulare; quia sine eo quod Deus assumat hominem in unitatem persone, est reperire modo supradicto concatenacionem quandam ab angelo usque ad hominem*'.

use of scientific learning in the service of exegetical understanding and pastoral or pedagogical analogy.

Both the *Dicta* and especially the *Hexaemeron* are, however, more than merely witnesses to a merging of Grosseteste's science and theology (Southern 1992). Not all aspects of his scientific thought are harnessed, and those that are, are carefully chosen. Rather, he places science within, and as an active part of, a grander enterprise, moving human intellect and spiritual understanding into all dimensions of Christian life. For all of the speculative, eclectic and extensive learning displayed, Grosseteste is careful to stress his fundamental purpose: the inculcation in the reader or listener of the frame of mind necessary for correct use of the imagination and subsequent interpretation. Reason and sin, Scripture and nature are part of one ordered, and ultimately, redemptive process: one that should be preached for the salvation of souls. Grosseteste looks to Basil and his hexaemeral homilies to emphasise this priority:

> So, since it is right to direct all things to their end, the art of expounding Scripture is to make everything found in it mean, in the end, something to do with the state of glory, or something which leads us directly to the state of glory, such as faith, hope and charity. That is why Basil says that the end of what is said in this teaching is not the praise but the salvation of those who learn (Grosseteste 1982/1996, I.V.1).[36]

Grosseteste brackets his *Hexaemeron* with two biblical verses, the first from John 17: 20–21: 'And not for them only do I pray, but for them also who through their word shall believe in me; that they may be one, as you, Father, in me, and I in thee; that they may also be one in us; that the world may believe that you have sent me' (Grosseteste 1982/1996, 1.I.1).[37] He ends with 1 Corinthians 15.22: 'As in Adam all died so in Christ are all justified'.[38] The *Hexaemeron* provides the theological ground and grammar for this pastoral vision, makes the intellectual space for his scientific work, and, in fulfilling both science and theology, offers both back to creation in a vision of joy and thanksgiving, both active and prayerful.

References

Anselm of Canterbury. (1946–1961a). *De veritate*. In *Opera omnia S. Anselmi Cantuariensis archiepiscopi*, 6 vols, F. S. Schmi Ed. [Vol. 1 printed at Seckau 1938; Vol. 2. at Rome 1940, all reset for the Nelson edn]. Edinburgh: Nelson.

---

[36] *'Unde, cum omnia ad finem oporteat dirigere, ars exposicionis huius scripture est ut totum quod in ea invenitur significet ultimo aliquid de statu glorie, aut aliquid directe deducens in statum glorie, velud est fides, spes et karitas. Unde Basilius ait quod sermonum huius doctrine finis est non dicencium laus, sed discencium salus'*. The Basil quotation is from I.1.2 (Basil 1958).

[37] *'Non pro eis rogo tantum, sed et pro eis qui credituri sunt per verbum eorum in me /Ut omnes unum sint, sicut tu Pater in me, et ego in te, ut et ipsi in nobis unum sint: ut credat mundus, quia tu me misisti'*.

[38] *'Et sicut in Adam omnes moriuntur, ita et in Christo omnes vivificabuntur.'*

Anselm of Canterbury. (1946–1961b). *De libertate arbitrii*. In *Opera omnia S. Anselmi Cantuariensis archiepiscopi*, 6 vols, F. S. Schmitt Ed. [Vol. 1 printed at Seckau 1938; Vol. 2. at Rome 1940, all reset for the Nelson edn]. Edinburgh: Nelson.

Anselm of Canterbury. (1946–1961c). *De Concordia*. In *Opera omnia S. Anselmi Cantuariensis archiepiscopi* 6 vols, F. S. Schmitt Ed. [Vol. 1 printed at Seckau 1938; Vol. 2. at Rome 1940, all reset for the Nelson edn]. Edinburgh: Nelson.

Anselm of Canterbury. (1976). *The complete philosophical and theological treatises of Anselm of Canterbury* (J. Hopkins & H. Richardson, Trans.). Toronto, ON: Mellen Press.

Basil. (1958). Eustathius. In Amand de Mendieta & S. Y. Rudbeger (Eds.), *Hexaëmeron S. Basilii Metaphrasis, Eustathius ancienne version latine des neuf homelies sur l'Hexaémeron de Basile de Césarée*. Berlin: Akademie-Verlag.

Bower, R. G., McLeish, T. C. B., Tanner, B. K., Smithson, H. E., Panti, C., & Lewis, N. (2014). A medieval multiverse?: Mathematical modelling of the Thirteenth Century universe of Robert Grosseteste. *Royal Society of London Proceedings Series A, 470*, 40025.

Callus, D. A. (1955). Robert Grosseteste as Scholar. In D. A. Callus (Ed.), *Robert Grosseteste: Scholar and Bishop* (pp. 1–69). Oxford: Oxford University Press.

Cooper, T. J. (2012). *One truth or many truths? Two medieval accounts of truth: Anselm of Canterbury and Robert Grosseteste*. Unpublished PhD diss., Catholic University of America.

Dales, R. C. (1961). Robert Grosseteste's Scientific works. *Isis, 52*, 381–402.

Dales, R. C., & Gieben, S. (1968). The proemium to Robert Grosseteste's *Hexaemeron*. *Speculum, 43*, 451–461.

Eastwood, B. (1988). Review of R.W Southern, Robert Grosseteste: The growth of an english mind in Medieval Europe. *Speculum, 63*, 233–237.

Freibergs, G. (1981). *The medieval Latin Hexameron from Bede to Grosseteste*. Unpublished PhD diss., University of Southern California.

Gasper, G. E. M. (2011). 'Oil upon the waters': On the creation of light from Basil to Peter Lombard. *Archa Verbi, 8*, 9–31.

Gasper, G. E. M. (2014). The beauty of creation. In G. Cipollone (Ed.), *Anselmo e la 'nuova' Europa* (pp. 41–56). Rome: Gregorian and Biblical Press.

Ginther, J. (2004). *Master of the sacred page: A study of the theology of Robert Grosseteste ca. 1229/30-1235.*. Aldershot: Ashgate.

Goering, J. W. (1995). When and where did Grosseteste study theology? In J. McEvoy (Ed.), *Robert Grosseteste: New perspectives on his thought and scholarship* (pp. 17–51). Turnhout: Brepols.

Goering, J. W. (2013). Robert Grosseteste's *Dicta*. The state of the question. In J. Flood, J. R. Ginther, & J. W. Goering (Eds.), *Robert Grosseteste and his intellectual milieu new editions and studies* (pp. 64–86). Toronto, ON: Pontifical Institute of Mediaeval Studies.

Gregory of Nyssa. (1855–1861). *De opificio hominis* (G. H. Forbes, Ed.). Burntisland: Pitsligo Press.

Grosseteste, R. (1861). *Roberti Grosseteste episcopi quondam Lincolniensis epistolae* (H. R. Luard, Ed.) Rolls Series 25. London: Longman, Green, Longman & Roberts.

Grosseteste, R. (1912). *De iride*. In L. Baur (Ed.), *Die Philosophischen Werke des Robert Grosseteste, Bischofs von Lincoln* (pp. 72–78). Münster: Aschendorff.

Grosseteste, R. (1982). *Hexaemeron* (R. C. Dales & S. Gieben, Eds.). Auctores Britannici Medii Aevi VI. Oxford: Oxford University Press.

Grosseteste, R. (1996). *On the six days of creation* (C. F. J. Martin, Trans.) Auctores Britannici Medii Aevi, VI (2). Oxford: Oxford University Press.

Grosseteste, R. (2010). *The letters of Robert Grosseteste, Bishop of Lincoln* (F. A. C. Mantello & J. Goering, Trans. & Eds.). Toronto, ON: University of Toronto Press.

Grosseteste, R. (2012). *On the cessation of the laws* (S. Hilderbrand, Trans.), The Fathers of the Church, Medieval Continuation. Washington, DC: Catholic University of America Press.

Grosseteste, R. (2013a). Grosseteste's *De luce*: A critical edition (C. Panti, Ed.). In Flood, Ginther, & Goering (Eds.), *Robert Grosseteste and his intellectual milieu* (pp. 193–247). Toronto, ON: Pontifical Institute of Mediaeval Studies.

Grosseteste, R. (2013b). Grosseteste's *on light*: An English translation (N. Lewis, Trans.). In Flood, Ginther, & Goering (Eds.), *Robert Grosseteste and his intellectual milieu* (pp. 239–247). Toronto, ON: Pontifical Institute of Mediaeval Studies.

Jerome. (1996). *Epistulae* (I. Hilberg &. M. Kamptner, Eds.). Corpus Scriptorum Ecclesiasticorum Latinorum, LIV–LVI. Vienna: Verlag der Österreichischen Akademie der Wissenschaften.

Lombard, P. (1971–1981). *Sententiae in IV libris distinctae*, 2 vols (I. Brady, Ed.). Grottaferrata: Editiones Collegii S. Bonaventurae ad Claras Aquas.

Lombard, P. (2008). *The sentences*, Book 2. *On creation* (G. Silano, Trans.). Toronto, ON: Pontifical Institute of Mediaeval Studies.

Marvin, L. W. (2008). *The Occitan war: A military and political history of the Albigensian Crusade*. Cambridge: Cambridge University Press.

McEvoy, J. (1983). The chronology of Robert Grosseteste's writings on nature and natural philosophy. *Speculum, 58*, 614–655.

McEvoy, J. (1995). Nature as light in Erigena and Grosseteste. In S. J. Ridyard & R. G. Benson (Eds.), *Man and nature in the middle ages* (pp. 49–53). Sewanee, TN: University of Sewanee Press.

McEvoy, J. (2000). *Robert Grosseteste*. Oxford: Oxford University Press.

Muckle, J. T. (1944). The Hexaemeron of Robert Grosseteste: The first twelve chapters of part seven. *Mediaeval Studies, 6*, 151–174.

Muckle, J. T. (1945). Robert Grosseteste's use of Greek sources in his Hexameron. *Medievalia et Humanistica, 3*, 33–48.

Muckle, J. T. (1951). Did Robert Grosseteste attribute the Hexameron of St. Venerable Bede to St. Jerome? *Mediaeval Studies, 13*, 242–244.

Panti, C. (2013). Robert Grosseteste and Adam of Exeter's physics of light, remarks on the transmission, authenticity, and chronology of Grosseteste's Scientific *Opuscula*. In Flood, Ginther, & Goering (Eds.), *Robert Grosseteste and his intellectual milieu* (pp. 165–190). Toronto, ON: Pontifical Institute of Mediaeval Studies.

Pegg, M. G. (2008). *A most holy war: The Albigensian crusade and the battle for Christendom*. Oxford: Oxford University Press.

Phelan, G. B. (1943). An unedited text of Robert Grosseteste on the subject-matter of theology. *Revue néoscolastique de philosophie, 36*, 172–179.

Power, D. (2013). Who went on the Albigensian crusade. *English Historical Review, 128*, 1047–1085.

Robbins, F. E. (1912). *The Hexaemeral literature: A study of the Greek and Latin commentaries on genesis*. Chicago: Chicago University Press.

Sharp, D. (1930). *Franciscan philosophy at Oxford in the thirteenth century*. Oxford: Oxford University Press.

Southern, R. W. (1992). *Robert Grosseteste, The growth of an English mind in medieval Europe* (2nd ed.). Oxford: Oxford University Press.

Wenham, G. J. (1987). *Word Biblical commentary, Vol. 1, Genesis 1-15*. Waco: Texas Word Books.

Part V
Science and Faith: Some Lessons from the Thirteenth Century?

Chapter 13
Intelligo ut credam, credo ut intelligam: Robert Grosseteste Between Faith and Reason

Angelo Silvestri

13.1 Faith and Reason

Robert Grosseteste, bishop of Lincoln from 1235 to 1253, was one of the most important bishops of England and probably of the whole of Europe; he therefore does not need much by way of introduction. He was a scholar, a theologian and a scientist. As he was one of the greatest scholars of his period, in this paper I have tried to examine whether or not it is possible to 'classify' the origins of his knowledge. That is to say I will try to comprehend whether Grosseteste belonged to one of the two main schools of thought of the middle ages: was he a philosopher whose great knowledge (*intelligo*) allowed him to understand the secrets behind faith and the mysteries of God (*ut credam*); or was he rather a strong believer (*credo*), probably a saint as many claimed, who simply needed faith to enhance his knowledge (*ut intelligam*) and to make progress on scientific understanding? In order to answer these questions I will consider some of Grosseteste's literary works, and the main philosophers and ideas influencing him.

Given the great output of Grosseteste in different fields, I make no attempt at providing a complete chronology of his works and thoughts. However, in order to produce a coherent analysis of the origins and the evolution of Grosseteste's ideas and his literary production, I will consider the works of those philosophers whom Grosseteste had studied and was familiar with, as well as the major philosophical, epistemological and theological works of the scholars of his time. Grosseteste began his 'career' not as a theologian, but rather as scientist.[1] Already during the

[1] McEvoy for instance wrote the following in the introduction to his book: 'I have suggested that without Grosseteste there might not have been mathematical-scientific tradition at Oxford' (James McEvoy 1982).

A. Silvestri (✉)
Cardiff University, Cardiff, UK
e-mail: SilvestriAM@cardiff.ac.uk

J.P. Cunningham, M. Hocknull (eds.), *Robert Grosseteste and the pursuit of Religious and Scientific Learning in the Middle Ages*, Studies in the History of Philosophy of Mind 18, DOI 10.1007/978-3-319-33468-4_13

first years of his adult life in Hereford, just after 1190, he was writing on chronology, astronomy and astrology (Southern 2004). Richard Bardney, the only direct source we have about Grosseteste's early years, tells us, he had the ability to cure people, *invenit voces quibus allevare solebat ægrorum morbos, subsidiumque dabat*.[2] Grosseteste's life during these early years is quite obscure and difficult to disentangle (Silvestri 2015), because of the paucity of the sources and the scarcity of direct and indirect evidence. In the recent past one of the main problems has lain in the attempt to classify his major works. The work and research of two (amongst others)[3] of the most important contemporary scholars of Grosseteste, Richard Southern and James McEvoy, seems to have produced a result with which everybody appears to agree (Callus 1969; Stevenson 1899). According to this perspective Grosseteste's works fall roughly into two distinct categories and periods; to the first period, which runs from his mastership in art to 1235, belong his commentaries on Aristotle and the Bible, while to the second period, from 1235 to his death in 1253, belong his translations from the Greek sources. However, Southern and McEvoy disagree about Grosseteste's scientific production. Southern argued that most of his scientific writing belonged to the years before 1225[4] and that thereafter he was devoted to theology, whereas McEvoy thinks that some of Grosseteste's important writing belongs to the period from 1225 to 1233.[5] This disagreement is not simply related to Grosseteste's production, but also to his alleged presence as a scholar in the University of Paris, another controversial point which has challenged the study of many other scholars in recent years.[6] I have briefly mentioned these studies due to their importance for the current academic debate and also because the chronological events unfolding in Grosseteste's life is a topic which cannot be overlooked by anybody who wants to understand his work. However, in this essay I will focus more on the message conveyed by his writings in order to try to understand not just where he was coming from, but also and especially at what he was aiming.

Grosseteste was a man of the thirteenth century and this century witnessed a major transformation in the form of scholarly thought. The development of medieval thoughts has never been linear, nor univocal and it has been subjected to major

[2] Wharton (1691) Richard of Bardney, like Gerald of Wales, clearly mentions Grosseteste's interests in medicine, but also his knowledge of and interest in the study of animals and horticulture. See (Goering 1995). See also (Southern 1986).

[3] Especially (Callus 1969). Harrison Thomson in 1940 had catalogued Grosseteste's works with great accuracy. According to his research, 129 sermons could be ascribed to Grosseteste. However, in a more recent work, Susan Paul has challenged Thompson's theory, advancing the hypothesis that only about 40 complete items, 19 summarized and 33 fragments or sermons notes could be clearly identified. See (Harrison 1940; Paul 2002).

[4] This would include *De Cometis* 1200, the commentary on *Posterior Analytics* 1220–5, and the commentary on *Physics* 1220–5. See (Southern 1986).

[5] Therefore according to his theory the commentary on *Posterior Analytics* is to be dated around 1228–30, and the commentary on *Physics* about 1228–32.

[6] The main supporters of Grosseteste's presence in Paris are: (Goering 1995; Schulman 1997).

fluctuations due to general as well as local contingencies. However, we can trace four different phases that have characterized the growth and the expansion of theological and philosophical understanding. The first phase, from the fifth to the ninth centuries, is commonly defined as an age of 'obscurantism' given the poor status of culture and cultural research. Indeed only two important philosophers emerged from this period, Boethius and Scotus Eriugena. Thereafter, until at least the twelfth century, the monastic reforms and the Crusades produced a new way of approaching the main questions concerning God, faith and humankind. A new scholarly thinking within and outside schools was epitomized by philosophers like Anselm, Abelard and those of the school of Chartres. This period with its growing tension between faith and reason and between dialectic and traditional theology was the harbinger of the golden age of theological thinking (Knowles 1963). There is, indeed, no doubt that the golden age of scholastic thinking was the following period, which started (and to some extent, also ended) in the thirteenth century. This period featured thinkers like Thomas Aquinas, Duns Scotus, Bonaventure and, of course, Robert Grosseteste whose work would lead into the last period of medieval thinking, the fourteenth century, which witnessed the crisis of the church and the Holy Roman Empire, intellectually embodied by the so called, 'razor of William of Ockham'. The development of medieval philosophy and theology spanned over six centuries during which philosophical and political ideas became intertwined with religious beliefs and dramatic social changes. Between St Augustine in the fourth century, who championed the union between reason and faith, and Ockham who divided faith and reason, Grosseteste lived in the golden age of scholastic thinking. Scholastic thinking embodied the philosophy and theology taught in the medieval schools; whether monastic, cathedral or palatine. The *Scholastica* investigated the relationship between faith and reason and in particular the use of philosophy to investigate and to interpret the Holy Scriptures. The masters who taught in the University were given the task by the established church to teach and implicitly to spread the knowledge of the Bible. This circumstance has been frequently dismissed simply as a scholastic-pedagogic change and relegated to within the University's cultural sphere, but this was not simply an academic variation; in fact it was much more than that. Indeed, this philosophical and theological development also had political and social consequences. On one side with the rise of the Universities a new power emerged alongside the *regnum* and *sacerdotium*, the power of *studium* with philosophy which had to 'serve' theology in the process of ascertaining the truth; on the other from St Augustine onward, believing was no longer enough, now one had also to understand. Grosseteste lived through these interwoven changes and was both a scientist and a theologian, but what path, if any, did he favour? Was he a philosopher who believed in order to understand (*credo ut intelligam*) or a scientist who needed to understand in order to believe (*Intelligo ut credam*)?

13.2 Intelligo ut credam

A clear feature of the thirteenth century is that scholars in the Universities 'committed themselves to Aristotle's model of science' (Marrone 1983) that is to say science based on logic. This chimes in with the idea proposed by Crombie (1953) that the modern scientific method of analysis was created by the thirteenth century's intellectuals and philosophers of Western Europe, who based their studies on ancient Greek geometers and logicians. Certainly, the thirteenth and fourteenth centuries were periods in which the philosophical and theological truths were investigated (and to some extent understood and taught) (Leff 1976) in light of a new methodological/scientific conceptual framework, but especially periods in which there was a concrete attempt to explain the natural world in natural terms rather than metaphysical ones. This implicitly created a philosophical problem about the nature of truth: what was truth and how could it be acquired? The traditional Augustinian thought of the Middle Ages featured a high level of philosophical speculation, together with the metaphysical idea of 'infused knowledge'. In particular Augustine had talked about the concept of Illumination, at work when a man acquired higher truth and higher knowledge through enlightenment by God. In Steven Marrone's interpretation for Augustine knowledge needed to lead everyone towards God. It is clear therefore that Augustine directed his thinking towards religion more than science, and that rather than identifying the nature of truth or the role of God, the Augustine philosophy simply linked men to God (Marrone 1983). This is because the philosophical and theological speculation of the late twelfth century tended to assume concepts more than explaining them; the nature of the universals as well as the concept of the illumination of the soul by Divine light (Leff 1976) were in themselves springboards to be used for further theological speculation rather than notions to be elucidated and clarified. However, scholars of the thirteenth century, diverging from Augustine, began to study reality and to see the world with different eyes, thanks especially to the great body of Latin translations from Arabic and Greek (Crombie 1953). In the thirteenth century, the West had been 'invaded' by Hebrew and Arabic philosophy as well as by the moral, metaphysical and scientific works of Aristotle and Al-Farabi in addition to the translations of Avicenna and Avicebron that had circulated since the twelfth century. What these two apparently contrasting tendencies had in common was the issue of universals; thirteenth century scholars began to think that it was necessary to study all the knowledge (or if one prefers all the scientific knowledge) available according to universal principles. At this point Aristotle's setting of rules became very important, especially his attempt to isolate and identify the small part of knowledge that could be called science (Marrone 1983). Indeed, one of the consequences brought about by the Jewish and Arabic philosophers who commented on Aristotle's works was that they frequently developed their own independent system of Aristotelian and Neoplatonic principles and theories, which very often explained the world in non-Christian terms (Leff 1976). In order to be scientific the thirteenth-century theories of truth had therefore to be theories detached from their original sources, that is to say from Augustine's thoughts. It is probably not a coincidence that William of Auvergne in his *Magisterium Divinale* maintained that he would defend the truth against all errors without making any appeal to Christian faith or the authority of Divine Revelation (Marrone 1983).

Scholars therefore turned to Aristotle for answers given that for him the aim of scientific inquiry had been to discover premises from which something already known as a fact could be deduced or demonstrated (Crombie 1953), because there was a difference between the knowledge of the fact (το οτι) and the reason for the fact (το διοτι) (Crombie 1969). However, although for Aristotle the entire process of cognitive investigation needs to take place in the domain of experience (Knowles 1963) (as for him, differently from Plato, the individual was fully real), it was soon clear to the thirteenth-century scholars that Aristotle did not provide exhaustive answers because he offered no universal solution for the problem of the nature of truth. In this context, therefore Grosseteste's scientific and epistemological works such as the *Commentary on Physics* can be seen as an attempt to explain the problem of truth, i.e. the real nature of truth, in relation to science. Despite the fact that Augustine's thoughts seemed to have been perceived as inadequate in relation to pure scientific standards, in his treatise *De Veritate* (probably written in the 1220s) (Marrone 1983) Grosseteste still attributed to the truth the concept of rightness, meaning that the simple Truth (first level of truth) could be defined in relation to its conformity to the eternal idea in the mind of God. The created truth (things, objects) revealed an existing substance, but only in the light of the first Truth (Ibid.). If one achieves simple truth—*adequationem rei ad intellectum*- one can understand the substance, the existence of the object, but only in the light of the first Truth. This theory was clearly still very much based on St Augustine's concept of Divine illumination but, in my opinion, also and especially on St Anselm's idea of God's perfection. Indeed with Anselm we are faced with one of the highest achievement of the Augustinian use of dialectic (Knowles 1963), an Augustinian thought without the Neoplatonism and especially without the idea of Divine illumination.

Anselm represents a sort of second step to understanding Grosseteste's thought. In the *Monologion* he gives us four proofs for the existence of God, evidence called *a posteriorem* (Hopkins and Richardson 1974), that is meaning based on things. He argues that if there are good things there must exist the Absolute Good that makes things good; also as there are different spaces so there must exist the absolute idea of space. However, it is Anselm's third and fourth evidences that are important as they directly influenced Grosseteste. The third one states that as there are multiple things existing in reality, therefore there must be the Supreme Being which makes all things exist. In the fourth, based on perfection, he maintains that as there are different levels of perfection there is therefore the absolute perfection. Grosseteste regards simple truth as Divine illumination as we have seen, but also, as in the *Posterior analytics*, as an element of reality, so knowledge of a simple truth becomes implicitly knowledge of whatever exists (Marrone 1983). Furthermore, he seems to regard the essence, the being of existing things, and the substance, as identical, all of these being placed on the same ontological level:

Being: ον—Substance: ουσια—(Matter: χωρα)[7]

Why did he place being and substance on the same ontological level? According to Steven Marrone (1983), this was because 'an object was, by its existence, true, and every true substance constituted a different simple truth.' The reason is twofold: on one side because of the Aristotelian idea that the essence forms the basis of all scientific knowledge,[8] and on the other because of Anselm's concept of grades of perfections. In the *Monologion* he established that 'I understand a nature to be the same thing as an essence.' (Hopkins and Richardson 1974), and also held that the two grades of perfection of all things were as follows: one thing needs to be itself, for the axiom that *things that are equal to the same thing are equal to one another* (Crombie 1953), and also fitted for the purposes for which it has been established. Therefore, the mere existence and essence of an object together constituted a simple state of reality, or the first state or reality. However, although this procedure was viable for simple truth, Grosseteste soon realised that it was not the truth he was looking for; this is because he was looking for scientific truth, and scientific knowledge by definition had to be at once universal, and necessary. Knowledge, properly speaking, is the comprehension of the truth of those things that always remain the same or that do not change; the reason is that if one thing remains the same or change in the same way every time it changes, the truth of these events can be understood and demonstrated. Therefore the demonstration of scientific or complex truth became the explanation of the elements included in one single phenomenon, as well as the description of the way in which those elements were combined and related to each other (Crombie 1969). As Marrone puts it, one way of addressing the problem of the truth was for instance the so called '*opinio*' (Marrone 1983; Ross 1957), which was simply the state of holding that something was true in the same way as different individuals have different opinions. It is easy to understand that this cannot be considered scientific knowledge because it is not universal, and not necessary.

Grosseteste began to devote himself to scientific investigation around the 1220s, investigating the problem of complex truth and how to attain it. In order to attain complex truth, which is scientific truth, one needs to know for certain that something is true. For instance if I want to know the truth about a pen, I need to know that the pen exists (Marrone 1983). How can I know it? The answer is through one fundamental thing: demonstration (Crombie 1969). This is Grosseteste's revolution in science despite the probable influence from Boethius (Herrera 1979), who was one of the first who fully applied Aristotelian logic (Knowles 1963). Demonstration

[7] Being is what makes things what they are: a man, a horse etc. Substance is, for instance, when one generates a son, he/she is going to have the same substance as the generator, whereas matter is when one fabricates things, one is going to use matter coming from this world.

[8] As underlined by David Knowles, '*A substance is the individual thing regarding which assertions can be made, but which cannot itself be asserted of a subject*' (1963).

was the sole basis for scientific understanding. The problem was that there cannot be demonstration of any random changing event, because scientific knowledge properly speaking is the comprehension of the truth of those things that always, or quite frequently, happen in one way (Crombie 1953). As scientific knowledge must be necessary, something must be true for everyone, not only as an opinion but also as universal, happening always in the same way. The conception of science which Grosseteste developed from Aristotle was one in which there was a double movement from theory to experience and from experience to theory, so that essentially scientific knowledge was demonstrative knowledge of things through their causes (Ibid.). So here lies the core of the problem: How can we demonstrate the causes? Grosseteste's approach to the entire issue has been considered revolutionary. He added to the traditional idea of the syllogism the idea of what in Latin is called *experimentum* (Marrone 1983), a mix of two main procedures, induction and deduction, and verification and falsification (Crombie 1953). In all probability he did not envisage his move as a development of a scientific method, for at the end of the day, as Anselm maintained, the truths, particularly the truths of faith, cannot be reached by demonstrative arguments (Knowles 1963). I am inclined to think that for experiment he probably did not mean simply everyday observation of the general phenomenon,[9] as Crombie has said, but rather something in relation to the verification of a conceptualized theory. Indeed, only in the fourteenth century would the West in general begin to treat natural phenomenon scientifically rather than philosophically. The real problem in the twelfth and thirteenth centuries was not in relation to the use of 'experiments' in a wrong way, but rather to the subordination of general phenomenon to non-scientific principles and to seeing science almost always in philosophical terms (Leff 1976). Having said that, it is certainly true, as stated by Riedl, that Grosseteste has the tendency 'to appeal to observation and experiment' (1942) and this is clearly proven by his numerous works on scientific topics.[10] However, Grosseteste tried to apply seemingly scientific principles in relation, for instance, to the movements of the planets. In this case Grosseteste was not satisfied by simply following the theories of his time. He 'used the Ptolemaic epicycles and eccentrics as devices to follow the movements of the planets as measured with instruments, to construct tables and to fix the length of the year' (Baur 1912). Certainly, and this is what is important for me, he introduced the West to the notion and the idea of experiment (or better, the idea of the necessity of having a concrete understanding of the natural phenomenon empirically) and related it to demonstrative science. What his study lacked, as has been underlined by Crombie and Marrone, was a systematic method by which the obtained results could be turned into an orderly scientific theory available to others; this would be the task and in some respects the achievement of fourteenth-century scholars. Indeed, we can note that some of his treaties were a mix of observation, the use

[9] See the example of horned animals in (Crombie 1953).

[10] Amongst others we can mention: *De Colore*, *De Lineis Angulis et Figuris*, *De Iride*, and *De Cometis*.

of authority, semi-scientific assumptions and/or observation and interpretation of facts. There is no doubt that the regularity of a phenomenon does not imply a general theory through which individual elements involved in the phenomenon under investigation could be explained, let alone the construction of a comprehensive theory.

However, on the basis of what I have just tried to explain, it is certainly possible to classify Grosseteste as a scientist or a natural philosopher, who believed strictly in the idea of science, so that he investigated the natural phenomenon in order to understand better God's creation. Based on this we can label him under the *intelligo ut credam*. However, can we also argue the contrary, that he believed in order to understand?

13.3 *Credo ut intelligam*

While it is certainly true that Grosseteste developed his own theology and an original idea of the necessity of the redemption, which led him to say in the *Cessatione Legalium* that the Incarnation would have happened even without the fall of man because it was already present in the mind of God from the very beginning, his philosophy and theology owed a massive debt to great thinkers, in particular Augustine, Anselm, Peter Lombard and of course Aristotle. Grosseteste needed to apply and to re-adapt his knowledge to the new theology, which in the thirteenth century was not a worshipping theology but rather a problem-solving theology which needed to find solutions to everyday problems: marriage, baptism, and the Eucharist. Grosseteste had lain great stress on the use of mathematics, the application of which, based on a physical theory of light, could produce a new form of knowledge (Leff 1976). In the work *De luce* Grosseteste attributed a very special role to the light, which he considered was the first thing God had created. Light therefore represented the beginning of everything, determining the creation of the universe and the world (Crombie 1953; Dales and Gieben 1982). The external limits or boundary is the firmament which reflects the light towards the centre of the sphere; this reflected light generates the nine spheres the lowest of which is the moon. The nine spheres are immutable, and below them are the four elements of fire, air, water and earth (Riedl 1942). The earth received all the actions of the superior spheres. This is very similar to what Aristotle had said about the 55 immutable and eternal spheres between the fixed starts and earth, each of them inferior to the next one closer to earth. There are clear similarities between Aristotle and Grosseteste, both enumerating the created spheres although they did so with different numbers. Aristotle did not explain the relation between the first engine (the first not moved) and the spheres and neither did Grosseteste. In order to overcome this impasse medieval philosophy would simply transform the 55 spheres into the angelical intelligences who would work as intermediate entities.

Of course, Grosseteste could not follow Aristotle as for the Greek philosopher, the world was not created by God but eternal, God did not and does not care about the world. God did not think about the world, because it would mean thinking about something inferior, mutable and imperfect; the only thing God thinks about is himself. Most important of all, God does not love, he is loved, but he does not love the world, let alone a single man. For Grosseteste of course there is, on the contrary, a relationship between God and men. However, the question was: before the incarnation, when God became man, how had God acted upon the world? Certainly not just through the angelical intelligences. Grosseteste resolved this deadlock by invoking the light. God acted upon the world through light for the creation and for the relationship between body and soul. Grosseteste held that 'every higher body in virtue of the light which proceeds from it is the form and perfection of the body that comes after it' (Ibid.). Where is this theory coming from? It is borrowed from St Augustine and possibly also from Plotinus (Crombie 1953). Augustine was looking for the truth and moved from 'science' to theology. This was exactly the shift that Grosseteste seems to have made and for the same reason. Science for Augustine meant Greek and Roman philosophy, in particular Cicero, who in the book *Ortensius* supported the idea of philosophy as wisdom and a way of living, which can give one happiness. As he was disappointed by this reading, Augustine moved to the Bible.

St Augustine was the first to create a synthesis between faith, philosophy and life as he deemed not only that faith could be helped and clarified by reason, but also that reason could be helped by faith: *credo ut intelligam* and *intelligo ut credam*. For St Augustine God is the truth, so that when he tried to demonstrate the existence of God he was simply looking for the demonstration of the universal truth. How can somebody know the truth? The soul cannot produce the criteria through which the truth is known. Augustine concluded therefore that there is something that he called Eternal Law which is above our mind and soul and this Eternal Law is the truth. This is because his only desire was to know God as the source of all knowledge and truth and 'to interpret the nature of God and of the soul' (Knowles 1963). God is a light that enlightens the human mind and for Augustine both the 'formal and efficient cause of the cognitive process' (Ibid.). This, of course, was not an original idea. It was taken from Plato[11] and revisited by Augustine on the basis of creationism. At the same time there are the intermediate ideas of the One, the supreme Good, considered as the Being existing above multiplicity, ideas that were taken from the Greek philosophers who followed Aristotle and preceded Plotinus. The step which was made between Aristotle to Plotinus was that Aristotle's remote Mind (or unmoved Being) was identified as Plotinus' Supreme Good and given the status of Creator' of the other beings. Therefore the object of a true life cannot be anything else than to purify oneself by finding the truth in order to achieve the Supreme Truth. This was the vision of the Supreme Being, partially achievable, and this is important, even in this life (Ibid.). Indeed Augustine himself, applying Aristotle, Plotinus, St Paul and the Scriptures, described God as immanent and

[11] In the Republic, Plato used the metaphor of the vessel which one uses to sail through a rough sea, for him the vessel is the equivalent of what we can call 'a divine revelation'. See (Bloom 1968).

transcendent at the same time, a Creator of the Universe, where everything revolved around the Creator himself. Augustine maintained that the process of understanding is similar to having a perception of an object struck by light. Knowledge at the lowest level of scientific understanding can be grasped through the understanding of immaterial numbers which are not visible but which are considered the basis of understanding the immutable. At the highest level, however, pure knowledge is represented by knowledge of the supreme truth (Ibid.). For Grosseteste in turn the light not only reverberates on the world, but through light the soul can act upon and control the body and through light one can come to know God. To Grosseteste, therefore, the light stands as the Universal-Soul stands to Plotinus. This is very much in line with the idea of *credo ut intelligam*, the opposite approach to the one he applied to science. Grosseteste however went beyond Augustine in borrowing concepts from St Anselm. Anselm himself had employed the concept of light,[12] reinforcing the idea that reason can clarify what we have already through faith: *Fides quaerens intellectum*, faith seeking to understand. Indeed St Anselm's motto had always been, 'I do believe it for unless I believe, I shall not understand' (Knowles 1963). In the *Proslogion* St Anselm elaborated the so-called, ontological[13] argument based on the idea of God.[14] God has all grades of perfection. Therefore he needs to exist because otherwise we could not think about him as perfect. He would be less perfect were he not to exist. Therefore for St Anselm, as for Grosseteste, God is, '*id quo maius cogitari nequit*' (Herrera 1979) something than which you cannot think of a greater. So for instance an atheist thinks about God as the greatest being, but then argues that in the real world He does not exist (Charlesworth 1965; Hopkins and Richardson 1974) so he implicitly admits that there might be something greater than God, locking himself into an irresolvable contradiction (Herrera 1979). We have a similar expression in Grosseteste who talked about God as the first thing, eternal, out of time, and especially, as he wrote on free will (*De Libero arbitrio*) (Lewis 2003), that God has the power 'to not know or not will, what he knows and what he will'.[15] Philosophically speaking this is a statement of omnipotence, which cannot have anything greater because when you want or not want, or will and not will at the same time it means that you are in everyone and that you can do everything and its contrary at the same time. It seems therefore that Grosseteste was also influenced by Anselm's reliance on faith in order to understand. Grosseteste borrowed concepts and theory from previous thinkers and philosophers, and we have seen that he can be considered as both a

[12] Surely You [God] dwell in light inaccessible [...]; How shall I approach unto a light inaccessible? See (Hopkins and Richardson 1974).

[13] Ontological as free or independent from experience, but rather based on concepts; in fact its proof requires nothing besides itself. See (Herrera 1979).

[14] The *Proslogion* is written, 'from the point of view of one trying to raise his mind to contemplate God and seeking to understand what he believes'. See (Herrera 1979).

[15] The expression is related to God's free will, understood as God's free decision. God exerts free will, but things are right anyway because He wants them to be like that (so that implicitly freedom and will coincide) and it cannot be different as God is all perfections and love.

scientist *intelligo ut credam*, and as a believer who used faith in order to understand, *credo ut intelligam*. In fact Grosseteste was both and neither, because he went beyond both concepts and represented a synthesis of both because he applied his scientific methodology to theology.

In Grosseteste we have a link between science and theology, because he applied the methodology of observing facts to both his theological writing and his scientific writing. The difference is that in the case of theology the observed facts were not experiments, but could only be gained from reading the Bible; that is why he had to go back and read the Greek and possibly Hebrew (Elliot 2012) to perfectly understand the text (although admittedly we do not have clear evidence of this). Although we do not have any catalogue which could testify to the books and/or manuscripts directly belonging to Robert Grosseteste,[16] we have indirect evidence from books annotated by him and other manuscript sources preserved in the Franciscan convent in Oxford (Hunt 1969). Further evidence of Grosseteste's clear link with Greek original manuscripts and sources came also from the analysis carried out by Harrison Thomson, who attributed to Grosseteste the translation of the whole Nicomachean Ethics as well as the compendium comments of Michael of Ephesus on the Ethics (Harrison Thomson 1933). More recently Ezio Franceschini, who studied Grosseteste's familiarity with Greek works and culture, has shared the same opinion about his ability and knowledge in translating directly from the Greek language. Franceschini stated that 'in some codices containing the works of the Pseudo-Dyonisius Areopagita in Grosseteste's translation are found marginal notes of varying length preceded by the notes, ex-greco' (Franceschini 1993). Less clear is the way in which Grosseteste used his Greek sources. Anna Dionisotti maintained that on one side Grosseteste made 'lavish use of his dictionary in his exegesis' (Dionisotti 1988), but on the other it is possible, judging from some of Grosseteste's expressions like, '*ut greci dicunt*' and/or '*secundum grecos*', that he had oral consultation with someone who was a native speaker or very advanced in the Greek language, possibly that Nicholas Graecus who is considered his main helper in the translation of Greek texts (Ibid.). Whatever is the case, his studying of the original sources is in my opinion the equivalent of the experiment in science. This is the link between *intelligo ut credam* and *credo ut intelligam* in his theological writings and in his actions, revealing at what his books were aiming. The link between science and theology is in his theological writings and in their aims. On one side Grosseteste relied on his faith and on the Bible, but on the other he applied his scientific method, which in this case involves departing from translation and going back to the original sources. A further link which is substantially new lies in the aim of his treaties. They were directed not just to academic speculation, as with Augustine, but rather to the people.

---

[16] Certainly in his commentary on the Psalms Grosseteste quoted a number of Greek sources which were not available in Latin at the time and this fact led M. R. James to write that most likely Grosseteste had a Greek Psalter before him when he was writing his commentary. See (Montague 1922).

De Cessatione Legalium, for instance, had this double role. Probably written around 1231–2 it is basically a manual of theology. In the text Grosseteste analysed natural law, determined by nature and positive law (of a given political community) in light of the Scriptures. There is a criticism of the ceremonial aspect of Mosaic Law practised by early Christians because sometime after the Council of Jerusalem the observation of these laws became heretical. However, it is not necessarily a text against the Jews nor a text which aims ultimately at their conversion to the Christian faith. The text seems to express Grosseteste's pastoral concerns. In this text there is also, according to my studies, a link with St Augustine (Elliot 2012), but not just in terms of sources (Dales and King 1986). Augustine at an early stage of his life adopted, for a period of time, the doctrine of Manicheism which entailed that Christ did not really have a human body, his birth, life and resurrection being apparent not real, and that Moses was not inspired by God, but by one of the evil princes of darkness. The Old Testament, therefore, must be rejected. Grosseteste did not go that far, but there is a possibility that he tried to analyse the laws/text critically in the light of what Augustine had already said. On the other hand the book also had a missionary purpose. As it was directed to help people it was theory put into practice, a theory followed if you like by suggestions on the best way to instruct and especially to help people. This text was followed by the translation of the *Testaments of the Twelve Patriarchs* which was intended for the refutation of the Jews (Stevenson 1899). According to Robert Henry Charles, quoted by De Jonge (1953), the original text underlined the importance of the Testament and the Scriptures. The original text presents some linguistic and historical difficulties in its interpretation and it is Charles' opinion that the original was a Jewish work and so for its content we must assume a Christian interpolation in a Jewish document (Ibid.). De Jonge is, however, more sceptical and he considers the text to be a Christian one based on an original Jewish text, which is impossible to reconstruct. Whatever is the case, what is really important is that the 'Testament' is not like other Grosseteste writings, directed to scholars and therefore to a small part of the population. This text was probably expected to be read by everyone who could read. Carlotta Dionisotti underlined this point clearly in her researches, stating that the translation has no notes or glosses and this is likely to mean that it was to be read 'as widely as possible' (1988). The reason why Grosseteste translated the text might be more difficult to ascertain. Matthew Paris reported that the text was written *ad majorem Judaeorum confusionem* (Paris 1872–83), suggesting that it may have been written in order to convert the Jews. However, it was not a missionary purpose which motivated Grosseteste, but most probably the strengthening of the Catholic faith or, as De Jonge writes, to make clear 'to all who opposed orthodoxy, that salvation can only be expected from Christ and the true Church which obeys Holy Scripture' (De Jonge 1991). The same can be said about his texts and writings concerning confessions and reconciliation. The *Deus est*, written to support the priest, and the *Perambulavit Iudas*, another writing on confession (Hassenauer 2012), are other examples of theory put into practice. A particular feature of his writings on confession is that he seems to want the penitents to understand how to examine themselves deeply, almost in a scientific way.[17] Thus it seems to have given penitents scientific instructions in relation to religious issues and of course the practical way to achieve the desired result. As reported by Hassenauer, when Grosseteste dealt with the best way to clean one's body and soul from sin, he used the example of cleaning the house, not from the ground floor up but rather from the top of the house down so not to spoil what has been already cleaned (Ibid.). It was an example clearly taken from observation of everyday life.

However, one of the best examples of faith and reason being combined is the *Templum Dei*. It was probably written in 1225; Goering called it the best example of how Grosseteste linked the art of medicine (science) and the role of the priest (religion), calling the priest the doctor of the soul. The book is written to educate priests charged with the care of souls. The thirteenth century was a good period for so-called practical theology, a theology aiming to educate priests or clergy in general in relation to their duties as ministers and preachers. Although many writers tried to make speculative theology available to the people, few of them really achieved this goal. Grosseteste was one who did and what is particularly striking, at least in my opinion, is the way in which he built his examples. For instance in the text he analysed the *Templum*, saying that the foundation is faith, the walls are hope and the roof is love. He then goes on to describe the role of faith, the role of hope and, when he reaches the roof, the role of love. The last, he says, protected the temple against the seven capital sins and against their three ministers; devil, world, flesh. The roof, moreover, is protection against the planetary influences of the days of the week and against the infirmities and wounds arising from astrological power (Goering and Mantello 1984). So it seems that there is a mix between belief—the idea of faith, love etc. as in the Christian religion—and elements of science as shown in the construction of the *Templum* and the idea of the planets and stars. Again in chapter XI of this book Grosseteste tells the priest to consider not only the sins but also the circumstances of the sinner and the possibility of sinning through an excess or a defect of virtue (Goering and Mantello 1985). The idea of excess or defect of virtue clearly comes from St Augustine who had talked about moral evil as a lack of good; a bad will might prefer an inferior good, instead of choosing the first good which is God. However, it also seems clear that Grosseteste has applied a sort of scientific procedure to this theological idea, considering sin as an 'unbalanced situation' determined by physical causes which can be scientifically analysed and resolved.

Another example of the link between science and theology is in the *Hexaemeron*, probably written in between 1228 and 1235 after he had completed his work on the *Posterior Analytics* (Dales and Gieben 1982). This text is connected with the *Cessatione Legalium*, as proven by the fact that the most important manuscript of the *Hexaemeron* (Bodleian MS lat. th. c. 17) is made up of two separate sections: fols 1^{A}-157^{D} contain *De universo spirituali et corporali* by William of Auvergne, whereas the second section includes *De Cessatione Legalium* on fols 158^{A}-189^{D}

[17] 'One may suppose that what is true for other fields of learning (law, medicine, science) is equally valid for theology...' (Hassenauer 2012).

and the *Hexaemeron* on fols 190^A-243^A (Dales and Gieben 1968). Because of his use of the sources we are on safe ground in affirming that when he wrote the *Hexaemeron* Grosseteste knew the Greek language sufficiently well. However, once again, given its popularity (Callus 1969), modern commentators view this work as a book not merely for scholars: *non enim sapientibus et perfectis ista scribimus* (Dales and Gieben 1982). Although Dales and Gieben correctly suggested that the *Hexaemeron* contained a number of significant arguments against pure scientific knowledge (Ibid.), and hostility towards astronomy,[18] I think that it contains a subtle defence of the suitability and correctness of scientific knowledge. In the text there is space for problems in relation to both philosophy and cosmology (Smalley 1969). It can be argued, therefore, that Grosseteste had not rejected science, but that he wanted to use and apply it for higher aims, one of which is the development of faith, as many of his theological treatises seem to prove. Beryl Smalley has stated how Grosseteste 'not only transcribed his sources, he added something on his own' (Ibid.), something original. However, what is new is not necessarily in terms of scientific discoveries, but rather in terms of methodological approach. Grosseteste did not try to reconcile Aristotle's view of the world as an eternal matter and the Christian doctrine of creation *ex-nihilo* (Ibid.), as many philosophers or scholars had tried to do before him. Rather, he presented his view as an uncompromising one (a Christian view if we prefer) where religion, or faith, has pre-eminence. Dales and Gieben in their analyses on the Prooemium of the *Hexaemeron* also underlined how Grosseteste, apart from being dependent upon the work of Augustine, 'brought the scientific knowledge of his day to bear on the biblical account of creation' (Dales and Gieben 1968), showing a subordinate link between the science of the day and the Catholic faith. As far as astrology is concerned, it seems that Grosseteste is not against it as science in itself, but rather against the superstitious use of it (Ibid.) as shown in the *Decem mandatis* (Smalley 1969) another text written for the help of the less educated clergy. Moreover in a passage in part five of the *Hexaemeron* Grosseteste confirms his view by comparing the physical world and the humankind, *'nullo tamen modo verum esset quod spiritus stelle superior esset natura spiritu hominis, cum homo secundum spiritum suum sit Trinitatis ymago'* (Dales and Gieben). In this passage we note that he confirms the superiority of religion/faith over nature, as nature is the creation of God and the creature cannot be superior to the creator.

In drawing some conclusions, I think it possible to prove that Grosseteste was both a scientist who needed to understand in order to believe (*Intelligo ut credam*), and a religious man who needed to believe in order to understand (*Credo ut intelligam*). Grosseteste went beyond classical theology because he applied his methodology of science to theology and, in my view, represented a synthesis of medieval thinking. It seems that Grosseteste used the scheme and pattern adopted by Peter Lombard's sentences. Indeed, Stevenson claimed that, 'Grosseteste's point

[18] Particula III seems to suggest the use of Avicenna's *De caelo et mundo* and Alpetrangius' *de motibus celorum*. See (Dales and Gieben 1982).

of view approximated rather to the positive than to the scholastic system of theology, and his training in that respect appears to have been largely guided by the sentences' (Stevenson 1899). This is partly true, despite some differences. Although Peter Lombard in *Libri quattuor sentetiarum* said that reason cannot be more important than faith, it is undeniable that the sentences were a clear attempt to reconcile reason with authority (which was very much what Grosseteste seems to have done), which in practical terms meant collecting under certain theological questions the statements contained in the Scriptures and in the Fathers. However, in doing so there was a tangible risk of treating Lombard's work as an authority in itself, instead of a means to study the original texts upon which it was based. This implicitly is a contradiction of what Grosseteste had always said and done, indeed he had repeatedly insisted on the necessity of going back to the original source without using a translation unless strictly necessary. So I do not deny the possibility that Grosseteste had used and known Lombard's sentences, but he cannot have been strictly guided by them because this would have been contrary to what he said about his work on theology. Moreover, Peter Lombard (Gilson 1994) with his sentences had summed up the philosophical comments of previous periods in order to present a speculative summary of Christian Doctrine, whereas Grosseteste had no such intention. He wanted his writings to have a practical use as we have seen, and therein lies one of his most relevant contributions.[19] Like Peter Lombard, Grosseteste represented a watershed in medieval ecclesiastical philosophy. Grosseteste represented (in a different way) one of the highest points of medieval knowledge and a synthesis of medieval thought both scientific and theological, and his fundamental contribution lies in his scientific reasoning as well as in his methodical approach to theology.

References

Baur, L. (1912). *Die Philosophischen Werke des Robert Grosseteste, Bischofs von Lincoln.* Münster: Aschendorff.

Bloom, A. (1968). *The republic of Plato* (A. Bloom, Ed. & Trans.). New York: Basic Book.

Callus, D. A. (1969). Robert Grosseteste as scholar. In D. A. Callus (Ed.), *Robert Grosseteste scholar and bishop. Essays in commemoration of the seventh centenary of his death* (pp. 1–69). Oxford: Oxford University Press.

Charlesworth, M. J. (1965). *St Anselm's proslogion.* Oxford: Oxford University Press.

Crombie, A. C. (1953). *Robert Grosseteste and the origins of experimental Science 1100-1700.* Oxford: Clarendon.

Crombie, A. C. (1969). Grosseteste's position in the history of Science. In D. A. Callus (Ed.), *Robert Grosseteste scholar and bishop* (pp. 98–120). Oxford: Clarendon Press.

[19] On this point Callus claimed he had discovered the evidence of Grosseteste's only surviving fragment of *Summa Theologiae* in the Manuscript MS 28, ff 306–307, in Exeter College Oxford, but Southern and others have raised serious doubts about his attribution. See (Callus 1969; Southern 1986).

Dales, R. C., & Gieben, S. (1968). The Prooemium to Robert Grosseteste's *Hexaemeron*. *Speculum, 43*(3), 451–461.

Dales, R. C., & Gieben, S. (Eds.). (1982). *Robert Grosseteste's Hexaemeron*. London: Oxford University Press for the British Academy.

Dales, R. C., & King, E. B. (Eds.). (1986). *De cessatione legalium*. Oxford: Clarendon.

De Jonge, M. (1953). *The testaments of the twelve patriarchs*. Assen: Van Gorcum.

De Jonge, M. (1991). Robert Grosseteste and the testaments of the twelve patriarchs. *Journal of Theological Studies, 42*(1), 115–125.

Dionisotti, A. C. (1988). On the Greek studies of Robert Grosseteste. In A. C. Dionisotti, A. Grafton, & J. Kraye (Eds.), *The uses of Greek and Latin, historical essays* (pp. 19–40). Dorchester: The Warburg Institute.

Elliot, M. (2012). Robert Grosseteste, the Jews and *De cessatione legalium*. In J. Cunningham (Ed.), *Robert Grosseteste, his thought and its impact*. Toronto, ON: Pontifical Institute of Medieval Studies.

Franceschini, E. (1993). Grosseteste's translation of the ΠΡΟΛΟΓΟΣ and ΣΧΟΛΙΑ of Maximus to the writings of the Pseudo-Dionysius Areopagita. *Journal of Theological Studies, 34*(136), 355–363.

Gilson, É. (1994). *La filosofia nel medioevo, dalle origini patristiche alla fine del XIV secolo*. Florence: La Nuova Italia.

Goering, J. (1995). When and where did Grosseteste study theology? In J. McEvoy (Ed.), *Robert Grosseteste: New perspectives on his thought and scholarship* (pp. 17–51). Turnhout: Brepols.

Goering, J., & Mantello, F. A. C. (1984). *The Templum Dei, edited from manuscript 27 of Emmanuel College, Cambridge*. Toronto, ON: Pontifical Institute for Medieval Studies.

Goering, J., & Mantello, F. A. C. (1985). The Meditaciones of Robert Grosseteste. *Journal of Theological Studies, 36*(1), 118–128.

Harrison, T. (1940). *The writings of Robert Grosseteste*. Cambridge: Cambridge University Press.

Hassenauer, M. (2012). For a larger audience, Grosseteste's Perambulavit Iudas in Anglo-Norman. In J. P. Cunningham (Ed.), *Robert Grosseteste, his thought and impact* (pp. 259–313). Toronto, ON: PIMS.

Herrera, R. A. (1979). *Anselm's Proslogion: An introduction*. Washington, DC: University Press of America.

Hopkins, J., & Richardson, H. (1974). *Anselm of Canterbury*, vol. I (J. Hopkins & H. Richardson, Ed. & Trans.). London: SCM.

Hunt, R. W. (1969). The library of Robert Grosseteste. In D. A. Callus (Ed.), *Robert Grosseteste Scholar and Bishop. Essays in commemoration of the Seventh Centenary of his Death* (pp. 121–145). Oxford: Oxford University Press.

Knowles, D. (1963). *The evolution of medieval thought*. London: Longmans.

Leff, G. (1976). *The dissolution of the medieval outlook*. New York: New York University Press.

Lewis, N. (2003). Robert Grosseteste. In J. J. E. Gracia & T. B. Noone (Eds.), *A companion to philosophy in the middle ages* (pp. 597–607). Padstow: Blackwell.

Marrone, S. P. (1983). *William of Auvergne and Robert Grosseteste*. Princeton, NJ: Princeton University Press.

McEvoy, J. (1982). *The philosophy of Robert Grosseteste*. Oxford: Clarendon.

Montague, R. J. (1922). Robert Grosseteste on the Psalms. *Journal of Theological Studies, 23*(90), 181–185.

Paris, M. (1872–83). *Matthæi Parisiensis Monachi Sancti Albani Chronica Majora*, vol. iv, RS, LVII (H. E. Laurd, Ed.). London: Longman.

Paul, S. (Ed.). (2002). *An edition and study of selected sermons of Robert Grosseteste*. Unpublished doctoral thesis, University of Leeds.

Riedl, C. C. (1942). *Robert Grosseteste on light*. Milwaukee, WI: Marquette University Press.

Ross, W. D. (1957). *Aristotle's prior and posterior analytics: A revised text*. Oxford: Clarendon.

Schulman, N. M. (1997). Husband, father, bishop? Grosseteste in Paris. *Speculum, 72*(2), 330–346.

Silvestri, A. M. (2015). *Power politics and episcopal authority. The bishops of Cremona and Lincoln in the middle ages*. Newcastle: Cambridge Scholars Publishing.

Smalley, B. (1969). The biblical scholar. In D. A. Callus (Ed.), *Robert Grosseteste scholar and bishop* (pp. 70–97). Oxford: Oxford University Press.

Southern, R. W. (1986). *Robert Grosseteste. The growth of an English mind in medieval Europe*. Oxford: Clarendon.

Southern, R. W. (2004). Robert of Grosseteste. In H. C. G. Matthew & B. Harrison (Eds.), *Oxford dictionary of national biography* (Vol. 24, pp. 79–86). Oxford: Oxford University Press.

Stevenson, F. S. (1899). *Robert Grosseteste, Bishop of Lincoln*. New York: Macmillan.

Thomson, H. S. (1933). A note of Grosseteste's work of translation. *Journal of Theological Studies, 34*(133), 48–52.

Wharton, H. (1691). *Anglia Sacra sive Collectio Historiarum. Antiquis Scriptarum de Archiepiscopis & Episcopis Angliæ. Pars Secunda*. London: Impensis Richardi Chiswel.

Chapter 14
Can Science and Religion Meet Over Their Subject-Matter? Some Thoughts on Thirteenth and Fourteenth-Century Discussions

Dónall McGinley

14.1 Natural Science and Religious Belief in the Thirteenth Century

When George Coyne, a Jesuit and former head of the Vatican Observatory, was asked how his work as a scientist related to his life as a Christian and a priest, he replied: 'They're not related at all!' This is a somewhat startling statement. Are natural science and religious belief really unrelated? If we go back to the great scholastics of the thirteenth century we certainly find enthusiasts for natural science who held that it could be of great aid to Christian belief. They adopted a modified Aristotelian picture of a hierarchy of sciences, with the physical sciences at the bottom and theology at the top, and adopted Aristotle's distinction between what was most knowable *to us* and what is most knowable *in itself*. In the order of acquisition of knowledge, we must start with sensible objects and eventually arrive at the purely intelligible, the order of inquiry being the opposite of the order of being. Aristotle's discussion at the very beginning of the *Physics* (Aristotle 1984a, *Physics* I, 1, 1184a17–b4) is the source of this account:

> The natural way of doing this is to start from the things which are more knowable and clear to us and proceed towards those which are clearer and more knowable by nature; for the same things are not knowable relatively to us and knowable without qualification. So we must follow this method and advance from what is more obscure by nature, but clearer to us, towards what is more clear and more knowable by nature.
>
> Now what is to us plain and clear at first is rather confused masses, the elements and principles of which become known to us later by analysis. Thus we must advance from universals to particulars; for it is a whole that is more knowable to sense-perception, and a universal is a kind of whole, comprehending many things within it, like parts (Aristotle 1984a, *Physics* I, 1, 1184a17–b4).

D. McGinley (✉)
Trinity College, Dublin, Ireland
e-mail: Donall_mcginley@yahoo.co.uk

J.P. Cunningham, M. Hocknull (eds.), *Robert Grosseteste and the pursuit of Religious and Scientific Learning in the Middle Ages*, Studies in the History of Philosophy of Mind 18, DOI 10.1007/978-3-319-33468-4_14

Philosophers like Robert Grosseteste, Roger Bacon and Albert the Great all held that natural science could be of benefit to religious belief. Roger Bacon went so far as to argue that experimental science was more important than theoretical knowledge because only through experiment can we reach *certainty* in our knowledge (Bacon 1962). Furthermore, Roger Bacon believed that access to empirical or experimental scientific knowledge was at least in part due to divine illumination and therefore relied on virtue as a precondition of scientific learning. In his monumental *Opus Majus*, much of which is devoted to optics and experimental science, he wrote:

> Moreover, there are seven stages of this internal knowledge, the first of which is reached through illuminations relating purely to the sciences. The second consists in the virtues [. . .] Moreover, Algazel says in his Logic that the soul disfigured by sins is like a rusty mirror, in which the species of objects cannot be seen clearly; but the soul adorned with virtues is like a well-polished mirror, in which the forms of objects are clearly seen [. . .] Virtue, therefore, clarifies the mind, so that a man comprehends more easily not only moral but scientific truths (Bacon 1962).

Bacon believed that experimental science could even be used to prove the articles of faith, could prolong life through medical understanding, could even be used to build flying machines, and that the technology derived from scientific learning (specifically optics) could be used to build terrifying machines that would frighten unbelievers into converting to Christianity (though he is unclear about how this persuasive power might work, or why onlookers would not simply marvel at the power of science). Bacon claimed, importantly, that experimental science could be used to uncover many dangerous and irrational beliefs and false science, injurious to faith and rationality.

> This science alone, therefore, knows how to test perfectly what can be done by nature, what by the effort of art, what by trickery, what the incantations, conjurations, invocations, deprecations, sacrifices, that belong to magic, mean and dream of, and what is in them, so that all falsity may be removed and the truth alone of art and nature may be retained (Bacon 1962).

So there are good reasons to suppose that reason and evidence are important for having a true picture of reality (primarily, in this instance, serving the negative function of avoiding serious error about the world), and therefore natural science can be of benefit to the correct religious view. However, is there a good reason to suppose that science can indeed come to the aid of religious belief in any way, or do they simply occupy distinct and unrelated realms? In this chapter I am going to consider the question of whether there is any overlap between the subject matters of religion and natural science, focusing on the thought of some thirteenth and fourteenth-century philosophers, particularly Thomas Aquinas and John Duns Scotus, who draws heavily on Aquinas in this instance. Among the things I will discuss are how sciences are distinguished, the relationship between the science of theology and the other sciences, and whether natural science could ever reach specifically religious conclusions, or reach conclusions in the realm of the supernatural. I shall argue that the distinctness of the sciences (as well as the general insistence among the orthodox scholastics that Aristotelian science is not capable of establishing everything we need to know about the proper end of human existence) shows that science and religion do not overlap in their respective subject-matters, and that one cannot derive religious or spiritual conclusions through natural science.

The thirteenth-century scholastic discussions of science accept Aristotle's division of sciences into three: physics, mathematics, and metaphysics (Aristotle 1984b, *Metaphysics* VI, 1, 1026a). In Duns Scotus's words,

> In addition, in *Metaphysics* 6, chapter 1 there is a distinction of theoretical habits into the mathematical, the physical, and the metaphysical; and from the proof of this in the same place it does not seem possible for there to be more theoretical habits, because in those habits the whole of being, both in itself and in its parts, is considered (Duns Scotus 2012, *Ordinatio*, Prologue, q.1, n.8; Duns Scotus 1950).

According to the Aristotelian division of the sciences, Physics (or natural science/natural philosophy) studies the natural world, material things, change and motion. Mathematics studies entities that while neither material nor changing are, nevertheless, not independent of material things. Metaphysics is the highest science and treats of eternal and immaterial things as such, the realm of the purely intelligible. The picture is made more complicated by the fact that Aristotle himself referred to metaphysics as 'theology' (Aristotle 1984b, *Metaphysics* VI, 1, 1026a19), whereas the scholastics generally distinguished theology from metaphysics, theology relying on divine revelation and not being subject to natural knowledge, and metaphysics being subject to natural knowledge. Natural theology ought to be considered to be a part of metaphysics, because its conclusions are derived from reason alone. According to Duns Scotus, theology refers to the study of what can be known about God via revelation; what can be known about God through reason itself is metaphysics rather than theology (Cross 1999). Revealed theology was regarded as a science in the true sense, despite its principles being revealed rather than self-evident, and was placed at the top of the hierarchy of sciences.

14.2 The Need for Revelation and the Insufficiency of Aristotelian Science

In arguing for the necessity of revealed doctrine and the insufficiency of Aristotelian science for acquiring the knowledge of man's ultimate end, which is necessary to man's ultimate happiness and to salvation, Thomas Aquinas argued thus:

> It was necessary for man's salvation that there should be a doctrine revealed by God, besides the philosophical disciplines investigated by human reason. First, because man is directed to God as to an end that surpasses the grasp of his reason. [...] But the end must first be known by men who are to direct their intentions and actions to the end. Hence it was necessary for the salvation of man that certain truths which exceed human reason should be made known to him by divine revelation. Even as regards those truths about God which human reason can investigate, it was necessary that man be taught by divine revelation.[...] It was therefore

> necessary that, besides the philosophical disciplines investigated by reason, there should be a sacred doctrine by way of revelation (Aquinas 1945, *Summa Theologiae*, Ia, q.1, a.1, co.).

Given that the doctrines for which we rely on revelation cannot be something which we have any access to via reason or any natural science, it looks as though our question is settled at the start: Science and religion can have nothing in common and must have totally unconnected realms of reality as their respective objects of inquiry. However, this response is too hasty. Even if we allow that there is a sharp distinction between what can be naturally known and what can only be accessed through divine revelation and must therefore be held on faith, all the sciences (the natural and the revealed) have in common the feature that they all aim at the *truth*. Furthermore, there are still several possibilities as to what the relation between theology and the natural sciences could be. We could take a purely Aristotelian position and argue that everything we need to know can be accessible through reason alone and reason is our only route to the truth, thereby discounting revelation at the start. Duns Scotus states this view thus:

> A philosopher might say, then, that no supernatural knowledge is necessary for man in this present life, but that he can acquire all knowledge necessary for himself from the activity of natural causes (Duns Scotus 2012, *Ordinatio*, Prologue, q.1, n.5, Ia; Duns Scotus 1950).

Or we could adopt the position that faith alone suffices and that rational sciences are of no consequence to the believer, or even the anti-rationalist position that reason itself will systematically mislead us as to the nature of reality. We might even adopt the position of the Averroists, according to whom pure reason, or philosophical dialectic, can reach conclusions radically at odds with the doctrines of faith, and yet claim that they are not really in conflict. (The Averroists were condemned for, though almost certainly never actually believed or taught, the absurd doctrine of 'double truth', whereby one thing can be true in theology but *false* in philosophy, and vice versa. However, the Averroists, and Averroes himself, imply that believing Scripture is fine for the unlearned, whereas philosophical inquiry will get you the real picture of reality. In more conciliatory mood though, they tend to opt for belief in the religious doctrine when a contradiction arises between religious doctrine and the findings of philosophical inquiry.) The remaining two alternative answers to the question of whether science and religion concern the same things or the same reality, I take to be the most plausible options for the religious believer.

1. Religion and science simply occupy utterly discrete realms, each being about distinct and mutually exclusive realities (religion concerning the immaterial and science the material, perhaps).
2. Religion and science have some overlap in their subject-matter, though the distinction between the two must remain.

What is the subject-matter of a science (either physical sciences, metaphysics, or theology), and how are distinct sciences demarcated in this Aristotelian system? A science investigates some genus, and sciences are differentiated by their distinct genera. This position is stated by Aristotle in *Posterior Analytics* I, Chap. 28:

> A single science is one whose domain is a single genus, *viz.* all the subjects constituted out of the primary entities of the genus—*i.e.* the parts of this total subject—and their essential properties.
>
> One science differs from another when their basic truths have neither a common source nor are derived those of the one science from those of the other. This is verified when we reach the indemonstrable premises of a science, for they must be within one genus with its conclusions: and this again is verified if the conclusions proved by means of them fall within one genus—*i.e.* are homogeneous (Aristotle 1925, *Posterior Analytics* I, 28, 87a38–b4).

So, in a Scotist vein, we might say that within a science one kind of unity (or set of unities) is studied, and the structure of this unity is established. But do distinct sciences necessarily demonstrate distinct truths? According to Thomas Aquinas, different sciences can establish the same truths, but by employing different methodology.

> Diverse conceptual characteristics *(ratio cognoscibilis)* make for diverse sciences. For instance, the astronomer and the natural philosopher demonstrate the same conclusion, *viz.*, that the earth is round. But the astronomer does this through a mathematical middle term ['e.g., through the shapes of eclipses or something else of this sort' (Aquinas 2014b, *Summa Theologiae*, Ia IIae, q.54, a.2, ad.2)]—*i.e.*, a middle term abstracted from matter—whereas the natural philosopher does it through a middle term considered materially ['e.g., through the movement of heavy things toward the middle of the earth or something else of this sort.' (Aquinas 2014b, *Summa Theologiae*, Ia IIae, q.54, a.2, ad.2)]. Hence, nothing prevents it from being the case that the same things that the philosophical disciplines treat insofar as they are knowable by the light of natural reason should be treated by another science insofar as they are known by the light of divine revelation. Hence, the theology associated with sacred doctrine differs in kind from the theology that is posited as a part of philosophy (Aquinas 2014b, *Summa Theologiae*, Ia, q.1, a.1, ad.2).

So if having a distinct method further distinguishes sciences, we can allow that distinct sciences can investigate the same genus. Metaphysics and theology have God among their shared subject-matter, metaphysics employing natural reason and theology using revelation as a principle to investigate the same reality,

It was a generally held opinion of the thirteenth-century scholastics that it is impossible to hold something on faith at the same time as possessing it as knowledge. Faith and knowledge are mutually exclusive states of mind. The definitions of *faith* and *knowledge* guarantee that one cannot believe something on faith while at the same time knowing it to be true. Consider a proposition *p,* (for instance, the proposition: 'The world has a beginning in time'). Now consider different kinds of belief I can have about the proposition *p*. I might know that *p* is true (or that it is false), I can have an opinion one way or the other, or I can be in a state of doubt (not tending either way). Furthermore, I could have faith that *p* is true. In the case of faith there is insufficient evidence for knowledge but the belief is held with the same degree of certainty as is a piece of knowledge. So faith is like opinion in terms of the evidence,[1] and faith is like knowledge in the certainty with which it is held.

[1] Among the scholastics, mere opinion was generally thought to require at least *some* justification or evidence.

The scholastics held that the mind has two principle powers, *intellect* and *will*. In Aquinas's *De Veritate* (*q*.14, a.1, 2) we find a description of the various ways in which the intellect can be related to the propositions *p* and not-*p*. In the case of 'doubt' the intellect is not inclined to hold either *p* or not-*p*. 'Opinion' is where 'the intellect is inclined more to one part than to the other' (Aquinas 2014a, *De Veritate*, q.14, a.1, 2), but not with certainty; the intellect does not totally discount the opposite being true. Aquinas continues:

> Now *sometimes* the possible intellect is determined in such a way that it adheres *totally* to one part. But it is determined in this way sometimes *by the intelligible object* and sometimes *by the will* (Aquinas 2014a).

In cases where the intellect is moved to assent by its object, the object can be 'known either by itself (as in the case of first principles, which are held by the habit of understanding), or through something else already known' (Aquinas 1945, *Summa Theologiae*, IIa IIae, q.1, a.4, co.). In the case of faith the object of the intellect is not sufficient to move the intellect to assent to the proposition, the intellect is moved to assent by an act of will, on the basis of elements sufficient to move the will but not the intellect. Furthermore, in the case of faith, even though the object of the intellect is insufficient to warrant the intellect's assent, the assent is nevertheless unhesitating; the proposition is held with the greatest degree of certainty. The non-overlapping *objects* of knowledge and faith, as distinct from the (less interesting) fact that faith and reason are *definitionally* distinct (i.e., they are mutually exclusive states of mind) give us an initial separation between what is held on faith and what can be known. But this result is superficial in the sense that it relies on the exclusive definitions of faith and knowledge.

It was nevertheless widely held that religious doctrines could be reached through human reason. There is a clear distinction between truths that could be known through natural reason and truths not accessible through reason, which can only be held on faith. Whereas this distinction was agreed upon by the scholastics, *which* truths belonged to which group was greatly disputed. For instance, both Aquinas and Scotus held that the existence of God could be established by reason alone, whereas the Trinity could not, and can only be an object of faith (in our present state). However, Aquinas believed that the immortality of the human soul could be demonstrated by reason, but Duns Scotus argued that while the immateriality of the soul could be demonstrated, the soul's immortality cannot.

14.3 The Eternity of the World

We are now in a better position to begin to answer our initial question, whether natural science and religion can coincide in their subject-matter. Let us consider one of the central disagreements between the Aristotelian world-view and that of Christianity: the eternity of the world. Where Aristotle, in common with many pagans, held that the world has always existed, Christians believed that the world had a beginning in time and was created by God out of nothing. The first great Christian offensive against the pagan idea of the eternity of the world was penned by the Christian Neoplatonist John Philoponus in 529 AD. This argument was widely endorsed in the Middle Ages and still has considerable appeal to some today. This is Bonaventure's version of the argument:

> It is impossible that infinitely many things have been gone through; but if the world did not begin, there have been infinitely many years. Therefore it is impossible to go through them. Therefore, it is impossible to get to this year (Bonaventure 1964, *Sentences*, book II, d.1, p.1, a.1, q.2, n.3).

So, given the supposed impossibility of passing through an infinite number of past years to get to this year, and since we *have* got to this year, it follows that the past years must be finite in number. Along with this argument, various other arguments were put forward by Bonaventure, Henry of Ghent and John Peckham, amongst others, as metaphysical demonstrations of the impossibility of an eternally existent world, claiming the impossibility of an actual infinite and the impossibility of an eternal world being created or being contingent. This argument and its close relatives play on various assumptions that were made about the nature of infinity in Aristotle's thought; that an actual infinite is impossible, that an infinite cannot be traversed (i.e., that we could never reach the far end of an infinity), and that infinity cannot be increased. Richard Sorabji believes that Philoponus's argument is successful against his opponents, i.e., Aristotle and the pagans, given their restricted notion of infinity. A solution to Philoponus's argument could only be reached when more subtle ideas about infinity were developed in the thirteenth and fourteenth centuries (Sorabji 1983). Thomas Aquinas, John Duns Scotus and William of Ockham all maintained that these arguments for the impossibility of an eternal world all fail. They believed that, although it was certain that the world *did* have a beginning in time, the creation of the world from eternity was not incompatible with *reason*, but only with revealed Christian doctrine. Here is Ockham's devastating response to Bonaventure's argument (above):

> ...It is true in general that an infinite that at some time is to be gone through never can be actually gone through; nor can there ever be a last [element] of such an infinite ... But an infinite that at *no* time was *to be* gone through but *always had been* gone through *can* be gone through despite its infinity. This is the reason why in virtue of the very fact that something *has been* gone through which at some time was *to be* gone through, it is finite. But if anything *has been* gone through which *never* was *to be* gone through, it need not be finite but can be infinite. Now, however, if the world existed from eternity, it *never* was the case that *all* past years were to be gone through, but because at no temporal instant would this proposition have been true: 'All these years (indicating all those [now] past) are to be gone through'. Therefore, the conclusion does not follow (William of Ockham 1997, *Quaestiones Variae*, q.3; Kretzmann 1985).

The argument against an infinite past relies on the assumption that some time in the past was infinitely remote from *now*. But there are no two designated parts of this infinite temporal continuum that are infinitely removed from each other. Richard Sorabji's statement of the apparent contradiction in an eternally existent world (and its solution) is so elegant that it warrants mention.

> One supposed difficulty is that, if an infinity of days had to pass before the arrival of today, then today would never arrive. This would certainly be so, if there was a first day, and then an infinity of days to cram in before today. But of course no first day is envisaged by those who postulate a beginningless universe; so there is ample room for a preceding infinity (Sorabji 1983).

So it looks as though neither the Christian nor the Aristotelian position on whether the World had a temporal beginning can be established by pure reason. Neither of the two alternatives is a logical necessity. This is something that must be held on faith. However, Duns Scotus comments that while it cannot be shown by reason alone that the world had a beginning in time, it may, nevertheless, be shown in some other way, such as by an understanding of natural causes and how things come into being; that is, by empirical science (Duns Scotus 1997b, *Questions on the Metaphysics of Aristotle*, book II, q.6, n.86; Geréby 1999). This appears to be an instance where science could be used to prove a religious doctrine. Science could be of aid to religion in an instance where metaphysics is not. However, we might reply that this is simply because Christianity, here, makes an empirical claim. That time had a beginning is a scientific empirical claim, and where religions make empirical claims, they open themselves up to empirical testing.

At this point we should clarify what follows from the supposition that the world never had a temporal beginning. It might be natural to think that if the universe had no temporal beginning, then this alone would suffice to show that there is no God, or at least that there is no warrant to postulate a creator. Our usual way of thinking of creation seems to lead us to this conclusion. When something comes into existence after not existing, we look for a cause of the thing's coming into existence. We think of creation as equivalent to making something begin to exist. However, Aquinas, Duns Scotus and Ockham argued that even if the universe has always existed it must nevertheless be something created; it cannot have been uncaused, and so there must be a first cause of its existence. In Thomas Aquinas's short treatise *On The Eternity of The World Against The Murmurers* ('the murmurers', here, being those theologians at Paris who held that it is nonsensical to say that the world has existed without beginning and yet was created by God, and yet refused to face Aquinas in disputation on the issue), we find an uncharacteristically caustic remark directed at those who held that the eternity of the world is incompatible with its being created:

> Anyone thinking seriously about it, then, must conclude that those who held that the world has always existed, but at the same time said that it was caused by God, are guilty of no conceptual incoherence. Those who detect this incoherence, therefore, must alone be men and wisdom must first have arisen with them! (Aquinas 1998).

Given a world without a temporal beginning, there is an infinite sequence of causes extending into the past, without beginning. Nevertheless, it is argued that there must be another kind of causal sequence that must terminate in a first cause. Here Scotus drew on the distinction between accidentally ordered causes and essentially ordered causes:

> Per se or essentially ordered causes differ from accidentally ordered causes in three respects. The first difference is that in essentially ordered causes, the second depends upon the first

precisely in the act of causing. In accidentally ordered causes this is not the case, although the second may depend upon the first for its existence or in some other way. The second difference is that in essentially ordered causes the causality is of another nature and order, inasmuch as the higher cause is the more perfect, which is not the case with accidentally ordered causes. This second difference is a consequence of the first, since no cause in the exercise of its causality is essentially dependent upon a cause of the same nature as itself, for to produce anything one cause of a given kind suffices. A third difference follows, *viz.* that all essentially ordered causes are simultaneously required to cause the effect, for otherwise some causality essential to the effect would be wanting. In accidentally ordered causes this simultaneity is not required (Duns Scotus 1982, *De Primo Principio*, 3.11).

So we can allow that an accidentally ordered series of causes can have an infinite number of members, because each cause in the sequence is of the same type, and does not constitute a more fundamental kind of causal explanation. But there must also be a more fundamental kind of causal series, and this must terminate in an ultimate cause, otherwise there will be not only explanation but also causality lacking in subsequent causes. Therefore, on this view, whether or not the world has a temporal beginning it will require explanation. So it looks as though the claim that the world did have a temporal beginning may not have much theological import, leaving the doctrine of creation as it is.

One might still insist that in the case of the world being created and having a beginning, and the case where the world is created but never began, we are using two quite different notions of creation, though it may well be that this is not relevant to the metaphysical dependency of the universe on God. Henry of Ghent argued that, even if sense can be made of a beginningless world having been created by God, such a universe could not be something contingent, and could not have been created freely by God (which contradicts Catholic doctrine). Henry writes:

If something always had being from eternity, there was never a preceding potential, neither belonging to the existing thing nor belonging to some efficient cause, by which its act of being could be stopped at some moment if we go backward in time. Therefore, it is absolutely necessary for it always to have been. If, then, the world creature is posited to have always had being from God and from eternity, it is absolutely necessary that it has been always and from eternity. And if this is the case, there was never from eternity—neither on God's part nor on the part of the thing—a potential by which it was able at some time not to have been (Henry of Ghent 1997, *Quodlibet* I, q.7 and 8, iii, 6, 1; Kretzmann 1985).

We cannot pursue the responses of Duns Scotus (*Ordinatio* II, d.1, q.3) and Ockham (*Quaestiones Variae*, q.3) here, but it is less clear that Henry's modal argument can be as easily defeated as the arguments relying on infinity. Odo Rigaldus (Eudes Rigaud), who was appointed Franciscan Regent Master at Paris in 1245, discussed the following objection to theology being considered a true science.

Also every science has to do with what is universal and imperishable, as the Philosopher points out. But theology is in great part concerned with singular events and with what passes away, e.g. with historical events which are deeds concerned with particulars. Therefore either theology is not a science or these things are not its concern. This second alternative however is clearly false (Rigaldus 1969).

We might argue, similarly, that when religions make historical claims they are thereby open to historical, and importantly archaeological, critique. So we might say that where religion encroaches on empirical science or history it becomes subject to testing, and becomes part of the subject-matter of these pre-established areas of natural inquiry, and therefore religious claims can be shown to be true or false by natural science. Insofar as religions make scientific claims, these religious claims become scientifically testable.

14.4 The Status of Scientific Claims in Religion

It may be worthwhile considering what place scientific or empirical claims have within religious belief. What function do they have within the specifically religious beliefs, and what relationship if any do they have to the supernatural elements of religion? It seems fair to say that the empirical claims made by religions started their lives not as scientific hypotheses but as revealed teachings about the structure and nature of the physical world. Nevertheless, the idea that claims about the physical world are capable of being known by natural means, without need of supernatural revelation, has a long history. Indeed, the majority opinion from Augustine to the late Middle Ages was that many religious doctrines could be known naturally, and therefore it should not be surprising that religious claims about the physical world are capable of being proved true or false by natural means.

As a preliminary way of investigating the relationship between empirical claims made by religions and the specifically religious claims, let us consider what follows from the truth or falsity of a scientific claim made by a religion. Even if one were to establish that any of the scientific claims made by a religion were true this would not act as epistemic warrant for believing that the religion itself is true. We can see that this must be so by considering that two contradictory belief systems might well make identical scientific claims. The two realms, the natural and the supernatural, seem to be clearly distinguished from each other, and claims about each do not appear to encroach on the other. Of course, while we can insist that the realms of the material and the immaterial are distinct in reality, the supernatural claims of religion and its empirical claims can be mutually dependent. If various historical and religious claims are true then the world must have a certain physical structure that allows these other things to be true. The truth of certain religious doctrines involves the physical world being a certain way. Likewise, if the world was created by God then there are empirical truths that are dependent on the supernatural and the immaterial. However, it looks as though the relationship between the two realms is not such that a deductive or causal relationship can be established between the two that would allow one to establish truths within the realm of the supernatural through investigation of the natural causal order. Consider the doctrine that God created the World freely and contingently. This doctrine has the logical consequence that the World might not have existed. Yet it is hard to see how, from within the realm of natural causes, anyone could ever be in a position to argue that there ever was a real potentiality for the world not to have existed, for this would be to have access to modal facts that are independent of the causal order to which we have access.

Why might anyone think that the truth of the empirical scientific claims of a religion is warrant for belief in the content of the religion as a whole? Given that two contradictory religions might well make identical scientific claims, the truth of the religion does not follow directly from the truth of its scientific claims. One might argue that these claims about the natural world that turn out to be certainly true point to the reliability of the source of the revelation. However, it is not so easy to derive this conclusion. One might argue that the believer must show the impossibility of making these scientific hypotheses without supernatural help, and all sorts of other things would need to be established as well, such as that the source of the revelation is reliable in other matters as well as the empirical, that it is in a position to have full knowledge, is not a deceiver, etc.

Now consider what follows for a religion if any of its empirical claims are shown to be false. Because the realm of natural causes and that of the supernatural are radically distinct the falseness of a religion's empirical claims will not impact on the truth or falsity of its specifically religious content (except where certain physical facts must obtain for a religious doctrine to be true). As in the case of the empirical claims of religions turning out to be true, the direct warrant for believing the spiritual or supernatural elements seems to be unaffected, but this is simply because the two realms are utterly distinct. For instance, if it turned out that we could prove that the World did not have a temporal beginning this would not obviously impact in any way on the belief that there exists something that is 'pure actuality', i.e., God as envisaged by Thomas Aquinas (*Summa Theologiae* Ia, q.2, a.3, co.) (It is not clear how any investigation into the Big Bang, for instance, could ever encroach onto the subject-matter of pure actuality, and, likewise, it is not clear how the claim that there exists some pure actuality could be construed as a scientific claim. These claims are metaphysical in nature.)

What effect might the truth or falsity of the empirical claims of a religion have on how reliable we consider the authority, i.e., the source of revelation, to be? If a religion makes empirical claims that turn out to be false, we can at least conclude that the religious authority is not infallible, and perhaps we ought to conclude that we should be suspicious of accepting something simply because it comes from this authority. It seems likely that were a religion to make detailed scientific hypotheses which systematically turn out to be true the believers would take it to be evidence for the truth of the religion as a whole. In reality, religions have not made such detailed empirical hypotheses. But even if they had we might argue that this only shows that they are reliable (albeit mysteriously so) when it comes to the realm of natural causes. The fact that the structure of the physical world is compatible with distinct and mutually incompatible religious views raises the worry that reliability in the realm of natural causes need not imply a similar reliability in the realm of the supernatural.

14.5 Can Natural Science Aspire to Spiritual Knowledge?

Natural science may at least be used for testing scientific claims made by religions, but can it reveal anything beyond the natural world? Can science be used to establish *purely* religious (that is, *non*-empirical) truths? Could science, for instance, give us reason to postulate some spiritual reality or an infinite being? Considering natural causes, we have some reason to doubt that this is possible. If we take any effect that we can observe in the world, could we ever be warranted in postulating a cause that is not similarly *finite*? Even if we accept the Neoplatonic axiom that the excellence or perfection of a cause is at least as great as its effect, it seems impossible that we might require an infinite cause for any finite effect or effects. Bonaventure did argue that creation itself demands an infinite cause, given that nothing and something are an infinity apart, but Scotus responded to this, arguing that nothing and something finite are separated by a finite distance (Cross 1998). Furthermore, Bonaventure's argument is metaphysical rather than proceeding from science. If the effect is finite, so might the cause be. Perhaps natural science can lead us to postulate a cause that is in a different *order of being* to the causes and effects we observe. We may postulate a first efficient cause, an unmoved mover, or particles that are not themselves composed of smaller particles. But here we have simply shown a dependency in our world, a dependency on some entity or entities of a different order of being. There seems to be no reason to suppose them infinite or even immaterial. Is it even *conceivable* that we might arrive at an infinite cause at the conclusion of a scientific investigation into finite effects? It looks as though science cannot make contributions to knowledge concerning infinite beings or their effects, for finite effects do not require infinite causes. So even if there *is* an infinite cause of these observed finite effects, the fact must remain hidden from natural science.

For the sake of argument, let us consider that we have encountered some real *infinite complexity* in something observed in our scientific inquiry (whatever an infinite complexity might mean). Or if an infinite complexity in the natural world is too wild an idea, consider the case of there existing an infinite number of animals (It is probable that in both these examples we might never be warranted in thinking that we were actually observing something infinite). What could we conclude from this? What we *can* conclude is that this infinite complexity or infinite population cannot have arisen naturally from a prior finitude. However, we cannot conclude immediately that the infinite was brought about by an infinite cause, for another alternative exists. The infinite might *always* have existed. So if there were an infinite number of creatures, all we could say is that they did not arise from a finite number by enumeration. The two alternatives are: (1) they were created, or (2) there have always been an infinite number of animals. The idea of there being an actual infinite number of entities in existence is quite counter-intuitive to us, and this may be because of how we know things actually come into being (e.g., starting with a small finite population, and then increasing.) Here we would need to prove a dependency, and it is not at all clear how one conceivably could prove it (through studying natural causes) in the case of any infinite phenomenon. In *Physics* book VIII, Chap. 10, Aristotle argued that the beginningless infinite motion of the cosmos must be caused by an infinite power (Aristotle 1984a, *Physics* VIII, 10, 266a12–24). Aristotle argued from the infinity of the motion to the infinity of the first mover, which must be completely metaphysically simple, not extended in space, and having no parts. A similar argument for an unmoved mover occurs again in book XII of the *Metaphysics* (XII, Chap. 7, 1072a21–1073b2). However, in this instance Aristotle's account is very far from what we might think of as a natural scientific account. Aristotle's physics is an a priori discipline investigating the nature of change and motion, and is much closer to metaphysics than to an empirical science, where observation and evidence are central. Aristotle's account also involves the stars moving because of their love of the unmoved mover. So the unmoved mover, as the first cause of motion, acts as a 'final cause'. In light of Galileo's attack on Aristotelian cosmology, final causes in nature, or the existence of directed purpose in the natural world, are now generally discounted in science (though contemporary philosophers such as Richard Swinburne and John Haldane have recently attempted to resurrect final causes in the natural world as motivating the existence of God). Even when Aristotle's first-mover argument is formulated in terms of efficient causes by Aquinas (*Summa Theologiae*, Ia, q.2, a.3, co.) it arguably fails because science has proved that moving causes do not operate in the way required for the argument to succeed (Kenny 1969).

There *may* be a conceptual possibility of generating an actual infinite from a finitude in a temporal succession, based on a case akin to the famous philosophical puzzle 'Thomson's Lamp' (Thomson 1954). In Thomson's thought-experiment, designed to show that there is a logical problem with infinite operations or infinite tasks, a lamp is switched on for one minute, then off for half a minute, then on for a quarter of a minute, off for an eighth of a minute, and so on *ad infinitum*. At two minutes the lamp has been switched on and off an infinite number of times. Thomson posed the question: 'is the lamp on or off after two minutes has elapsed?' Given that there is no last member in the series, because any given member is succeeded by another (given the infinite series), both possible answers appear to be impossible. This led Thomson to believe that completing an infinite task was a logical impossibility. However, it should not surprise us that there cannot be an answer one way or the other, because it is specified that there are an infinite number of switches. There is simply no truth to the matter of what the last member of the series is (at two minutes), because a last member is precluded by the parameters of the thought-experiment. While it is true that an infinite number of tasks has been completed by the time two minutes has elapsed, it is also the case that the parameters of the thought-experiment do not include any specifications for what state the lamp will be in once the two minutes have elapsed; indeed the stated parameters preclude the lamp being in *any* state at all after two minutes.

Along similar lines to the preceding thought-experiment, we might consider it a possibility that an actual infinite could indeed come into being successively, though, of course, not by natural means. In considering another argument against the possibility of a beginningless world, the argument from an actual infinity of souls, Scotus argued that it is possible that an actual infinity could be created successively by God. The problem of an actual infinity of souls, discussed by Avicenna, Algazel and Averroes, amongst others, and in turn drawn upon by the scholastics, is that, given that the existence of a beginningless world would mean that there have been an infinite number of past days, God could have created one soul per day meaning that there would now exist an actual infinity of souls (which was thought impossible). Duns Scotus replied to this, arguing that what God could do given an infinite time he could also do in a finite time. Souls can be created at successive discrete instants in time at an infinitely fast rate, and by the end of one day God could have created an actual infinity of souls.

> To the other point, about an infinity of souls, I reply that anything which cannot be made by God in one day 'because it involves a contradiction' cannot, for the same reason, be made by him in an infinite past time (if there had been an infinite past time). For in this one day there are infinite instants (nay, in one hour of this day), in each of which he could create a soul just as he could in one day of the whole of infinite time, if there were such infinite time (for it is not necessary that God rest from one day to the next in order to create one soul after another), and so if in the infinite instants of this day he cannot create infinite souls (because this cannot be done), neither could he have created infinite souls in the infinite days of the whole of past time.[...] the instants of this day—or of this hour—seem to have an infinity equal to the infinite instants of the infinite days, and so the proposed conclusion seems to follow (Duns Scotus 2014, *Ordinatio* II, d.1, q.3, 3, n.168–9; Duns Scotus 1973; Duns Scotus 1997a).

In a similar way, we can consider a successive generation of creatures in the physical world, which we might think of as 'Thomson's animals'. We obviously regard this as a natural impossibility, because in the case of natural processes there are natural constraints as to what is possible. In the case we are considering here animals would be generated at an ever increasing rate, quickly occupying space, and in motion at a speed greater than light. Once we allow that there are natural constraints as to what can happen naturally in the physical world (i.e., that there are laws of nature) we can dismiss any *real* possibility of an infinite being naturally generated in reality from a prior finitude. Again, if an infinity were to exist in the physical world it could only be because either it already existed (from eternity), or it was created by non-natural means.

In our world it may be possible to prove a dependency on some different order of being, but it seems impossible to show there is any kind of infinite dependency, or that there exists in the material world a dependency upon immaterial or spiritual realities. (Again, it seems impossible to reach these conclusions through the pursuit of physics or natural science, though it may well be possible through the pursuit of metaphysics.) If these do exist they cannot be shown by natural science, that is, by studying natural causes.[2] Duns Scotus agrees that we cannot through natural science reach religious conclusions about the supernatural or immaterial realities. Following Aquinas, Scotus writes:

[2] Again, metaphysical arguments may well conclude that there are immaterial realities, such as minds.

> Note: it cannot be shown by natural reason that something supernatural exists in the wayfarer, nor that it is required necessarily for his perfection. Therefore it is impossible to use natural reason here against Aristotle; if one argues from things believed, it is not a reason against the philosopher, because he will not concede the believed premise. Hence the reasons given against him here possess one or other premise as something believed, or as proved from something believed; therefore they are only persuasive theologically, from things believed to a thing believed (Duns Scotus 2012, *Ordinatio*, Prologue, q.1, n.12; Duns Scotus 1950).

So where does this leave the harmony of science and religion in mediaeval thought? We might say that they are in harmony because they do not contradict each other, but this is because they cannot contradict each other, for they concern different realms of reality. The arguments of these thirteenth and fourteenth-century scholastics for the insufficiency of Aristotelian science and the necessity of supernaturally revealed doctrine show that there must always be some discipline beyond what is susceptible to natural reason, *viz*. theology. Duns Scotus writes:

> However at least in respect of the wayfarer in this present life the said knowledge [of man's nature and the proper end of that nature] is supernatural, because it exceeds his natural faculty; natural, I say, in the sense of in accord with the state of fallen nature (Duns Scotus 2012, *Ordinatio*, Prologue, q.1, n.37; Duns Scotus 1950).

The objects of religious faith seem to occupy a different realm to those of the natural sciences. Given considerations about cause and effect, we must conclude that natural science cannot establish any truths about God or any infinite, about any immaterial causes, or spiritual realities. Can natural science encroach on religion? I answer that insofar as religions make empirical claims, they become scientifically testable. Religious beliefs and doctrines extend across the realms of metaphysics, history and physics, and in some cases even into biology, chemistry and botany. Therefore, it is only when religious beliefs include empirical beliefs that religion and science share in a common subject-matter.

14.6 Conclusion: The Importance of Metaphysics

We have reached the conclusion that natural science and religion can only share a common subject-matter, and indeed only come into conflict, when religions make empirical claims, claims about the structure of the natural world. Further to this, there appears to be a lack scientific warrant for purely religious beliefs. Are the specifically religious beliefs so remote from the world of experience as to be devoid of meaning, might religious belief be an irrational state of mind, or are we perhaps led to the uncomfortable position of fideism, the view that matters of faith have simply nothing whatever to do with reason?

On a traditional realist account of religious belief, religious faith is at least in part constituted by beliefs about reality, about the nature of the extra-mental world. Religious beliefs are true or false depending on what is the case in the world. The character of religious faith is such that, despite there being insufficient evidence to warrant certainty of belief, the belief is nevertheless held unhesitatingly. The intellect is not moved to assent by the evidence (as it is in the case of ordinary empirical knowledge), rather the will is moved by the good (as an object of desire), and the intellect assents to what is believed because it is moved by the will.[3] However, if we take religious faith to be virtuous it cannot be virtuous simply due to the epistemic character of the belief, (i.e., its being believed despite the absence of evidence). If faith is virtuous it must be so because it is a belief in what is *true*. Believing a falsehood while lacking evidence for the belief cannot be considered virtuous. So faith (as an epistemic state) is not a per se virtue; its virtue must, at least in part, be due to the content of what is believed and, importantly, on the content being true.

So far we have seen that both Aquinas and Duns Scotus emphasised that natural science is insufficient to give us access to specifically religious truths, including those essential to human happiness and salvation. They argued for the necessity of supernaturally revealed doctrine, which would give us access to truths not naturally accessible through reason or empirical investigation. Natural science and religion appear to have completely discrete realms as their objects of inquiry. Any role that natural science might have in relation to religion would seem to be very slim (perhaps acting negatively in uprooting false views about the physical world), but it cannot lead to supernatural or infinite hypotheses.

What could bridge this apparent gap between the observable world and the world of the immaterial and supernatural? The source of knowledge of the immaterial and the supernatural is the discipline of metaphysics. So the insufficiency of natural science leads these mediaeval thinkers not only to the need for supernatural revelation, but also acts as a plea for the importance of metaphysics to theology. We can see a clear distinction between the evident progress in scientific knowledge and what appears to be the inherently uncertain nature of religious doctrines. If there is to be any form of dialogue between these apparently disparate disciplines it must take place in the realm of philosophy, where reality is discussed in terms general enough to bridge the gap between the observed world of natural causes and effects and the world of the immaterial, and of ultimate and infinite causes. In particular, the scholastics emphasised the importance of metaphysics in understanding religious doctrine and in the articulation and interpretation of theological ideas. An additional concern is that our beliefs be rational, and the scholastics gave metaphysics a central role in defending the rationality of religious beliefs.

According to the mediaeval scholastics, the existence of God can be established by unaided reason through metaphysics. Natural theology, then, is strictly speaking a branch of metaphysics. Metaphysics is the realm in which the wayfarer can bridge the divide between the world of natural causes and the realm of revealed religion. Few theologians these days would go as far in their endorsement of philosophy as

---

[3] 'And it is also in this way that we are moved to believe what someone says because the reward of eternal life is promised to us if we believe; and the will is moved by this reward to assent to the things that are said, even though the intellect is not moved by what is understood' (Aquinas 2014a, *De Veritate*, q.14, a.1, co. 2).

the Jesuit theologian and philosopher Francisco Suárez, who, in the foreword to his monumental *Metaphysical Disputations* of 1597, wrote: 'It is impossible for anyone to become a competent theologian unless he builds upon a solid metaphysical foundation' (Vollert 1947). However, it can be argued that metaphysics is important to defending the intelligibility of religious doctrines, as well as understanding the potential limits of human thought. Historically, philosophy has been used not only to try to prove certain religious doctrines (e.g., that God exists, or that God created the universe), but also, where such proof was not thought possible, to show at least that the doctrines are logically coherent and do not give rise to contradictions (e.g., in the case of the Trinity being compatible with the simplicity of God). According to Duns Scotus, we can only know what terms we can intelligibly apply to God through metaphysics.

> But we do not immediately know whether any proper conceivable notion about God exists. Therefore no knowledge acquired naturally in this life represents any characteristic of God that is proper to him. The minor premise is evident, for the first proper notion we have about God is that he is the first being. 'First being,' however, is not something initially known from the senses, for we must first ascertain that the combination of these two terms makes sense. Before we can know that this combination represents something possible, we need to demonstrate that some being is first (Duns Scotus 2004, *Reportatio* 1-A, Prologue, q.3, a.1).

Following Augustine, the scholastic philosophers emphasised that, in addition to metaphysics, revelation is necessary. Metaphysics is restricted to what can be discerned through natural reason, whereas theology has access to other truths which are above natural knowledge, coming through a supernatural revelation. It should not come as a surprise that the subject-matter of theology occupies a realm distinct from that of natural science. I think it highly plausible that if Christianity is true then naturalism must be taken to be false, that is, there must be truths that are not accounted for by spatio-temporal entities. In addition to the insistence on the necessity of supernatural revelation, the scholastics relied on metaphysics in evaluating and articulating religious doctrines, and it is relatively uncontroversial that many religious doctrines are metaphysical claims about the nature of the world. The final outcome of our discussion here is that, in addition to the necessity of supernaturally revealed doctrine, philosophy is important to theology in being clear about what religious doctrines actually claim to be true of the world, and in defending them, and furthermore that natural science cannot make any serious contribution to religious knowledge.

References

Aquinas, T. (1945). In A. C. Pegis (Ed. & Trans.), *Basic writings of Saint Thomas Aquinas* (Vol. 1). New York: Random House.

Aquinas, T. (1998). On the eternity of the world against the murmurers. In R. McInerny (Ed. & Trans.), *Thomas Aquinas: Selected writings*. London: Penguin Books.

Aquinas, T. (2014a). *Quaestiones disputatae de veritate*, q. 14 (A. Freddoso, Trans.). Accessed February 1, 2016, from http://www3.nd.edu/~afreddos/translat/aquinas3.htm

Aquinas, T. (2014b). *Summa Theologiae* (A. Freddoso, Trans.). Accessed February 1, 2016, from http://www3.nd.edu/~afreddos/summa-translation/TOC.htm

Aristotle. (1925). *Posterior analytics* (G. R. G. Mure, Trans.). Oxford: Clarendon Press.

Aristotle. (1984a). *Physics* (R. P. Hardie & R. K. Gaye, Trans.). In J. Barnes (Ed.), *The complete works of Aristotle* (Vol. 1, pp. 315–446). Princeton, NJ: Princeton University Press.

Aristotle. (1984b). *Metaphysics* (W. D. Ross, Trans.) In J. Barnes (Ed.), *The complete works of Aristotle* (Vol. 2, pp. 1552–1728). Princeton, NJ: Princeton University Press.

Bacon, R. (1962). *Opus Majus*, vol. 2 (R. B. Burke, Trans.). New York: Russell & Russell.

Bonaventure. (1964). *In II Sententiarum* d.1, p.1, a.1, q.2. In *On the eternity of the world: St. Thomas Aquinas, Siger of Brabant, St. Bonaventure* (C. Vollert, L. Kendzierski, & P. Byrne, Trans., pp. 106–114). Milwaukee, WI: Marquette University Press.

Cross, R. (1998). *The physics of Duns Scotus: The scientific context of a theological vision*. Oxford: Clarendon Press.

Cross, R. (1999). *Duns Scotus*. Oxford: Oxford University Press.

Duns Scotus, J. (1950). *Opera Omnia, vol. 1, Ordinatio, Prologus* (P. C. Balic, Ed.). Vatican: Polyglot Press.

Duns Scotus, J. (1973). *Opera Omnia, vol. 7, Ordinatio, Liber Secundus, Distinctiones 1–3* (P. C. Balic, Ed.). Vatican: Polyglot Press.

Duns Scotus, J. (1982). *De Primo Principio: A treatise on God as first principle* (A. Wolter, Trans.). Chicago: Franciscan Herald Press.

Duns Scotus, J. (1997a). *Ordinatio*,II, d.1 q.3. (M. Tweedale, Trans.) In R. N. Bosley & M. Tweedale (Eds.), *Basic issues in medieval philosophy* (pp. 215–230). Ontario: Broadview Press.

Duns Scotus, J. (1997b). *Questions on the metaphysics of Aristotle*, vol. 1 (G. Etzkorn & A. Wolter, Trans.). New York: Franciscan Institute Publications.

Duns Scotus, J. (2004). *The examined report of the Paris Lecture, Reportatio 1-A*, vol. 1 (A. Wolter & O. Bychkov, Ed. & Trans.). New York: Franciscan Institute Publications.

Duns Scotus, J. (2012). *Ordinatio, prologue* (*Opera Omnia, vol. 1*) (P. L. P. Simpson, Trans.). Accessed February 1, 2016, from http://aristotelophile.com/Books/Translations/Scotus%20Prologue.pdf

Duns Scotus, J. (2014). *Ordinatio II, Distinctions 1–3* (*Opera Omnia, vol. 7*) (P. L. P. Simpson, Trans.). Accessed February 1, 2016, from http://www.aristotelophile.com/Books/Translations/Scotus%20Ordinatio%202%20dd.1-3.pdf

Geréby, G. (1999). Eternal allegiances, Duns Scotus' place in the debate about the possibility of an eternally created world. In B. Nagy & M. Sebök (Eds.), *The man of many devices, who wandered full many ways: Festschrift in honor of János M. Bak* (pp. 372–383). Budapest: Central European University Press.

Henry of Ghent. (1997). *Quodlibet* I, q. 7 & 8 (M. Tweedale, Trans.). In R. N. Bosley & M. Tweedale (Eds.), *Basic issues in medieval philosophy* (pp. 207–214). Ontario: Broadview Press.

Kenny, A. (1969). *The five ways: St Thomas Aquinas's proofs of God's existence*. London: Routledge and Kegan Paul.

Kretzmann, N. (1985). Ockham and the creation of a beginningless universe. *Franciscan Studies, 54*, pp. 1–31.

Rigaldus, O. (1969). Is theology a science? (A. Wolter, Trans.). In J. Wippel & A. Wolter (Eds.), *Medieval philosophy: From St. Augustine to Nicholas of Cusa* (pp. 265–272). New York: Macmillan.

Sorabji, R. (1983). *Time, creation and the continuum*. Ithaca: Cornell University Press.

Thomson, J. F. (1954). Tasks and super-tasks. *Analysis, 15*(1), pp. 265–274.

Vollert, C. (1947). Translator's introduction. In Francis Suárez, *On the various kinds of distinctions: Disputationes Metaphysicae, Disputatio VII, de variis distinctionum generibus* (C. Vollert, Ed. & Trans., pp. 1–15). Milwaukee, WI: Marquette University Press.

William of Ockham. (1997). *Quaestiones Variae*, q.3 (M. Tweedale, Trans.). In R. N. Bosley & M. Tweedale (Eds.) *Basic issues in medieval philosophy* (pp. 231–248). Ontario: Broadview Press.

Chapter 15
Medieval Lessons for the Modern Science/Religion Debate

Tom McLeish

15.1 Cultural Narratives for Science

The medieval intellectual world is fascinating, its cultures colourful, the greatest number of its lives soberingly short and hard (life expectancy was about 30 years) (Lancaster 1990), and its emerging political maps intriguing. However that may be, we do not usually turn to the thirteenth century for guidance or 'lessons', as the title of this chapter suggests we might. We read from the medieval world with interest, but rarely look it for advice. We enjoy thinking through the contrasts between the medieval schools and our universities, the power struggles between barons and kings, and our contemporary questions over decentralisation of political power, even between medieval Aristotelian natural philosophy and contemporary science. But we rarely seek to apply the knowledge so gained to twenty-first century life. To a modern, let alone post-modern, reader it appears strange to suggest that a thinker such as Robert Grosseteste, however powerful a mind he possessed, might helpfully instruct us as he did those early Oxford Franciscans, in such a modern and media-fuelled confrontation as that of science with religion.

Our suspicions will be justified if we believe that the current 'science and religion' debate is indeed to be framed as the clash of two incommensurable worldviews, as claimed for example in Dawkins' *The God Delusion* or Dennett's *Breaking the Spell* (2007). If science, as these writers would have it, represents the dominant force propelling us out of centuries of dogmatic religious thought-control into a future of enlightened and freethinking materialism, then nothing can be learned to advantage from a medieval thinker deeply committed to such outdated Christian philosophy and praxis, other than just how intellectually dark was the

T. McLeish (✉)
Durham University, Durham, UK
e-mail: t.c.b.mcleish@durham.ac.uk

J.P. Cunningham, M. Hocknull (eds.), *Robert Grosseteste and the pursuit of Religious and Scientific Learning in the Middle Ages*, Studies in the History of Philosophy of Mind 18, DOI 10.1007/978-3-319-33468-4_15

world from which we are fleeing. As serious historians of science have repeatedly and carefully shown however, such a view of intellectual history is not supported by the evidence (Principe 2011). Furthermore, as I have argued at length elsewhere (McLeish 2014), this falsely-projected confrontation is not even the most urgent challenge, nor productive of the most interesting set of questions, that concern our current tangled public narratives around science and religion.

Other signs - less obvious, but more consequential - indicate that our thinking has taken a wrong turn. For example, although we now deploy unprecedented technical power and possess once undreamed-of knowledge of the hidden subatomic and cosmic worlds, our public and political discourses around both science and technology are dismally shallow and argumentative. Why is it that we cannot seem to sustain an adult debate in our public spheres on science-driven questions such as genetically modified organisms (GMOs), climate change mitigation, nanotechnologies, fracking—the 'troubled technologies'? In place of a critical engagement with evidence and goals, in the light of a publicly-owned set of values, we witness repeated restatements from entrenched positions. Worse, as Phil Macnachten (Davies et al. 2009) and Jean-Pierre DePuy (2010) have pointed out, although the public debates are ostensibly framed as evaluations of risks in new technologies, the discourse is fuelled in reality by deeply-lying and ancient narratives. DePuy labels them: the narrative of desire ('be careful what you wish for'), of the sacred ('don't mess with sacred Nature') and of evil ('open Pandora's box at your peril'). If his analysis is correct, then science is currently without a *cultural narrative of purpose* that provides a guide to navigating the possibilities it opens up.

The only alternative to these negative and risk-averse framings of scientific knowledge is the shrill positivism of the 'New atheists', whose approach to the needful categories of purpose and meaning it to deny them, rather than supply them. Although welcomed by a few, their position has been discredited philosophically (Flew and Varghese 2007), and historically (Principe 2011). Bruno Latour has pointed out that the insufficiency of either of these polarised positions in regard to epistemology is reflected in another polarised deadlock—the impasse in environmentalism. Pointing out the different forms of contradiction in both the modernist and naturalist positions, he writes,

> Everything happens as if modernists were unable to reconcile their idea of Science and Nature—which, remember, according to their narrative, is supposed to be farther and farther removed, as time passes, from law, subjectivity, politics and religion—with the alternative reality that the connections of science and technologies are more pressing every day, more confusing, requiring even more intervention, more assemblies, more scrutiny, more stewardship (2008).

Science and technology are rendering our relationship with the natural world more, not less, complex. The negotiation of these complexities calls for a richer cultural narrative for science, not a simpler one. Our problem is a lack of resource from where to draw such a narrative—we have nowhere in modern or post-modern thinking to look for it. Neither DePuy's ancient (and incidentally pagan) myths of warning and threat, still alive and stifling effective dialogue, nor the myopic scientism of ardent materialism, have anything to offer other than their own bipolar deadlock. We are perhaps reminded of the humorous 'search for Wisdom' in Job:

28. In answer to the urgent question 'But where can Wisdom be found?' neither the foundations of the earth, nor the depths of the sea, can find it hidden within their recesses, though 'Death and Destruction' have 'heard a rumour of it'. Nor, appositely, is Wisdom to be found in the marketplace, soaked as it is in riches (no fewer than six different words for 'gold' are used in as many verses as the writer travels to the centres of commerce in jewellery and other luxuries). The 'science and religion' question that matters is not so much an intellectual exercise of reconciling epistemologies—it is a search for wisdom to guide and to frame our astonishing power to discover and to change the material world around us.

If on the one hand we accept that the commonly accepted public historical narrative of science and its religious context is deeply flawed, and on the other that science and its public framing is in serious trouble, then a look into the 'distant mirror' of the thirteenth century might provide some needed perspective on our current difficulties. More than this—we might well find ingredients there with which to construct a healthy narrative support for our engagement with nature. It is surely here that such cultural roots must lie, when the Aristotelian transmission from Muslim Spain into northern Europe galvanised the formulation of new questions of what we might come to know of the ordered universe and its workings. This milieu contains the search for questions themselves—what are the fruitful avenues of investigation that might lead to an understanding of nature, and which unprofitable? Is there a theological mandate to search for order in the material world, and to re-imagine it? What is the role of mathematics in description of the world, if any? Might an investigation of nature call on experimental manipulations as well as observation? What constitutes a complete understanding of a phenomenon? When this level of question is on the table, fundamental issues of teleology are inescapable—in stark contrast to our contemporary intellectual scientific world, in which they are hardly ever raised. For these are questions of vital importance to science itself, yet which cannot be answered within scientific methodologies. The thirteenth century reminds us that at great turning points in science, we need to go beyond its disciplinary boundaries for resources to re-frame its direction of travel (Kuhn 1962).

For these reasons, it is after all not such a strange idea to ask what we might learn, or at least what questions we might ask, by visiting the nascent scientific world of Grosseteste and his sources. I think that there are five chief ways in which this thirteenth-century master, and his intellectual and theological milieu, can assist in escaping our current impasse. I have called these: (1) the disruption of damaging myths, (2) the long history of science, (3) a cultural narrative for science, (4) a unified vision and (5) a relational and incarnational metaphysics. We next discuss each thread in more detail.

15.2 Disruption of Damaging Myths

As has already been noted, a common meta-narrative of the history of science in both public media and (at the least) school education, is that nothing remotely resembling science existed before the early modern period (or the late sixteenth century). According to this story, before Galileo and Newton any philosophy of the natural world was clouded with magic, alchemy, superstition, and—worse of all—the dogma of theology (Numbers 2010). There are other sub-narratives that emerge—that the scientific method is entirely modern, that medieval thinkers' chief goal was in any case to recapitulate the thoughts of the classical philosophers and not to move beyond them, that the medieval church repeatedly suppressed innovative thinking in general, and that 'theology' and 'science' were indistinguishable in the medieval world of scholasticism. Grosseteste's scientific corpus serves as an immediate gust of fresh air to remove such flimsy cobwebs of reconstructed history.

The shortest of the scientific treatises, the *De colore* (On colour) is enough on its own to remove credence in such a fiction. As I and others have explored in depth elsewhere (Dinkova-Bruun et al. 2013), the *De colore* represents a piece of work that a modern scientist would recognise as being in continuity with, though naturally distant from, questions posed and methods pursued today. Grosseteste does not allegorise or mystify colour; he does not accord any supernatural powers of transformation to it; he writes no explicitly theological material in his treatment at any point. On the contrary he treats colour as a perceived property of the natural world.

Color est lux incorporata perspicuo– (Colour is light incorporated in a diaphanous medium) the opening line of the treatise—introduces the conjecture that colour is an emergent property of light and matter (op. cit.). Readers familiar with his more substantial work on the physics and cosmology of light, the *De luce*, will recognise from the outset that Grosseteste is working with colour as a corollary of his more general theory of light. If material extension of all bodies (including the largest body of all—the cosmos itself) depends on an active indwelling of continuously self-multiplying light within material body, then one might expect the eye to detect visible effects beyond the fact of substantiality itself. And so it is—he identifies the different colours of objects as betraying the activity of different lights (characterised by the variation of two quantities of greatness—*multa/pauca*—and clarity—*clara/obscura* within materials characterised along a third dimension of purity—*purum/impurum* (op. cit.). There is to this day an unsolved problem in cognitive psychology of the apparent ordering, continuity and perceptive proximity of colours (Wuerger et al. 1995). Grosseteste prepares the ground for an approach to this issue by creating an abstract theatre of colour space. He is also working in a highly mathematical way (though this has not always been recognised in the secondary literature on the *De colore*- even by Crombie (1953). The numbers of possible colours and their contingencies are calculated in terms of the combinatorics of his three bipolar qualities. Never explicit, but strikingly obvious to mathematically equipped readers of his and Aristotle's theories of colour (*De sensu et sensate*), is that in developing a three-dimensional colour space between the opposing poles of black and white, he is going far beyond the Philosopher.[1] For Aristotle, the ascending series of colour from black to white is

[1] Aristotle, *De sensu et sensatu* available in translation at http://classics.mit.edu/Aristotle/sense.html

linear, or one-dimensional. All colours are met with at some point on a single pathway from one pole to the other. But the *De colore* describes in combinatorial clarity the higher dimensionality of the space which ascending and descending series of colours inhabit. We can deduce that the entire space is three-dimensional, and that the central meeting place of ascending and descending colours is a two-dimensional subspace. So, the treatise can be read as a constructive criticism of Aristotle's one-dimensional ascending series of colours as, by implication, an inadequate account of the phenomenon. Grosseteste insists that *per experimentum* (whether by thought or in action is beside the point here) one only reaches all possible colours by the variation of three independent quantities. The treatise does not represent a mere recapitulation of ancient thought, but goes far beyond it in imaginative theory as well as in mathematical complexity and observational relationship.

Within this short text of 400 Latin words we find, in this reading, a recognisably scientific approach to the mathematical modelling of an observed physical phenomenon. Naturally it is of its own time, not of ours—we now understand the origin of the three-dimensionality of colour to have its origins in the three types of photosensitive cone cells in the human retina, not directly in the properties of light or materials. But the core characteristic of science is not to be found in the answers it holds *pro tem*, but in the questions it poses, the way it goes about answering them and in the direction of its intellectual travel. In this sense, the questions and methods in colour science today are in continuity with Grosseteste's thought. If that were not true, it would be hard to explain why a team of scientists encountering this work in detail, and the related treatise on the rainbow, the *De iride*, were immediately inspired to create some new science. They recast the physical optics of the rainbow, and the perceptual framework of human colour vision, to show that even in contemporary terms, Grosseteste was correct in asserting that colour space can be both spanned and mapped by 'the space of all possible rainbows' (Smithson et al. 2014). Remarkably, this analytic work, required originally to establish whether the colour space of the *De colore* was indeed equivalent to the perceptual space used today, led to the discovery of a new mapping for colour space in which the coordinate system is inspired by the spectral characteristics of rainbows.

By the same token, this single work refutes the commonly held but misguided notion that early science was uniformly suppressed by the church. We read a Christian thinker in the thirteenth century developing pagan philosophy from the fourth century BCE transmitted to him via the Islamic tradition of the early medieval period. In the case of the *De colore* he drew explicitly from the Cordoban Muslim scholar Averroes (Ibn Rushd). Grosseteste was one of the first western masters to read and employ Averroes's *Commentary on the Metaphysics* in his own work. Such a confident and open use of sources from radically different and theologically incommensurate traditions by one charged, a little later in his career, with the care of Franciscan students, does not speak of a repressive ecclesiastical milieu. This is not to ignore or downplay acts such as the papal prohibitions of sets of Aristotelian teachings during the same century, but to point out that these were exceptions rather than the norm, and in any case did not have an adverse effect on a thinker such as Grosseteste either in terms of the sources he called upon or the conclusions he came to. To allow, in addition, a later instance to illustrate a general point, Pierre Duhem proposed that the 1277 Étienne Tempier condemnations may have even stimulated conjectures that the Earth, rather than the sun, might be in motion (1906–13). One of the condemned propositions was Aristotle's teaching that 'the earth is in the centre of the universe and necessarily at rest'. To draw attention to an idea, especially by means of the bright light of official disapproval, constitutes an irresistible encouragement for the academic thinker to toy with it.

This summarised case study illustrates, finally, the invalidity of an attempt to conflate the scientific and theological disciplines even in the thirteenth century. In all the treatises on light, Grosseteste is self-consciously engaging in work that is not theology. His motivation to explore scientific topics might be consequent to a theologically derived ethic or teleology (see Sect. 15.5 below), but it remains nevertheless quite distinct from it. His logic is tested, at least in thought, against observation and demonstration, not against doctrine. He derives, likewise, no direct consequences for theology from his conceptualisations of colour, his geometric optics of the rainbow, or his physical theory of the cosmogony of the celestial spheres. He is perfectly capable of doing this, but does so only in his theological works. So, for example, in the *Hexaemeron* he draws on the physical properties of light to make a theological point—'Among corporal things it is light which provides the most evident demonstration, through example, of the Most High Trinity' (referring to the triple property of luminosity, splendour and heat) (Grosseteste 1996). In the scientific works he achieves detailed conceptualisations of hidden dynamics and structures that satisfy his desire for an explanation of observed phenomena (colour, the rainbow, the motions of the stars and planets), but nowhere makes explicit allusion to theological ideas such as the Trinity. Again, this is by no means to suggest that he disconnects his scientific work from all theological motivation and framing, as commonly even believing scientists do today, as we shall see in the following, but it is to assert that he is perfectly clear on when he is doing science, when theology, and how to employ distinct methodologies in the two endeavours.

15.3 A Long History of Science

A second aspect of our deconstruction of the 'modern science' myth requires some comment: it is one thing to show that Grosseteste and his contemporaries were working in a potential *logical* continuity with science today; to show that this is also an actual *historical* continuity with it is another. It may never be possible to retrace the full pattern of reception of his scientific corpus. These treatises, remarkable as they are, are not as widely referred-to as the *Hexaemeron* (op. cit.) and Psalm commentaries, for example (Ball 2012). Yet nearly two generations after their probable first writing, Roger Bacon had grounds to acclaim Grosseteste as the greatest mathematical genius of the century (Bacon *Opus Maius* I, 108). The conceptual continuity of his geometric optics and work on the rainbow, with those of Bacon, Theodoric of Freiburg, the Prague school of the fifteenth century, and onwards to Newton's own *Optics*, strongly suggests a historical transmission of his science. His years as master to the Oxford Franciscans, a role dedicated to the formation of young scholarly minds, closely aligned with the period of production of his scientific works, makes it inconceivable that the excitement of these new ideas were not communicated with those cohorts, and adopted by those of their number who later went on to teach others. The brightest minds among them (we know of at least Adam Marsh, and possibly Bacon) would not have failed to be inspired and to think about their rich conceptual content themselves (Felder 1904; Panti 2012).

But whatever the detail and extent of their later adoption and development, Grosseteste's scientific works are testament to the longer continuity of a human intellectual story that we now call 'science', but which went by other names in earlier ages (McLeish). It might better be termed 'natural philosophy' in the eighteenth and nineteenth centuries or even 'natural wisdom' in antiquity. A vital thread is that of a developing story—natural philosophers are consciously drawing from ideas of the past, but building on and correcting them. Our evolving understanding of nature has a history, with a more occluded past, a present mixture of partial understanding and of open questions, and a hoped-for future of clearer insight.

Grosseteste's own methodology within such history of science has already emerged in the way that the *De colore* works with Aristotle's and Averroes theories of colour. It is also worth recalling that he would also have known Bede's compact catalogue of natural phenomena, the *De natura rerum* from the early eighth century.[2] This remarkable monastic instructional text adapts the successive works (under the same title) of Pliny and of Isidore of Seville and was widely copied and read in the succeeding five centuries. Bede demonstrates by example how, even in the early Middle Ages, science was not simply transferred, but could be critically assessed against observation and reason. A good example is found in his discussion of the saltiness of seawater. The problem is a longstanding one from antiquity: how is it that the seas remain salty when fresh water from rivers the world over flows into them unremittingly, and for centuries? Pliny's answer is that the fresh river water sinks on meeting the ocean and is recycled via underground culverts to rise again from springs. But Bede points out that this is inconsistent with the observation that fresh water is lighter (we would say 'less dense' today) than salt water, so that if it did not mix on meeting seawater then it would float upon it as a surface layer rather than sink. Bede claims (*contra* Pliny) support for the alternative hydrological cycle that returns the fresh water via the atmosphere. If Grosseteste had any need for

[2] Grosseteste's access to and knowledge of this seminal work of Bede is discussed in (Southern 1986).

authority that permitted him to correct authorities, then in reading Bede closely, he would have absorbed the notion that received natural philosophy is not determinative of current thought, but should be re-evaluated against others' ideas, direct observation, and reason (Kendall and Wallis 2010).

Although the two strong characteristic elements of current scientific methodology: experimental testing and mathematical modelling, are of course far less well developed in either eighth or thirteenth centuries than today, this does not mean that the work of Bede or Grosseteste is out of continuity with them. Nor should we expect scientific method and goals to develop in sudden transformational leaps when a gradual account will suffice to explain the historical evidence, and is consistent with the written record. So, although Crombie's claim that Grosseteste was the 'first to set out a systematic and coherent theory of experimental investigation…' (Crombie 1953) is usually considered an overstatement today, our example of the *De colore* illustrates a history of science demonstrably at the dawn of experimental thinking. It certainly embodies an early account of explicit mathematical modelling in its three-dimensional colour space, together with explicit suggestions that this mathematical approach can in principle be verified by manipulations of light and materials.

If the thirteenth century is marked by the dawn of experimental method, then in Grosseteste it also represents a clear new departure in the ubiquitous application of mathematical thought to natural science. From our modern perspective, it is hard to imagine an intellectual milieu in which this would not seem natural. But that is because we do not share the same sharp dualism of the perfect and imperfect inherited philosophically from Plato and cosmologically from Aristotle. Grosseteste himself comments on the *Posterior Analytics* that we are able to do with mathematics that which God is able to do with physics—that is to deduce conclusions from axioms within a closed system. We do have access to the fundamental axioms of mathematics, but only the Creator has that access in regard to nature. Our task is to arrive at nature's axioms inductively from observations of their consequences. Such human predicament of incompleteness is a consequence of our dwelling in the sublunary world of imperfection. Now, while is it uncontroversial that (perfect) mathematics applies to the structure and motion of the (perfected) spheres above that of the moon, it is by no means clear that it will be as commensurate with the (imperfect) realm of the elements. To assay a mathematical analysis of sublunary nature is therefore not only a critical, but a bold, step. Yet it is one that Grosseteste takes in each of his scientific treatises. In spite of the unavailability of advanced algebraic notation of any kind, he is able to compute, for example, abstract vectors combinatorially in his three-dimensional colour space. Perhaps more impressive is the continuation of his discussion of colour in the *De iride*, in which he considers the conceptual space of all possible rainbows. Though not immediately apparent as such, this high degree of abstract and structured thinking is highly mathematical.

In re-thinking Aristotle in critical ways, and in advancing mathematical tools to conceptualise the structures that lie behind the superficial perception of phenomena such as colour, Grosseteste partakes in both the reception and advancement of a much longer story of science than typically frames discussions of religion and science today.

15.4 A Cultural Narrative for Science

Perhaps the most striking contrast between Grosseteste's intellectual world and ours can be found in our differing teleology. Cultural narratives are able to generate purpose, or equally, to proclaim purposelessness. So, as I suggested in the introduction, while he knows why he is exploring the natural world, and develops a strong sense of purpose in doing so, we have in our own time lost any such propelling meta-narrative. In late modernism a faint echo of a human reason that we do science remained, but only in an instrumental narrative of national economic prosperity. In a post-modern atmosphere of suspicion around all overarching stories, that too (possibly healthily) has withered.

There are both simple and more sophisticated strands within Grosseteste's motivations to engage in natural science. On a delightfully childlike level, at one point in his commentaries on the Psalms, he reflects that, if the Bible chooses to convey truth to its readers through the illustrations of natural objects (trees, clouds, falling leaves etc.) then it behoves us to discover as much as we are able concerning them, simply in order that we might better understand the Scriptures[3] An application of this very direct thinking appears in an explanatory note accompanying his edition of John Damascene's *De Fide Orthodoxa*. Two chapters in the earliest manuscripts at his disposal concerned scientific topics that ostensibly had no contact with the theological substance of the work as a whole. Earlier editors had sometimes omitted them for that reason. But Grosseteste reinstates both, explaining that:

> These two chapters, namely the 24th about seas and the 25th about winds, are omitted in some Greek manuscripts; perhaps because they did not seem to contain a theological subject. But according to truly wise men, every notice of truth is useful in the explanation and understanding of theology (Cf. Rome, Bibl. Vat., MS Chigi A. VIII. 245, f. 16va).

We see immediately the impressively connected philosophy of knowledge that drives his studies. Although he is perfectly able to distinguish theology and science (again, there was no age—certainly not the thirteenth century—in which they were 'indistinguishable'), he takes the two as mutually dependent in at least illustrative ways. He maintains a clear distinction between theological and scientific writing, but within an implicit and deep connectivity. So although we find no explicit theological introductions or conclusions to the scientific works, this is because their theological task speaks for itself. For an explanation of deeper connection between the silent theological framing of his natural philosophy, and the science

[3] See the chapter on the *Hexaemeron* by Giles Gasper in part IV this volume for Grosseteste's views on the all-encompassing canvas of Scripture.

itself, we need to turn to the philosophical works. In the *Commentary on the Posterior Analytics* (of Aristotle) Grosseteste places a more sophisticated theological philosophy of science within the overarching Christian narrative of Creation, Fall and Redemption. Employing a Boethian metaphor for the effect of the Fall on the higher intellectual and spiritual powers (in descending hierarchy those of understanding, memory, imagination) as a 'lulling to sleep' by the weight of fallen flesh, he maintains that the lower faculties, including critically the senses, are less affected by fallen human nature than the higher (Crombie 1953). Human understanding (*aspectus*) is now inseparable from human emotion and loves (*affectus*); the inward turning of the latter now dulls the former. However, there is an avenue of hope that the once-fallen higher faculties might be re-awakened: engaging the *affectus*, through the still-operable lower senses, in the created external things of nature allows it to be met by a remainder (*vestigium*) of other, outer *light*. So a process of re-illumination can begin once more with the lowest faculties and successively re-enlighten the higher:

> Since sense perception, the weakest[4] of all human powers, apprehending only corruptible individual things, survives, imagination stands, memory stands, and finally understanding,[5] which is the noblest of human powers capable of apprehending the incorruptible, universal, first essences, stands!

Human engagement with the external world through the senses, necessary because of our fallen nature, becomes a participation in the *theological* project of salvation. Furthermore, the reason that this is possible is because this relationship with the created world is also the nexus at which human seeking is met by divine illumination. As a central example, the 'physics of light' grounded in the cosmogony of the *De luce* informs a 'metaphysics of light' as a vehicle to become a 'theology of light'. The *De impression elementorum* makes explicit the theological action of light that remains implicit in the *De luce*—light (following the epistle of James 1:19) is a symbol of the perfect gift that descends from the Father of Lights. The implied restorative process that begins with an alertness to nature through our senses becomes another of Grosseteste's 'critical Aristotelian' moves (Crombie 1953). With Aristotle he insists that all knowledge of particulars and universals comes through the senses, but against Aristotle he allows this to be met with divine illumination. This double move even suggests a theological motivation for novel combination of experiment and mathematics implied in his scientific works—in every case it is at the meeting-point of observed phenomena and mathematical reasoning that understanding is born. The teleological employment of scientific investigation as an instrument of human participation in a reversal of the effects of

[4] We recall Paul's categories in 1 Cor. 1: 7.

[5] This may be an abbreviation of a five-step 'ladder of intelligence' detailed by Isaac of Stella in his Sermon 4 on the Feast of All Saints (1977): 'For the soul too, while on pilgrimage in the world of its body, there are five steps towards wisdom: sense-perception, imagination, reason, intelligence and understanding.'

sin in the fall, is an idea that itself reawakens in the early modern period, especially (but by no means exclusively) in Francis Bacon (Harrison 2009).

We might expect that, since light is for Grosseteste both the supreme physical and theological form, so among the senses his preferred metaphorical example to illustrate the result of a meeting between sense and revelation would be sight. And so it proves to be. As the higher senses become sharpened by their infusion of illumination through the lower, so a higher penetrating power, *sollertia* (Grosseteste borrows the use of the word from James of Venice) awakens. In the *Commentary on the Posterior Analytics* he writes:

> Sollertia, then, is a penetrative power by which the vision of the mind does not rest on the surface of the thing seen, but penetrates it [the thing seen, rem visam] until it reaches a thing naturally linked to itself [sibi naturaliter coniunctam]. In the same way as [sicut si] corporal vision, falling on a coloured object, does not rest there, but penetrates into the internal connectivity and integrity of the coloured object, from which connectivity its colour emerges, and again penetrates this connectivity until it reaches the elementary qualities from which the connectivity proceeds (Rossi 1981).[6]

This is his great articulation of the restorative effect of the divinely-assisted contemplation of nature, but it is also a striking articulation of the experience that epitomises the work of science in any age. Any contemporary scientist would recognise the meaning as a felt experience. An enhanced form of seeing, by which not only the outer appearances of things, but their inner logic and workings also are perceived, is arguably the prime metaphor for scientific understanding. 'Ah—I see it now!' is not an arbitrary exclamation. This exposition on *sollertia* fills out in more detail the scale of restored human understanding. The penetrative power of the 'vision of the mind' as a connectivity with the object understood is preceded by a similarly-patterned connectivity of the 'corporal vision' with the inner integrity of the object perceived. From the *De colore* we know that Grosseteste develops his theory of colour as the natural consequence of form (light) giving extension and integrity to matter. So once more we see his metaphysics of light developing from his reading and thinking through Aristotle, towards a theory for scientific method itself. What is remarkable is how he manages to work in an isomorphism between the physics of light and matter, and the perception of the same light and matter by the human intellect.[7] Colour vision becomes the vehicle by which he explains that a subjective 'connectivity', or perhaps better, 'complex structure' is generated in our mind through sense-connection with an objective connectivity within materiality of the world.

---

[6] Trans. Sigbjørn Sønnesyn (personal communication); Aristotle's Post. An. II.19 is also in the background here, where the emergence of general understanding from particulars of sense-perception is described: 'It is like a rout in battle stopped by first one man making a stand and then another, until the original formation has been restored'.

[7] Remarkably, the visual perception of depth in materials beneath a translucent surface, is currently an active topic in vision research, see e.g. (Motoyoshi 2010) and the chapter by Hannah Smithson (Part I this volume).

Grosseteste does not have to invent the idea of an extended sense of vision in regard to the human relationship with nature—it occurs both in the Church Fathers and in Biblical Wisdom literature with which he is familiar. Gregory of Nyssa -whose writings we know were familiar to him through the Psalm commentaries and the *Hexaemeron* (Southern 1986) -records a remarkable death-bed dialogue with his sister Macrina ('The Teacher') in *On the Soul and the Resurrection*. The greater part of the discussion is a debate on the reality of 'the soul' (in context the notion might better be translated for contemporary readers as 'mind' for the purposes of that treatise). Macrina's final and decisive move against Gregory's position (assumed for the sake of the argument—that mind is merely an epiphenomenon of matter) is, almost verbatim, that it 'penetrates to something below the visual image'. She chooses the example of the phases of the moon: we do not assume that the appearance of a waxing and waning object is sufficient to describe the reality, but understand that the Moon is a sphere passing through successively different angles of illumination by sunlight as seen from the Earth. It is the mind that performs this task—'the mind that sees', seeing below the surface of phenomena, or in Macrina's words:

> You see what the eye does teach; and yet it would never of itself have afforded this insight, without something that looks through the eyes and uses the data of the senses as mere guides to penetrate from the apparent to the unseen. It is needless to add the methods of geometry that lead us step by step through visible delineations to truths that lie out of sight, and countless other instances which all prove that apprehension is the work of an intellectual essence deeply seated in our nature, acting through the operation of our bodily senses (http://www.newadvent.org/fathers/2915.htm).

A closely parallel Biblical source is found in the Hymn to Wisdom of Job 28. Grosseteste refers to this ichneutic search for Wisdom, humorously described in the Hymn, in developing a discussion of theology itself in the opening of the *Hexaemeron*. James McEvoy points out that in this context he is distinguishing theology from the sciences by emphasising the place of divine revelation—it possesses a necessary totality that the 'wise of the world' are not able to discover, but that must be received by faith (McEvoy 2000). Yet, as we have already seen, he derives theological motivation for his work in the liberal arts in general, and cannot have been unmoved by the reason given at the close of the Hymn, that God himself knows the way to wisdom:

> But God understands the way to it; it is he who knows its place.
>
> For he looked to the ends of the earth, and beheld everything under the heavens, so as to assign a weight to the wind, and determine the waters by measure (Job 28: 23–24; Clines 2006).

Here once more is the special, enhanced way of 'seeing' that recruits other aspects of mind than perception alone, including quantitative reasoning, to the task of beholding all of creation. Furthermore, although the Hymn concludes with this description of divine beholding, no student could miss the structural sense in which this conclusion balances its opening verses, equally powerfully descriptive of the unique view of the Earth from below afforded to the eyes of human miners dangling by ropes in their deep-cut shafts. Not even the sharp-eyed falcon can claim their vision of the earth 'from beneath, transformed by fire' (28: 5). It is not only God who has access to the deep perception of creation which is the Way to Wisdom—the invitation is extended to humans as well.[8]

So Grosseteste has plenty of Biblical and Patristic material to work with in developing a cultural (and in his case, necessarily theological) narrative of science. However, the central place within his thought that he accords to his own metaphysics of light, and the detailed example of the 'physical structure' underlying colour that he develops in the *De colore* and the *De iride* give him material to expand and develop *sollertia* as a running teleological metaphor. He is explicit in his demonstration that sense perceptions can awake the higher senses into a grasp of underlying reality (the two qualities of light itself and their intersection with a third quality of the indwelt matter) when mathematics and geometry are also summoned to the task of deeper seeing. Finally, all this is set within an overarching Biblical narrative of Creation, Fall and Redemption in which humankind is invited to participate in the process of recreation.

15.5 A Unified Vision

Reading Grosseteste from a scientific perspective excites resonances with a class of thinkers for whom a unified map of the world has the highest value. Einstein is perhaps the most celebrated modern example. The prime motivation for his Nobel Prizewinning work on the photoelectric effect was not a central attack on that problem—it is in any case only a corollary to the paper (Einstein 1905)—but a desire to develop a thermodynamic account of light. Similarly, relativity arises, not from a direct analysis of time and motion, but from an attempt to overcome an uncomfortable incommensurability between late nineteenth century electromagnetism and mechanics.

A similar passion for a single vision has already emerged in our examination of the *De colore* and *De iride*—taken together with the *De luce* these works replace a fragmented universe of coloured objects by a unified theory of the activity of light within body to generate the phenomenon of material extension that in turn produces the phenomenon of colour (Dinkova-Bruun et al. 2013). Furthermore, the abstract geometry of colour itself works as a unifying mathematical framework in all of its occurrences, arising from the product of internal properties of materials and of light.

Perhaps more remarkable is the completely original unification that Grosseteste makes, at least by implication, in the *De luce*, of the superlunary and sublunary cosmic regions. For Aristotle, as we have seen, the universe contains two incommensurate and separate realms in which, for all time, both nature and physics are

[8] This special sort of 'seeing' which is Wisdom- and also the great metaphor for scientific insight- is also picked up strongly by Oxford theologian and philosopher Paul Fiddes (2014).

different.[9] The imperfect spheres of the elements sustain vertical motion, mixing, disease, while above the moon all matter is perfect, crystalline and all motion circular. There is not even a temporal connection between the two regions, since this separation has been the case for all time. The cosmogony of the *De luce* is not only a remarkable application of Aristotelian physics taken in a critical vein to overthrow an Aristotelian cosmos without beginning, nor is it just an impressively clever theory of origins. It also demonstrates how the same creative force of light (in its two forms of *lux* and *lumen*) and its action of rarefaction and compression on matter, can give rise to both superlunary and sublunary regions within a single process of structure development, itself determined by a uniform set of properties. Grosseteste explains that the inward progression of *lumen*, together with its successive perfection of the spheres, is eventually weakened through distance from the firmament and through the work it needs to do in passing through all the underlying spheres. Below the orbit of the Moon, there is insufficient power within the field of *lumen* to form any further perfected spheres, so what materials remain—the elements—are compacted but left unperfected. Today we would term this process a 'symmetry breaking': the operations of a uniform physical process on a system that originally possesses a state of symmetry, breaks that symmetry by creating two regions in different states. A detailed computational study of the physics in *De luce* has confirmed that such a programme can be taken further than the text alone is able to, using tools unavailable before the invention of the calculus (Bower et al. 2014), but translating only Grosseteste's own physics into computational mathematics.

The *De luce* succeeds in demonstrating that the apparent *heterogeneous* structure of the cosmos can arise via the working of *homogeneous* local physical processes. There is no need to postulate different physics at work within different regions of the world. Instead the later development of distinctly structured regions is implicit in the original 'laws' of interaction between matter and light, and importantly in the temporal boundary conditions of the cosmos.

The unifying theme within Grosseteste's thinking informs both his science and his theology. It is within the theologically-motivated desire to seek out unifying causes (since all things originate from the same Creator) that we can most clearly perceive his alertness to pattern, and especially numerical pattern. No number is more significant than three in this regard. If ever he looks to nature for the signature of God (and, as we noted above, he is far more shy of doing this than the exponents of natural theology in the eighteenth and nineteenth centuries), then it is in signs of the Trinity. Not only does light serve as a unifying physical substance so that (*Hexaemeron*) every object is *aliquod genus lucis* (some kind of light) but that, 'Among corporeal things it is light which provides the most evident demonstration, through example, of the Most High Trinity' (Hexaemeron part viii).

Grosseteste goes on to explain that this is because light has three properties—lightness, splendour and heat. In the *Commentary on the Posterior Analytics* he likewise sees the Trinitarian imprint in the material constitution of matter, form and

[9] See for example (Bowen and Viltberg 2009).

the combination and realisation of the two. Even in this subtle way, it seems strange to modern ideas that a theological framing might be expected to pattern science. But there is one crucial step in the scientific process that, especially since Popper (1934), has been overlooked methodologically, or certainly understudied. In the shadows cast by the long debate on validation and refutation of scientific hypotheses is the vexed question (quite unvisited by his *Conjectures and Refutations*) of the source of scientific ideas in the first place. This is where imagination has to run without limit—there is no logical path to the creation of hypotheses. A scientific imagination inspired by theological ideas is, as history shows, likely to be a fruitful one. A notion that light is fundamental as a form giving body to matter at all scales may have its root in an enlightened reading of *Genesis* but it is tested against its explanatory power to account for cosmic origins in the physics of the early universe, whether that is within a medieval geocentric structure or a twentieth-century relativistic Einsteinian one. Perhaps science today needs to rediscover unfamiliar places in which to draw inspiration for the mighty acts of imagination it needs to reconceptualise nature.

15.6 A Relational and Incarnational Metaphysics

There is another purpose evident in Grosseteste's thought behind the re-engagement of the human mind with the inner structures of the cosmos, one that is independent from the post-lapsarian invitation to re-awaken fallen minds. This second strand is important to him, for one of his great theological questions concerns an alternative history—one in which there is no Fall from grace. In the *De cessatione legalium* he asks famously *An Deus esset homo etiam si non esset lapsus homo?*- Would God would become man had man not fallen? (Grosseteste 2012). The question of the incarnation in such an unfallen world has corollaries—in particular would we be doing 'science' in such a world? Is there, in other words, a motivation for natural philosophy that goes beyond the restoration of a mind once perceiving nature clearly, but now clouded and dulled? Although the text does not address this question directly, it points in very strong directions that parallel Grosseteste's conclusion that there would indeed have been an incarnation of God in an unfallen world, and that his relationship with human and non-human creation maintains a directional narrative even without its disastrous first turn.

Grosseteste points out, once again driven by the primacy of his unifying principle of light, that the human body communicates with all corporeal natures ('*communicat in natura*') because of the way light is incorporated into all elements by its reflection from the heavenly bodies. All of the rational soul of humans, the sensitive souls of animals and the vegetative souls of plants share both the same indwelling of constitutive light, and the composition of the elements. He entertains a very early insight into the material way in which humankind is, literally, earthed into creation. An even more impressive account of such material connectedness across the cosmos is found towards the end of the *De luce*, and is worth quoting in full:

> And it is clear that every higher body in respect of the luminosity begotten from it is the species and perfection of the following body. And just as unity is potentially every following number, so the first body by the multiplication of its luminosity is every following body. Earth, in contrast, is all higher bodies by the collection in it of the higher luminosities. Thus, the poets call it 'Pan' [that is,'All'] and it is named Cybele as if cubele from the cube [that is, from solidity]; because it is the most compressed and dense of all bodies, it is Cybele and mother of all the gods, for although all higher luminosities are brought together [in earth], they have not come forth in it through their operations, but it is possible that the luminosity of any celestial sphere you please be drawn out from earth into act and operation, and so from earth, as if from a kind of mother, any god will be procreated (Grosseteste 2011).

A modern version of this sentiment was made famous by the scientist and communicator Carl Sagan, drawing a material communication between human and cosmic materiality not from light, but from the atomic generative properties of stars, 'The nitrogen in our DNA, the calcium in our teeth, the iron in our blood, the carbon in our apple pies were made in the interiors of collapsing stars. We are made of star-stuff' (Sagan 1980).

For both writers there are real, material reasons that connect us to even the most distant objects in the universe. The difference is in the material detail: Grosseteste deduces them from the structuring properties of light, Sagan from the unique environments within the cores of stars, where alone heavy elements can be manufactured. In spite of the efforts of thinking such as this, almost poetic in the connective and emotive force of its idea, the deeply relational cultural context that it suggests for science has not taken root.

The structure of Grosseteste's theological underpinning, evident by implication in his natural philosophy, and explicitly in the theological works, meets some of the needs of our own age in surprisingly fresh ways. In the light of the problematic cultural narratives of our time, identified by DePuy from the evidence of contemporary debates around science and technology, and following the suggestion from Latour to mine theological strata for material that could lead to a resolution of their recurrent impasses, we have located a rich seam. Taking his theology in suitably transformed context, and drawing also on his sources, Biblical and Patristic, I have suggested that a cultural narrative generated from a 'Theology of Science' for today would recognise and incorporate a set of seven foundational principles (McLeish 2014).

(i) It would recognise a long, complex history of relationship between human and non-human. The ancient beginnings of the story recognise the Biblical (and other ancient wisdom traditions) embedding of the need to be reconciled to a world that puzzles and threatens. It removes the damaging analysis that science is exclusively modern, or represents an awakening from inappropriate, superstitious or ignorant shackles of thought. Rather it assists in relocating the deeper seeing, the imagination constrained by observation, the recreation of nature—all that we now call 'science' as part of a longer and deeply human story. In this view, science belongs with art, story, drama, music in the collective creative, therapeutic and constitutive human endeavours.

(ii) It entertains a high view of human aptitude to re-imagine nature. The special theological position of humankind, made *in imagio Dei* and under divine command as caretakers of the Earth, encourages, rather than suppresses, the adventure of discovery that re imagines nature and illuminates the world's inner structures. The simplicity, the naturalness, with which Grosseteste explores scientific questions, aligns with his theological anthropology when this is explicit. Both warn against a view that doing science is 'unnatural' or disconnected with our ancient human origins.

(iii) It balances and integrates a science of illumination with the wisdom of cautious intervention in the world. The dual structure of Wisdom and Knowledge, *sapienta* and *scientia* is present in Grosseteste's Aristotelian source material, but becomes transformed through his Christian theology, drawing as it does on the wisdom of Job and the Psalms. This is not to say that we should retreat from nature—as Latour has pointed out, humankind has long passed the point at which this was a realistic option, even if it were ever appropriate. But it does insist that our technical transformations, our co-creations, should place the value of sustaining a fruitful natural world before our own profit.

(iv) Drawing on the traditions of the consequences of the Fall, and the repeated reminders in Proverbs, Psalms and Job of the 'thorns and briars' that characterise engagement with the earth, and with Paul's letter to the Romans of the groaning of creation (as if in childbirth), a theological narrative of science would recognise that engaging with nature is ambiguous and painful. Again, this is not to signify that it is inappropriate. But it is to recognise, with Job, that the search for wisdom is a difficult and long one, and with Grosseteste that our starting point is one in which our understanding and our wisdom have been lulled to sleep. In the task of reconciling a natural world that hides itself and threatens, our unavoidable foolishness is likely to lead to painful mistakes.

(v) The experience of pain and difficulty in rediscovering a relationship with nature is due in part to the balance of order and chaos that constitutes a world in flux. Learning to live with uncertainty is as certain an experience in science as in any other lived experience. It is especially true of a world where the in-built predisposition of inanimate matter to explore new potential in structures on all levels from the molecular to the macroscopic can only evolve on a substratum of microscopic random motion. Although there could have been no notion of the essential thermal ('Brownian') motion in the thirteenth century, (although Lucretius has the atomic constituents of matter in constant motion),[10] thinkers like Grosseteste were grappling with the science of the imperfect (sublunary) world. As we noted, it takes intellectual courage to

[10] Lucretius *De rerum natura*.

suppose that the logic and grammar of mathematics might bring some order, and not inappropriately, to the realm of the elements.

(vi) The astonishing fruitfulness and centrality of the *creative question* we read everywhere in Grosseteste's work addresses one aspect of the meagre current public narratives for science—that it is not the place to expect a role for creativity and imagination. In re-reading the intellectual history of the thirteenth century, we are struck repeatedly by how difficult it is to identify which questions will open doors to understanding, and at what time, and which are premature or even ill-defined. To read into a time when it was by no means obvious what science could become, humbles us today. It sharpens the realisation that our disciplinary methodologies have become narrowly defined, and dangerously closed to new thinking.

(vii) Finally it would recognise the role and work of love in rediscovering a participative and reconciliatory project of the human relationship with nature. One of Grosseteste's most striking and moving works is written, not in Latin, but in Anglo-Norman French—the *Château d'Amour* (Mackie 2003). Full of physical structure, light and colour, this 'Hymn to the Virgin' is as faithful to a structural *sollertia* of nature as it is to the theology of faithfulness and love. A project that successfully engages a wider participation in the playful joys of science, as well as its necessarily painful task, will also be explicit in celebrating the centrality of love—both within the community that undertakes the task, and also of the object of our gaze, created in the first place from the same creative *caritas*.

The long story of natural wisdom, a high expectation of human ability and responsibility, a balance of practical and intellectual wisdom, the enduring of difficulty, an accommodation with uncertainly, a celebration of the question, and the exercise of love—these are some of the lessons we can begin to draw from a deep engagement with medieval science. They are far from irrelevant to our time. Very much more than a fascinating period in the early history of science, the thirteenth century and its thinkers, of whom Grosseteste is the prime example, speak with wisdom we urgently need to rediscover.

References

Ball, R. M. (2012). Robert Grosseteste on the Psalms. In J. P. Cunningham (Ed.), *Robert Grosseteste: His thought and its impact* (pp. 79–108). Toronto, ON: Pontifical Institute of Medieval Studies.

Bowen, A. C., & Viltberg, C. (Eds.). (2009). *New perspectives on Aristotle's De Caelo*. Leiden: Brill.

Bower, R. G., McLeish, T. C. B., Tanner, B. K., Smithson, H. E., Lewis, N., & Gasper, G. (2014). *A medieval multiverse?* Mathematical modelling of the Thirteenth Century Universe of Robert Grosseteste. Royal Society of London Proceedings, Series A470, 20140025. doi:10.1098/rsp.2014.0025.

Clines, D. J. A. (2006). *World biblical commentary*. London: Thomas Nelson.

Crombie, A. C. (1953). *Robert Grosseteste and the origins of experimental science 1100-1700*. Oxford: Oxford University Press.

Davies, S., Macnaghten, P., & Kearnes, M. (2009). *Reconfiguring responsibility: Deepening debate on nanotechnology*. Durham: Durham University.

Dennett, D. (2007). *Breaking the spell. Religion as a natural phenomenon*. London: Penguin.

Dinkova-Bruun, G., Gasper, G. E. M., Huxtable, M., McLeish, T. C. B., Panti, C., & Smithson, H. (2013). *Dimensions of colour: Robert Grosseteste's De Colore; Edition, translation and interdisciplinary analysis*. Toronto, ON: Pontifical Institute of Medieval Studies.

Duhem, P. (1906–13). *Etudes sur Leonard de Vinci* (3 vols). Paris: Hermann.

DuPuy, J.-P. (2010). The narratology of lay ethics. *Nanoethics, 4*, 153–170. doi:10.1007/s11569-010-0097-4.

Einstein, A. (1905). *Annalen der Physik, 17*(6), 132–148.

Felder, H. (1904). *Geschichte der wissenschaftlichen Studieren im Franziskanerorden*. Freiburg: Herder.

Fiddes, P. (2014). *Seeing the world and knowing god*. Oxford: Oxford University Press.

Flew, A., & Varghese, R. A. (2007). *There is a god: How the World's most notorious atheist changed his mind*. New York: Harper One.

Grosseteste, R. (1996). *On the six days of creation* (C. F. J. Martin, Trans.). Oxford: Oxford University Press.

Grosseteste, R. (2011). *La luce: Introduzione, Testo Latino, Traduzione E Commento* (C. Panti, Ed.). Pisa: Edizioni Plus.

Grosseteste, R. (2012). *On the cessation of the law* (S. M. Hilderbrand, Trans.). Washington, DC: Catholic University Press.

Harrison, P. (2009). *The fall of man and the foundations of science*. Cambridge: Cambridge University Press.

Isaac of Stella. (1977). *Sermon 4 on the feast of all saints* (B. McGinn, Trans.). Minnesota, MN: Cistercian Press.

Kendall, C., & Wallis, F. (2010). *Bede: On the nature of things and on times*. Liverpool: Liverpool University Press.

Kuhn, T. S. (1962). *The structure of the scientific revolution*. Chicago: University of Chicago Press.

Lancaster, H. O. (1990). *Life expectancy*. New York: Springer.

Latour, B. (2008). 'It's development, stupid!' or: How to modernize modernization. In J. Procter (Ed.), *Postenvironmentalism* (pp. 17–25). Cambridge, MA: MIT Press.

Mackie, E. A. (2003). Robert Grosseteste's Anglo-Norman treatise on the loss and restoration of creation, commonly known as Le Château d'amour. In M. O'Carrol (Ed.), *Robert Grosseteste and the beginnings of a British tradition* (pp. 151–179). Rome: Istituto Storico dei Cappucini.

McEvoy, J. (2000). *Robert Grosseteste*. Oxford: Oxford University Press.

McLeish, T. (2014). *Faith and wisdom in science*. Oxford: Oxford University Press.

Motoyoshi, I. (2010). Highlight- shading relationship as a cue for the perception of translucent and transparent materials. *Journal of Vision, 10*(9), 6. 1–11.

Numbers, R. (2010). *Galileo goes to jail and other myths about science and religion*. Cambridge, MA: Harvard University Press.

Panti, C. (2012). Scienza E Teologia Agli Esordi Della Scuola Dei Minori Oxford: Roberto Grossatesta, Adamo Marsh E Adamo Di Exeter. In E. Menestò (Ed.), *I Francescani E Le Scienze Atti Del 39 Convegno Internazionale Assisi*, 6–8 Ottobre 2011 (pp. 311–351). Perugia: CISAM.

Popper, K. (1934). *The logic of scientific discovery*. London: Routledge.

Principe, L. (2011). *The scientific revolution: A very short introduction*. Oxford: Oxford University Press.

Rossi, P. (Ed.). (1981). Robertus Grosseteste, *Commentarius in Posteriorum Analyticorum Libros*. Unione Accademica Nazionale Corpus Philosophorum Medii Aevi, Testi e Studi.

Sagan, C. (1980). *Cosmos: The story of cosmic evolution, science and civilisation*. London: Random House.
Smithson, H. E., et al. (2014). *Journal of the Optical Society of America, 31*, A341–A349.
Southern, R. (1986). *The growth of an English mind in medieval Europe*. Oxford: Clarendon.
Wuerger, S. M., Maloney, L. T., & Krauskopf, J. (1995). *Vision research, 35*, 827–835.

Index

A

Abelard, Peter, 45–47, 206, 229, 230, 247
  The Sentences, 230
Abū al-'Alā'ibn Sahl, 31
Abu Ma'shar al-Balkhi (Albumasar), 14
Adam of Exeter, 161, 234
Aertesen, Jan, 189
Alain de Libera, 123, 129
Alan of Lille, 193, 195, 202
Albertus, Magnus, 48, 95, 216
Albertus of Cologne (Coloniensis), 124
Alexander of Hales, 114, 123, 148, 205
Alexander's Dark Band, 81
Al-Fārābī (Alpharabius), 13, 248
Al-Ghazali, 131
Al-Kindi, 23, 131, 174
Alpetragius, 229
Amalricians, 49–52, 54
Amaury of Bêne (Amalric), 49, 50, 52
Angels, 148, 149, 205, 212, 231, 239
Anselm of Canterbury, 224, 226
Aquinas, Thomas, 88, 136, 189, 210, 219, 247, 264, 265, 267, 269, 270, 273
Arab, 90, 95, 124, 191, 199
Arabic, 14, 24, 29, 30, 32, 41, 48, 92–94, 122, 222, 248
Arcs, 34, 70, 71
Aristotle
  De anima, 23, 32
  metaphysics, 13, 22, 49, 54, 130, 131, 166, 169, 191, 200, 206, 265, 270, 291
  meteorologica, 24, 26, 135, 136, 146, 150
  physics, 7, 13, 21, 54, 65, 152, 169, 194, 200, 222, 229, 263, 265, 275, 293
  posterior analytics, 3, 7, 121, 191, 197–198, 201, 222, 266, 267, 290, 291
Ashworth, Jennifer, 201
Astrology, 94, 130, 131, 246, 258
Astronomy, 110, 130, 131, 246, 258
Astronomy-Astrology, 130
Atom, 8, 15, 212
Atomist, 23
Augustine, St., 177, 272, 279
Averroës (Ibn Rushd), 22, 23, 123, 131, 170, 183, 266, 276, 285
Avicebron (Solomon Ibn Gabirol), 41–43, 46, 170, 174, 216, 248
Avicenna (Ibn Sīnā), 13, 23, 24, 34, 88, 89, 91, 123, 124, 131, 152, 158, 170, 174, 199, 229, 233, 276

B

Bacon, Francis, 145, 291
Bacon, Nicholas, 124, 125
Bacon, Robert, 122, 145
Bacon, Roger
  Communia Mathematica, 147
  Communia naturalium, 128, 180, 181
  Compendium studii philosophie, 151, 155, 157
  De multiplicatione specierum, 29, 128, 169, 174, 178, 180
  Opus maius, 29, 123, 124, 127–132, 134, 135, 137, 139, 146, 148, 149, 151–161, 287
  Opus minus, 151, 153
  Opus tertium, 120, 121, 123, 124, 128, 132, 138–140, 146–149, 151, 154, 155, 160

J.P. Cunningham, M. Hocknull (eds.), *Robert Grosseteste and the pursuit of Religious and Scientific Learning in the Middle Ages*, Studies in the History of Philosophy of Mind 18, DOI 10.1007/978-3-319-33468-4

Baeumker, Clemens, 45
Bandoun, John, 147
Bartholomew the English Man, 111
Baryon, 17, 18
Basil the Great, 228
Bede, St., 287, 288
Bernard of Chartres, 43, 121
Bérubé, Camille, 199
Big Bang, 3, 273
Boethius, 16, 21, 44, 129, 192–194, 201, 203, 204, 247, 250
Boethius of Dacia, 124
Bonaventure, St., 177, 269, 274
Boyer, Carl, 158
Brams, Joseph, 135
Brethren of Purity (Ikwān al-Safā), 34

C
Campanus of Novaria, 147
Cartwright, Nancy, 166
Cathar, 228
Cathars, 43, 51
Catoptrics, 25, 31–34
Cause
efficient, 47, 151–161, 180, 214, 253, 271, 274, 275
final, 151–156, 158–161, 166, 205, 275
CERN, 18
Charles, Robert Henry, 256
Christ, 42, 87, 89–91, 97–99, 170, 177, 226, 232, 236, 238, 240, 256
Christ's incarnation, 238
CIELAB, 77–82
Cistercians, 290
Clarembald of Arras, 43, 44, 46
Clausius, Rudolf, 9
Clayton, Philip, 42
Clegg, Brian, 119
Cloud, 4, 26, 27, 32–34, 136, 152, 155–160, 289
Corbiel, Peter, 51
Corporeity, 8, 22, 23, 42, 54, 165–183, 216
Corpus, 10, 26, 53, 65, 90, 91, 190, 192, 195, 202–204, 221–223, 284, 286
Cosmogony, 3, 286, 290, 294
Cosmology, 3, 47, 50, 54, 224, 239, 258, 284
Courçon, Robert, 51, 52
Coyne, George, 263
Creation, 9, 43, 44, 47, 54, 104, 152, 154, 155, 161, 172–174, 202, 210, 213, 215, 217, 219, 221–240, 252, 253, 258, 269–271, 274, 290, 292, 293, 295, 297
Crombie, A.C., 3, 4, 6, 7, 23, 28, 29, 91, 158, 167, 169, 172, 248–253, 288, 290
Crowley, Theodore, 120, 121

D
D'Onofrio, Giulio, 43, 44
Dales, Richard C., 222, 223, 225, 252, 256–258
Daniel of Morley, 14
David of Dinant, 47–49, 51–54, 210
De cessatione legalium, 219, 238
Debye series, 70
DePuy, Jean-Pierre, 282, 296
Descartes, Rene, 36, 67, 68
Dickson, Gary, 49, 51
Dionisotti, Carlotta, 255–257
Dionysius Exiguus, 196, 202–204, 206, 231
Dioptrics, 23–26, 31, 34
Disease, 87–91, 97, 98, 103, 105, 106, 111, 112, 294
Docking, Thomas, 143, 144
Dominicans, 30, 129, 145, 212, 213, 218, 219
Donati, Silvia, 119, 120, 200
Duhem, Pierre, 286
Duns Scotus, John, 134, 170, 210, 211, 219, 247, 264–266, 268, 270, 271, 276–279

E
Eclipses, 267
Eco, Umberto, 191, 204, 205
Edmund of Canterbury, 121
Eidola (phantasms), 23
Einstein, Albert, 293
Eleatic tradition, 48, 201
Elford, Dorothy, 46
Emden, A.B., 124, 125
Emission, 10, 23, 25, 30, 32, 157
Essence, 22, 28, 31, 32, 41, 45, 46, 49, 65, 166, 171, 172, 176, 178, 179, 181, 182, 197, 210–212, 218, 250, 290, 292
Euclid, 23, 132
Eve, 88
Experimental method, 34, 83, 144, 151, 288

F
Fall, 12, 55, 87–89, 159, 169, 196, 198, 228, 238, 246, 252, 290, 291, 293, 295, 297
Fermat's principle, 6
Fideism, 277

Finite, 8, 23, 182, 211–213, 216, 217, 219, 269, 274
Finitude, 210, 274–276
Fishacre, Richard, 210, 211
Form (*eidos*), 22
Forrest, William, 92
Franceschini, 255
Franciscans Oxford School, 161
Fulbert of Chartres, 87, 99

G
Galen, 88, 91
Galileo, Galilei, 275, 284
Garnier of Rochefort, 50
Gell-Mann, Murray, 17, 18
Genesis, 9, 43, 44, 152, 155, 203, 222, 223, 227–231, 233, 238, 239, 295
Genesis creation, 228
Geoffrey of Aspall, 200
Geometry, 7, 23, 30, 34, 69, 123, 153, 156, 157, 160, 161, 172, 292, 293
Gerard of Huy, 119
Ghisalberti, Alessandro, 192
Gieben, Servus, 3, 28, 199, 225
Gilbert of Tournai, 199
Gilbert the Englishman, 111
Giles of Rome (Aegidius Romanum), 131
Giles of Spain, 41
Gilson, Étienne, 44, 170, 209, 259
Ginther, James, 147, 209, 211, 216, 222, 223, 232
Goering, Joseph, 147, 223
Gregory of Nyssa (Nyssen), 228, 231, 292
Gregory of Tours, 149
Grosseteste, Pietro B., 157
Grosseteste, Robert
  Chateau d'Amour, 298
  Commentary of ecclesiastical Heriarchy, 203
  Commentary on Divine Names, 190, 202
  Commentary on the celestial hierarchy, 47, 202
  Commentary on the Mystical Theology, 195–197
  Commentary on the physics, 200–201, 215
  Computes correctorius, 150, 222
  De Artibus Liberalibus, 17
  De cessatione legalium, 115, 182, 218, 223, 224, 256, 295
  De Colore, 28, 60, 61, 64–67, 72, 82, 287
  De Cometis, 125
  De Impressionibus elementorum, 290
  De Iride, 4, 6, 24, 26, 27, 60, 66, 81, 83
  De Libero Arbitorio, 198, 203, 254
  De lineis angulis et Figuris, 156
  De lineis et Figuris, 23, 26, 138, 251
  De Luce, 3–19, 22, 171, 174, 214, 215, 235, 236, 238, 252, 293
  De Motu corpoali, 22, 47, 173, 214
  De Motu Supercaelestium, 47
  De Natura Locorum, 28, 29, 154
  De Operationibus, 47, 54, 173, 175
  De Sphera, 12, 47
  De Statu Causraum, 54, 167
  De Unica Forma Omnium, 54, 218
  De Veritate, 223, 238, 249, 268
  Dictum, 211, 213, 214, 218
  Perambulavit Iudas, 256
  Templum Dei, 98, 99, 257
  Testament of the Twelve Patriarchs, 256
Group Theory, 19
Guy le Gros de Foulque (Guido Fucoldi), 122, 126–128

H
Haartrup, Dieter, 137
Hauréau, Barthélemy, 45
Health, 87–92, 99, 113, 130, 167
Henry III, 150
Henry of Ghent, 211, 269, 271
Henry of Suse (Ostiensis), 50
Hezekiah, 227
Hierarchy
  celestial, 47, 202
  ecclesiastical, 203
Hippocrates, 88, 90, 91
Hippocratic, 91
Hirsch, S.A., 126, 140
Hispalensis, Johannes, 94, 95
Hoccleve, Thomas, 92
Holy Spirit, 45–47, 50, 133, 226
Hugh of Amiens, 229
Hugh of St. Cher, 109
Hume, David, 165
Humoral, 91, 96–98
Hylomorphism, 41, 43

I
Ibn al-Haytham (Alhazen), 21, 23, 24, 28–34, 36, 125, 132–134
Ibn Sahl, 5, 24, 31, 36
Impurity, ritual, 103–115
Incarnation, 51, 226, 238, 239, 252, 253
Incarnational, 283, 295–298

Infinity, 34, 209–212, 215–217, 219, 269–271, 274–276
Isidore of Seville, 228, 287

J
Jacobi, Klaus, 190, 192
James, Letter of, 128
James, M.R., 255
Jeauneau, Édouard, 44, 45
Jesus. *See* Christ
Job, 297
Job, book of, 232, 233, 282
John Damascene, 47, 221, 235, 289
John of Basingstoke, 124
John of La Rochelle, 114
Johnson, Timothy, 120, 137
Jourdain, Charles, 120

K
Kamāl al-Dīn al-Fārisī, 21, 28, 29
King, Peter, 134
Kings
  David, 148
  Henry III, 145
Knorr, Wilbur R., 147
Knowledge, 31, 44, 90, 105, 120, 143, 166, 189, 212, 223, 245, 263–268, 273–279, 281, 282, 289, 290, 297
Krause, Karl Freidrich, 42

L
Langton, Stephen, 114
Laonese historian, 49
Lateran Council IV, 53
Latin, 21–24, 41, 43, 47, 49, 61, 81, 92, 94–96, 104, 122, 126, 135, 144, 158, 170, 176, 177, 222, 228, 229, 248, 251, 255, 285, 298
Le Blon, Jacob Christoph, 76
Lee diagram, 77, 80, 81
Leff, Gordon, 54, 55, 248
Leprosy, 103–114
Leviticus, 104–111, 113, 114
Liberal arts, 17, 222, 225
Lincoln, 23, 42, 55, 99, 119–141, 148, 150, 154, 155, 209, 211, 213–215, 217–220, 222, 245
Lincolniensem, 139
Lincolniensis, 55, 143
Lincolnshire, 125
Lindberg, David C., 4, 88, 90, 120, 132, 133, 137
Little, A.G., 122, 125, 139
Lombard, Peter, 229, 230, 252, 258, 259
Lupus, William, 124, 150
Luscombe, David, 48
Lydgate, John, 92

M
Maccagnolo, Enzon, 48–50
Macnachten, Phil, 282
Macrobius, 47
Magic, 131, 264, 284
Maimonides, Moses, 113, 114, 131
Maloney, Thomas, 120, 121, 123, 129, 138, 139
Manichaeans, 203
Mantello, F.A.C., 54, 114, 234, 257
Marenbon, John, 206
Marrone, Steven, 248–251
Marsh, Adam, 121–125, 137, 143–161, 287
Martin of Tours, 149
Mathematics, 7, 13, 55, 62, 121, 123, 126, 129–131, 135, 143, 147, 153–155, 157, 158, 161, 167, 168, 172, 232, 252, 265, 283, 288, 290, 293, 294, 298
McCleery, Iona, 88, 99
McEvoy, James, 47, 125, 146, 168, 173, 174, 196, 202, 204, 215–217, 221–223, 229, 233, 236, 245, 246, 292
Medicine, 87–99, 105, 201, 246, 257
Metamerism, 82
Meteorological, 24
Meteorology, 23, 24, 33, 34, 36, 146
Mie theory, 70
Molland, George, 120
Moon, 9, 10, 134, 146, 153, 160, 176, 239, 252, 288, 292, 294
Motion, 9, 11, 12, 17, 21, 22, 47, 159, 161, 165, 176, 180, 182, 196, 203, 265, 275, 276, 286, 288, 293, 294, 297
Music, 16, 17, 297

N
Nagel, Sylvia, 135, 153
Neckham (Nequam), Alexander, 229
Neoplatonism, 41, 42, 249
Neumich, 110
Neutrinos, 8
Newton, Isaac, 70
North, John, 119, 126, 127, 130

O

Ockham, William, 6, 134, 247, 269
Optics, 4, 6, 21–37, 65, 70, 125, 132–136, 143, 144, 154, 156, 160, 167, 237, 264, 285–287
Ordered Universe Project, 224
Origin, 12, 42, 44, 59, 74, 76, 87, 165, 169, 170, 174, 176, 177, 181, 213, 218, 245, 285, 294, 295, 297
Original Sin, 88, 89, 98
Oxford, University of, 132

P

Panentheists, 42
Pantheist, 42, 46, 47
Panti, Cecilia, 3, 8, 9, 11–13, 22, 28, 29, 54, 60, 143–161, 222, 287
Papal
  bull, 53
  legate, 51, 52
  letters, 99
Parens scientiarum, 53
Paris
  city, 49, 51, 54, 122, 123, 125, 132, 137, 146–149, 151
  University of, 48, 49, 52, 55, 129, 148, 212, 246
Paris, Matthew, 122, 256
Paul, St., 253
Pecham, John, 123, 132, 133, 137
Peripatetic School, 61
  De coloribus, 61
Peter of Limoges, 129, 130, 132, 133, 140
Peter of Maricourt (Maharncura), 131, 153, 160
Peter the Chanter, 109
Philip of Tripoli, 95
Philip the Chancellor, 189, 192, 194, 195, 198, 202–204
Philo of Alexandria, 228
Philoponus, John, 269
Physicalists, 48
Physics, 54, 275–277, 284, 288, 290, 291, 294, 295
Physiology, 111
Pierre de Rivleaux, 122
Pierre des Roches, 122
Pittenger, Norman, 42
Planets, 9, 17, 251, 257, 286
Plato, 8, 44, 45, 47, 174, 227, 229, 249, 253, 288
Platonist, 23, 25
Plotinus, 174, 253, 254
Pope
  Clement IV, 120, 122, 126–129, 131, 132, 137, 144
  Gregory I (the Great), 227, 230
  Gregory IX, 52
  Honorius III, 53, 95
  Innocent III, 48, 50, 53
  Innocent IV, 53
Poullion, 192
Psalms, 223, 224, 239, 289, 297
Pseudo-Aristotle, 131
Pseudo-dionysius
  Divine Names, 190
  Mystical Theology, 197, 203
Pseudo-Ptolemy, 131
Ptolemy, 4–6, 9, 12, 30, 129, 135
  Optica, 23, 26
Pythagoras, 129, 227
Pythagorean, 16

Q

Quadrivium, 16, 123, 132, 232

R

Rainbow, 21–37, 59–83, 135–137, 143–161, 172, 237, 285–288
Ralph of Flaix, 109
Reason, 12, 21, 43, 62, 81, 107, 111–113, 115, 123, 125, 126, 129, 132, 133, 148, 149, 152, 158, 159, 177–179, 189, 191, 204, 205, 211, 213, 215, 217, 225, 226, 231–233, 235, 245–259, 264–270, 274, 276–279, 283, 287 290, 292, 296
Reason, sufficient, 108
Reduction and impossible, 172
Reflection, 6, 22, 24–27, 29–32, 34, 35, 67, 68, 70, 74, 81, 109, 152, 156–160, 168, 169, 193, 205, 214, 236, 237, 295
Refraction, 4–7, 22, 24–27, 29–32, 34–37, 67–70, 74, 81, 152, 156–161, 168, 169, 179, 237
Regimen sanitatis Salerni, 92, 95
Renaissance, 23, 130, 189
Rome, 91, 289
  Curia, 55
Rosemann, Philipp W., 51
Rossi, Pietro B., 7, 223

Rufus, Adam, 218, 236
Rufus, Richard (of Cornwall), 123, 125

S
Sapientes antiqui/moderni, 119–141
School of Chartres, 43–47, 247
Scientia divina, 144
Scientia experimentalis, 130, 131, 135, 153
Scientist, 9, 10, 77, 91, 95, 119, 129, 137, 245, 247, 252, 254, 258, 263, 284–286, 291, 296
Secreta secretorum, 92
Sens, Council of, 49, 51, 52
Septuagint, 229
Siger of Brabant, 124
Simon of Tournai, 193, 195, 202
Simplicius of Cilicia, 170
Sin, 88–90, 97–99, 103, 109, 111, 112, 238, 240, 257, 290
Smalley, Beryl, 108, 109, 113, 258
Smith, A.M., 4, 132
Snell's law, 5, 6, 31
Soissons, Council of, 45
Sollertia, 291, 293, 298
Sorabji, Richard, 269, 270
Sorbonne, 133
Sound, 17, 90, 175
Southern, Richard, 3, 13, 89, 121, 124, 145, 146, 222, 223, 225, 233, 240, 246, 259, 287, 292
Space, 10, 23, 59–83, 178, 182, 191, 213, 224, 228, 234, 240, 249, 258, 276, 284, 285, 288
Species, theory of, 153
Sphere, 9–15, 17, 19, 23, 28, 29, 34, 36, 66, 169, 174, 176, 214, 216, 217, 224, 235, 247, 252, 282, 288, 292, 294, 296
Spheres, celestial, 9–17, 22, 217, 239, 286
St.Ambrose, 149, 228–230
St Augustine, 21, 87–89, 198, 204, 219, 227–232, 235, 247–249, 252–257
St. Bede, 228, 230
St. Bonaventure, 123, 127, 131, 137, 199, 210, 211, 247
Stability, 3, 7–9
Stars, 9, 10, 12, 24, 130, 131, 136, 176, 179, 217, 237, 257, 275, 286, 296
Suárez, Francisco, 279
Sub-lunar, 11, 12, 14, 15
Substance, 8, 13, 14, 22, 25, 41, 42, 49, 107, 153, 168, 170, 171, 176–179, 181, 196, 200, 201, 218, 249, 250, 294
Sweeney, Leo, 210–212, 215, 218, 219
Symmetry, 3–19, 26, 294

T
Tempier, 286
Theodoric of Cervia, 111
Theodoric of Freiburg, 26, 28, 30, 160, 287
Theophilus, 228
Thierry of Chartres (Theodoric), 43, 45, 46
Thomas of Chobham, 114
Thomson's Lamp, 275
Tide, 161, 172
Transcendentals, 189–206
Trichromacy, 74, 77, 79, 82
Trinity, 143, 179, 196, 213, 215, 268, 279, 286, 294

U
Uniformity, Principle of, 3, 173
Unity, 3–19, 165, 174, 176, 181, 182, 196, 200, 201, 203, 206, 213, 223, 224, 232, 238–239, 267, 296

V
Valente, Luisa, 190, 192, 193
Verger, Jaques, 52
Vescovini, Graziella Federici, 130
Vicenna (Ibn Sīnā), 248
Victorinus, Marius, 177
Vuillemin-Diem, Gudrun, 135, 136
Vulgate, 95, 227, 229

W
Wallensis, Thomas, 124
Wavelength, 5, 68–76, 80–82
William of Auvergne, 55, 103, 109, 114, 148, 192, 194, 249, 257
William of Auxerre, 192, 194, 203, 232
William of Conches, 44–46
William of Middleton, 109
William of Moerbecke, 135–137
William of Ockham, 6, 134, 247, 269
William of Sherwood, 124, 150, 151
William of Thierry, 44, 46
William the Breton, 49
Wilson, Catherine, 165
Witelo, 4, 23, 132

Y
Yahyä ibn ul-Bitrīq, 93
Young, Spencer, 52

Z
Zupko, Jack, 192

Printed in the United States
By Bookmasters